PRAISE FOR *ALONE TOGETHER*

"Savvy and insightful." —*New York Times*

"Nobody has ever articulated so passionately and intelligently what we're doing to ourselves by substituting technologically mediated social interaction. . . . Equipped with penetrating intelligence and a sense of humor, Turkle surveys the frontlines of the social-digital transformation." —Lev Grossman, *Time*

"In this beautifully written, provocative, and worrying book, Turkle, a professor at MIT, a clinical psychologist, and, perhaps, the world's leading expert on the social and psychological effects of technology, argues that internet use has as much power to isolate and destroy relationships as it has to bring us together." —*Financial Times*

"What [Turkle] brings to the topic that is new is more than a decade of interviews with teens and college students in which she plumbs the psychological effect of our brave new devices on the generation that seems most comfortable with them." —*Wall Street Journal*

"A fascinating portrait of our changing relationship with technology." —Newsweek.com

"[Turkle] summarizes her new view of things with typical eloquence . . . fascinating [and] readable."
—*New York Times Book Review*

"Important . . . [a]dmirably personal. . . . [Turkle's] book will spark useful debate."
—*Boston Globe*

"Turkle is a sensitive interviewer and an elegant writer."—Slate.com

"Turkle is not a luddite, nor is *Alone Together* a salvo in some analogue counter-reformation. But it does add to a growing body of cyber-sceptic literature."
—*Observer* (UK)

"Subtle and interesting."
—*Guardian* (UK)

"Perhaps the world's leading authority on the impact of gadgetry on our lives . . . Turkle is brilliant—and brilliantly disturbing."
—*Sunday Times* (UK)

"Disturbing. Compelling. Powerful."
—*Seattle Times*

"A fascinating, insightful, and disquieting 'intimate ethnography' of our digital, robotic moment in history."
—*Natural History Magazine*

"Turkle is a gifted and imaginative writer . . . [who] pushes interesting arguments with an engaging style."
—*American Prospect*

"Enlightening."
—*Globe and Mail* (Canada)

"[Turkle's] interviews with teens and young adults who've grown up online are fascinating and her conclusions are provocative."
—*Oregonian*

"A must-read."
—*Albany Times Union*

"Worth talking about." —*History News Network*

"Sherry Turkle has distinguished herself as an astute observer of the subjective side of the relationship between people and technology. . . . Turkle does an excellent job of bringing out the important questions behind the technologies with which she deals. . . . Turkle is an engaging writer, and her stories are interesting, accessible, and thought-provoking." —*Journal of Technology, Theology, and Religion*

"[Turkle's] unplug-before-it's-too-late message is stark enough to make you reach for a landline and call someone you love."
  —*Las Vegas Business Press*

"A carefully researched and well-thought book." —*Deseret News*

"[Turkle's] decades of teaching technology and daily living add authority to her fine survey!" —*Bookwatch*

"No one has thought more deeply nor researched more thoroughly human-computer relations [than Turkle]. Her book *Alone Together* is an immensely satisfying read." —*Sherbrooke Record*

"A thorough look at how devices affect our relationships by a psychologist with decades in the field. . . . Lucid, well-written analysis."
  —*Bloomberg Businessweek*, four-star review

"There is such a wealth of information in *Alone Together* that it is impossible to cover all the insights and epiphanies in it that spark the mind and stir the soul." —*Spirituality and Practice*

"No one has a better handle on how we are using material technology to transform our immaterial 'self' than Sherry Turkle. She is our techno-Freud, illuminating our inner transformation long before we are able see it. This immensely satisfying book is a deep journey to our future selves." —Kevin Kelly, author of *What Technology Wants*

"Sherry Turkle has observed more widely and thought more deeply about human-computer relations than any other scholar. Her book is essential reading for all who hope to understand our changing relation to technology."

> —Howard E. Gardner, Hobbs Professor of Cognition and Education, Harvard Graduate School of Education

"*Alone Together* is a brilliant, profound, stirring, and often disturbing portrait of the future by America's leading expert on how computers affect us as humans. She reveals the secrets of 'Walden 2.0' and tells us that we deserve better than caring robots. Grab this book, then turn off your smart phones and absorb Sherry Turkle's powerful message."

> —Rosabeth Moss Kanter, Harvard Business School professor and author of *Evolve!*, *Confidence*, and *SuperCorp*

"Based on an ambitious research program, and written in a clear and beguiling style, this book will captivate both scholar and general reader and it will be a landmark in the study of the impact of social media."

> —Jill Conway, president emerita, Smith College, and author of *The Road from Coorain*

"Sherry Turkle is the Margaret Mead of digital culture. Parents and teachers: If you want to understand (and support) your children as they navigate the emotional undercurrents in today's technological world, this is the book you need to read. Every chapter is full of great insights and great writing."

> —Mitchel Resnick, LEGO Papert Professor of Learning Research and head of the Lifelong Kindergarten group at the MIT Media Laboratory

"*Alone Together* is a deep yet accessible, bold yet gentle, frightening yet reassuring account of how people continue to find one another in an increasingly mediated landscape. If the net and humanity could have a couples therapist, it would be Sherry Turkle."

> —Douglas Rushkoff, author of *Program or Be Programmed*

Alone Together

alone together

Why We Expect
More from Technology
and
Less from Each Other

THIRD EDITION

Sherry Turkle

BASIC BOOKS

New York

Basic Books
Hachette Book Group
1290 Avenue of the Americas, New York, NY 10104
www.basicbooks.com

Printed in the United States of America
Revised trade paperback edition: November 2017
Published by Basic Books, an imprint of Perseus Books, LLC,
a subsidiary of Hachette Book Group, Inc.
Originally published in hardcover and ebook by Basic Books in 2011

The Hachette Speakers Bureau provides a wide range of authors for speaking
events. To find out more, go to hachettespeakersbureau.com
or call 866-376-6591.

The publisher is not responsible for websites (or their content) that are not
owned by the publisher.

The Library of Congress has catalogued the hardcover as follows:
Turkle, Sherry.
   Alone together : why we expect more from technology and less from each
other / Sherry Turkle.
      p. cm.
  ISBN 978-0-465-01021-9 (alk. paper)
      1. Information technology—Social aspects. 2. Interpersonal relations.
3. Human-computer interaction. I. Title.
      HM851.T86 2010
      303.48'33—dc22

                                                          2010030614

ISBN 978-0-465-03146-7 (2012 paperback)
ISBN 978-0-465-09365-6 (2017 paperback)
ISBN 978-0-465-09366-3 (e-book)

LSC-C

10 9 8 7 6 5 4 3 2 1

TO REBECCA

My letter to you, with love

"Everything that deceives may be said to enchant."

—Plato, *The Republic*

"I'm done with smart machines. I want a machine that's attentive to my needs. Where are the sensitive machines?"

—Tweet available at dig_natRT @tigoe via @ramonapringle

CONTENTS

turning points

thirty years ago, when I joined the faculty at MIT to study computer culture, the world retained a certain innocence. Children played tic-tac-toe with their electronic toys, video game missiles took on invading asteroids, and "intelligent" programs could hold up their end of a serious chess match. The first home computers were being bought by people called hobbyists. The people who bought or built them experimented with programming, often making their own simple games. No one knew to what further uses home computers might be put. The intellectual buzz in the still-young field of artificial intelligence was over programs that could recognize simple shapes and manipulate blocks. AI scientists debated whether machines of the future would have their smarts programmed into them or whether intelligence might emerge from simple instructions written into machine hardware, just as neurobiologists currently imagine that intelligence and reflective self-consciousness emerge from the relatively simple architecture and activity of the human brain.

Now I was among them and, like any anthropologist, something of a stranger in a strange land. I had just spent several years in Paris studying how psychoanalytic ideas had spread into everyday life in France—how people were picking up and trying on this new language for thinking about the self. I had come to MIT because I sensed that something similar was happening with the language of computers. Computational metaphors, such as "debugging" and "programming," were starting to be used to think about politics, education, social life, and—most central to the analogy with psychoanalysis—about the self. While my computer science colleagues were immersed in getting computers to do ingenious things, I had other concerns. How were computers changing us as people? My colleagues often objected, insisting that computers were "just tools." But I was certain that the "just" in that sentence was deceiving. We are shaped by our tools. And now, the computer, a machine on the border of becoming a mind, was changing and shaping us.

As a psychoanalytically trained psychologist, I wanted to explore what I have called the "inner history of devices."[1] Discovering an inner history requires listening—and often not to the first story told. Much is learned from the tossed-off aside, the comment made when the interview is "officially" over. To do my work, I adopted an ethnographic and clinical style of research as I lived in worlds new to me. But instead of spending hundreds of hours in simple dwellings, as an anthropologist in a traditional setting would do, listening to the local lore, I lurked around computer science departments, home computer hobbyist clubs, and junior high school computer laboratories. I asked questions of scientists, home computer owners, and children, but mostly I listened to how they talked and watched how they behaved among their new "thinking" machines.

I heard computers provoke erudite conversations. Perhaps, people wondered, the human mind is just a programmed machine, much like a computer. Perhaps if the mind is a program, free will is an illusion. Most strikingly, these conversations occurred not just in seminar rooms. They were taking place around kitchen tables and in playrooms. Computers brought philosophy into everyday life; in particular, they turned children into philosophers. In the presence of their simple electronic games—games that played tic-tac-toe or challenged them in spelling—children asked if computers were alive, if they had different ways of thinking from people, and what, in the age of smart machines, was special about being a person.

In the late 1970s and early 1980s, I witnessed a moment when we were confronted with machines that invited us to think differently about *human* thought, memory, and understanding. The computer was an evocative object that provoked self-reflection. For me, this was captured in a conversation I had with thirteen-year-old Deborah in the early 1980s. After a year of studying programming, Deborah said that, when working with the computer, "there's a little piece of your mind and now it's a little piece of the computer's mind." Once this was achieved, you could see yourself "differently."[2] Face-to-"face" with a computer, people reflected on who they were in the mirror of the machine. In 1984, thinking about Deborah (and in homage as well to Simone de Beauvoir), I called my first book on computers and people *The Second Self.*

That date, 1984, is of course iconic in Western intellectual thinking, tethered as it is to George Orwell's novel. *Nineteen Eighty-Four* describes a society that subjects people to constant government surveillance, public mind control, and loss of individual rights. I find it ironic that my own 1984 book, about the technology that in many a science fiction novel makes possible such a dystopian world, was by contrast full of hope and optimism. I had concerns about the "holding power" of the new technology: some people found computers so compelling that they did not want to be separated from them. And I worried whether losing oneself in worlds within the machine would distract us from facing our problems in the real—both personal and political. But, in this first work, I focused on how evocative computers fostered new reflection about the self.

In the decade following the publication of *The Second Self*, people's relationships with computers changed. Whereas in the 1980s that relationship was almost always one-on-one, a person alone with a machine, in the 1990s, this was no longer the case. By then, the computer had become a portal that enabled people to lead parallel lives in virtual worlds. People joined networks such as America Online and discovered a new sense of "place." These were heady times: we were no longer limited to handfuls of close friends and contacts. Now we could have hundreds, even thousands, a dazzling breadth of connection. My focus shifted from the one-on-one with a computer to the relationships people formed with each other using the computer as an intermediary.

I began throwing weekly pizza parties in the Boston area to meet people who could tell me the stories of their lives in the new virtual worlds. They described the erosion of boundaries between the real and virtual as they moved in and out of their lives on the screen. Views of self became less unitary, more protean. I again felt witness, through the prism of technology, to a shift in how we create and experience our own identities.

I reported on this work in my 1995 *Life on the Screen*, which offered, on balance, a positive view of new opportunities for exploring identity online. But by then, my optimism of 1984 had been challenged. I was meeting people, many people, who found online life more satisfying than what some derisively called "RL," that is, real life. Doug, a Midwestern college student, played four avatars, distributed across three different online worlds. He always had these worlds open as windows on his computer screen along with his schoolwork, e-mail program, and favorite games. He cycled easily through them. He told me that RL "is just one more window." And, he added, "it's not usually my best one."[3] Where was this leading?

Two avenues forward became apparent by the mid-1990s. The first was the development of a fully networked life. Access to the network no longer required that we know our destination. With browsers and search engines—Mosaic, Netscape, Internet Explorer, Google—one had the sense of traversing an infinite landscape always there to be discovered. And as connections to the Internet went mobile, we no longer "logged on" from a desktop, tethered by cables to an object called a "computer." The network was with us, on us, all the time. So, we could be with each other all the time. Second, there was an evolution in robotics. Now, instead of simply taking on difficult or dangerous jobs for us, robots would try to be our friends. The fruits of such research made their way into children's playrooms: by the late 1990s, children were presented with digital "creatures" that made demands for attention and seemed to pay attention to them.

Alone Together picks up these two strands in the story of digital culture over the past fifteen years, with a focus on the young, those from five through their early twenties—"digital natives" growing up with cell phones and toys that ask for love. If, by the end of researching *Life on the Screen*, I was troubled about the costs of life with simulation, in the course of researching this book, my concerns have grown. These days, insecure in our relationships and anxious about intimacy, we look to technology for ways to be in relationships and protect ourselves from them at the same time. This can happen when one is finding one's way through a blizzard of text messages; it can happen when interacting with a robot. I feel witness for a third time to a turning point in our expectations of technology and ourselves. We bend to the inanimate with new solicitude. We fear the risks and disappointments of relationships with our fellow humans. We expect more from technology and less from each other.

In this book I concentrate on observations during the past fifteen years, but I also reach back to the prehistory of recent developments. To tell the story of artifacts that encourage relationship, I begin with the ELIZA program in the 1970s and take the story through to the "sociable" humanoid robots, such as Domo and Mertz, built at MIT in the 2000s. Along the way there have been many other digital "creatures," including Tamagotchis, Furbies, AIBOs, My Real Babies, Kismet, Cog, and Paros, these last, robot baby seals designed specifically to provide companionship for the elderly. I thank the more than 250 people involved in my robot studies. Some who met robots came to MIT; other times I brought robots to schools, after-school centers, and nursing homes. When working with children, whenever possible, I provided them with a robot to take home for several weeks. Children and their families were asked to keep "robot diaries," accounts of home life with an AIBO, My Real Baby, or Furby.

In the story of computer-mediated communication, I began my investigations in the 1980s and early 1990s with e-mail, bulletin boards, Internet Relay Chat, and America Online and went on from there to the first virtual communities and multiuser online role-playing games. Over the past decade, as the network dramatically changed its contours, I broadened my investigation to include mobile devices, texts, instant messages, social networks, Twitter, and massively multiplayer online games. My work also included studies of virtual communities where three-dimensional avatars inhabit photorealistic spaces.

The focus of my research on networking was the young, and so I did most of my observations in high schools and on college campuses. But I also spoke with adults who gave me insight into how the network is changing parenting and communications patterns in fields from architecture to management consulting. Over 450 people have participated in my studies of connectivity, roughly 300 children and 150 adults. I thank everyone who lent their voices to this work over the past fifteen years. I am grateful for their generosity and good will.

The work reported on here, as all of my work, includes field research and clinical studies. In field research, one goes to where people and their technologies meet to observe interactions, sometimes ask questions, and take detailed notes. Depending on the nature of the field setting, casual conversations may take place over coffee or over snacks of milk and cookies. I teach courses about the computer culture and the psychology of computation, and some of my material comes from the give-and-take of the classroom. In the clinical component of my work, I pursue more detailed interviews, usually in an office or other quiet setting. I call these studies clinical, but of course my role in them is as a researcher, not a therapist. My interest in the "inner history" of technology means that I try to bring together the sensibility of ethnographer and clinician in all my work. A sensitive ethnographer is always open to the slip, to a tear, to an unexpected association. I think of the product as an intimate ethnography.

In my studies of robots, I provided the artifacts (from primitive Tamagotchis and Furbies to sophisticated robots such as Kismet and Cog). This meant that I was able to study children and seniors from a range of social and economic backgrounds. In the research on the networked life, I did not distribute any technology. I spoke to children, adolescents, and adults who already had Web access and mobile phones. Necessarily, my claims about new connectivity devices and the self apply to those who can afford such things. This turned out to be a larger group than I had originally supposed. For example, in a public high school study in the spring of 2008, every student, across a wide range of economic and cultural situations, had a mobile phone that could support texting. Most students had phones that could put them on the Web. I am studying a moving target. In January 2010, a Nielson study reported that the average teen sends over three thousand text messages a month.[4] My data suggests that this number is steadily increasing. What I report here is nothing less than the future unfolding.*

My investigations continue. These days, parents wait in line to buy their children interactive Zhu Zhu robotic pet hamsters, advertised as "living to feel the love." And one of the hottest online programs is Chatroulette, with 1.5 million users, which randomly connects you to other users all over the world. You see each other on live video. You can talk or write notes. People mostly hit "next" after about two seconds to bring another person up on their screens. It seems right that Zhu Zhu pets and Chatroulette are the final "objects" I report on in this book: the Zhu Zhus are designed to be loved; in Chatroulette, people are objectified and quickly discarded. I leave my story at a point of disturbing symmetry: we seem determined to give human qualities to objects and content to treat each other as things.

I preserve my subjects' anonymity by changing identifying details, except where I cite scientists and researchers on the public record or those who have asked to be cited by name. Without mentioning "real" names and places, I express appreciation to everyone who has spoken with me and to the school directors and principals, teachers, and nursing home directors and staff who made my work possible. I studied robots in two nursing homes and have data from students in seven high schools (two public and coeducational; five private, one for girls, two for boys, one coeducational; and one coeducational Catholic high school). In some cases I have been able to follow children who grew up with Tamagotchis and Furbies through their adolescence and young adulthood as they entered the networked culture to become fluent with texting, Twitter,

---

* In this book I use the terms *the Net*, *the network*, and *connectivity* to refer to our new world of online connections—from the experience of surfing the Web, to e-mail, texting, gaming, and social networking. And I use the term *cell phone* to describe a range of connectivity devices such as BlackBerries and iPhones that do a lot more than make "calls." They provide access to instant messaging, texting, e-mail, and the Web.

MySpace, Facebook, and the world of iPhone apps. I thank these young adults for their patience with me and this project.

I did much of the work reported here under the auspices of the MIT Initiative on Technology and Self. I thank all of my colleagues and students who worked with the initiative and in the Program for Science, Technology, and Society, which is its academic home. I have profited from their support and good ideas.

Collegial relationships across MIT have enriched my thinking and been sources of much appreciated practical assistance. Rodney Brooks provided me with an office at the MIT Artificial Intelligence Laboratory to help me get the lay of the land. He gave me the best possible start. Cynthia Breazeal and Brian Scassellati, the principal developers of Kismet and Cog, worked with me on the first-encounters study that introduced sixty children to these robots. These two generous colleagues helped me to think through so many of the issues in this book. On this study, I worked with research assistants Anita Say Chan, Rebecca Hurwitz, and Tamara Knutsen, and later with Robert Briscoe and Olivia Dasté. The Kismet and Cog support team, including Lijin Aryananda, Aaron Edsinger, Paul Fitzpatrick, Matthew Marjanavic, and Paulina Varchavskaia, provided much needed assistance. At the very beginning of my research on virtual worlds, I worked with Amy Bruckman. For me, it was a touchstone collaboration. Jennifer Audley, Joanna Barnes, Robert Briscoe, Olivia Dasté, Alice Driscoll, Cory Kidd, Anne Pollack, Rachel Prentice, Jocelyn Scheirer, T.L. Taylor, and William Taggart all made precious contributions during the years of interviews with children, families, and elders. I worked with Federico Castelegno at MIT on a study of online gaming; I thank him for his insights.

In this diverse and talented group, four colleagues deserve special recognition: Jennifer Audley worked on this project from the earliest studies of Tamagotchis and Furbies through the work on the robots Kismet and Cog. Olivia Dasté joined the project in 2001, working closely with me in nursing homes and schools and on the analysis of the "first encounters" of Kismet and Cog. William Taggart and Cory Kidd worked in nursing homes, primarily with the Paro robot. Each of them has my deepest thanks.

I also am grateful to Professors Caroline Jones, Seymour Papert, Mitchel Resnick, William Mitchell, Rosalind Picard, and William Porter. Conversations with each of them brought new ideas. For my thinking about Domo and Mertz, thanks to Pia Lindman, Aaron Edsinger, and Lijin Aryananda of MIT's Computer Science and Artificial Intelligence Laboratory (the Artificial Intelligence Laboratory's successor) who shared their experiences and their robots with me.

Conversations with five psychoanalytic colleagues were particularly important in shaping my thinking on children and the culture of simulation, both online and robotic: Dr. Ellen Dolnansky, Dr. James Frosch, Dr. Monica Horovitz, Dr. David Mann, and Dr. Patrick Miller.

My MIT colleague Hal Abelson sent me an e-mail in 1997, suggesting that I "study those dolls," and I always take his advice. In the late 1970s, he was the first to introduce me to the special hopes of personal computer owners who were not content until they understood the "innards" of their machines. In the late 1980s, he introduced me to the first generation of virtual communities, known at the time as "MUDs." Following his leads has always led me to my life's work. I can only repay my debt to Hal Abelson by following up on his wonderful tips. I thank him and hope I have done him proud.

Colleagues at Harvard and presentations at that institution have consistently broadened my perspective. In particular I thank Professors Homi Baba, Mario Biagioli, Svetlana Bohm, Vanessa Conley, Peter Galison, Howard Gardner, Sheila Jasonoff, Nancy Rosenblum, Michael Sandel, and Susan Sulieman for individual conversations and opportunities to meet with groups.

There are other debts: Thad Kull tirelessly tracked down sources. Ada Brustein, William Friedberg, Katie Hafner, Roger Lewin, David McIntosh, Katinka Matson, Margaret Morris, Clifford Nass, Susan Pollak, Ellen Poss, Catherine Rea, and Meredith Traquina gave excellent advice at key moments. Jill Ker Conway's reading of my first full draft provided encouragement and direction. Thomas Kelleher at Basic Books contributed organizational ideas and a much-appreciated line editing; Jennifer Kelland Fagan copyedited this manuscript with great care. Any infelicities of language are surely the result of my not taking their good advice. Grace Costa and Judith Spitzer provided the administrative support that freed my time so I could interview, think, and write.

I have worked with Kelly Gray on six book projects. In each one, her dedication, intelligence, and love of language have been sustaining. In *Alone Together*, whose primary data spans thirty years of life in the computer culture, it was Kelly who helped me find the narrative for the book I wanted to write. Additionally, some of my favorite turns of phrase in this book are ones that Kelly introduced into our many conversations. I wanted to list them; she told me not to, but her modesty should not deceive my readers about her profound contribution.

My work on robotics has been funded by the Intel Corporation, the Mitchell Kapor Foundation, the Kurzweil Foundation, and the National Science Foundation (NSF Grant # SES-0115668, "Relational Artifacts"). Takanori Shibata, the inventor of Paro, provided me with the baby seal robots to use in my studies. The Sony Corporation donated one of their very first AIBOs. My work on adolescents has been funded by the Intel Corporation, the Mitchell Kapor Foundation, and the Spencer Foundation. Among all this generosity, the contribution of Mitchell Kapor must be singled out. He understood what I was trying to accomplish with an Initiative on Technology and Self and gave it his full support. In all cases, the findings and opinions expressed here are mine and do not reflect the positions of the organizations and individuals who have helped me.

I have worked on the themes of this book for decades. It is certain that I have many unacknowledged debts. I take this opportunity to say thank you.

There is a final debt to my daughter Rebecca. Since she was six, she has patiently made friends with the talkative robots—simple and fancy—that I have brought into our home. I have asked her to take care of Tamagotchis, to play with Kismet and Cog, to befriend our own stay-at-home Paro. The My Real Babies frightened her, but she made a good effort to tell me why. Rebecca calls our basement storage room "the robot cemetery" and doesn't much like to go down there. I thank Rebecca for her forbearance, for her insightful and decisive editorial support, and for giving me permission to quote her. She refused to friend me on Facebook, but she taught me how to text. The story of digital culture has been the story of Rebecca's life. The book is written as a letter to her about how her mother sees the conversations in her future.

Now Rebecca is nineteen, and I know that, out of love for me, she is glad this book is finished. As for me, I'm not so sure. Thinking about robots, as I argue in these pages, is a way of thinking about the essence of personhood. Thinking about connectivity is a way to think about what we mean to each other. This book project is over; my preoccupation with its themes stays with me.

Sherry Turkle
BOSTON, MASSACHUSETTS
AUGUST 2010

PREFACE TO THE 2017 EDITION

I write this at the cottage where in the summer of 1999 I first challenged teams of eight- and nine-year-olds to play with Furbies, a first-generation robotic pet. It was pay and play. Children who put in time with my Furby studies got to take home a Furby, and I got to interview them more!

The Furbies came out of their boxes as furry visitors from another planet who arrived speaking Furbish but wanted to learn English. They were primitive sociable robots, objects designed to win us over not with their smarts but with their sociability. Furbies illustrated a basic lesson about how humans attach to artifacts: nurturance is the killer app. We nurture what we love, but we also love what we nurture. In this case, if you can get a child to teach something, anything, to a robot, you will forge a deep but one-sided attachment.

In fact, Furbies do not learn English from the children who play with them. A Furby's English-speaking "ability" unfolds depending on how many hours it is active. The eight-year-olds mostly didn't figure this out, however; they thought they were teaching the Furbies and loved the Furbies they had personally trained. When a Furby broke, the child who owned it didn't necessarily want it replaced. The Furby had died, and the child mourned it.

The Furby, although just a toy, was part of a turning point in the story of artificial intelligence. Since the 1960s, when the term was coined, "artificial intelligence" (AI) had, for the most part, focused on the intelligence that might be embedded in artifacts. Its primary virtue would be to inform. In the 1990s, a group of AI researchers began to design robots to exhibit other virtues. If people needed information, well, the nascent Internet would take care of that. People needed robots that could offer tending. People were lonely; robots would pay attention. Sociable robots could be our companions and our friends.

Before I began the research for *Alone Together*, I was, if anything, an enthusiast of digital culture. My introduction to sociable robotics put me on another path. Indeed, I was recently asked if *Alone Together* had an origin story, a moment when I turned from looking at technology optimistically to take a more cautionary position. My interviewer was taken aback when I not only said, yes, I had a story, but named the day it took place.

I began this book on the day I met Cog, a robotic head attached to a neck, torso, and arms, that was socially responsive to the humans it interacted with. When you paid Cog a visit, it could make eye contact, reach toward you, and mimic your motion. I first went to see Cog at the MIT Artificial Intelligence Laboratory with a colleague. We were both at a conference on the future of artificial life. Together, we walked into Cog's room and tried to get its attention. Only my colleague succeeded. He and Cog locked eyes and moved in synchrony. I was left out of the game. Cog showed no interest in my presence at all. Intellectually, I knew there was some technical explanation—for example, Cog might be programmed to attend to the color red, whereas I was dressed in black—but at the time this knowledge seemed of little use. I felt competitive and jealous. I wanted Cog to notice me.

That moment of pique—when I fully experienced my own vulnerability to the seductions of technology—was the origin of this book. I realized that I wanted a technology that had no emotional care to give to emotionally care for me, to make me feel important. My reaction was not unique. The more I studied responses to sociable robots, the more common I found it to be—and the more I understood what these new objects knew how to do: push people's Darwinian buttons. Making eye contact, tracking and mimicking motion, saying our name: these make people experience a robot as not just clever and thus deserving of respect but as deserving of care and attention—and beyond this, as having care and attention to give to us.

Why do I say Darwinian? Because as we evolved, people were the only other creatures who responded to us with suggestions of empathy. But now the robots showed us "as if" empathy, and we were, you might say, cheap dates. We proved willing to talk to robots about personal matters, to take them as confidants, even as psychotherapists, if they could give us the feeling that they understood (regardless of whether they had any comprehension of the loves and lives we were describing).

In *Alone Together*, I report on more than a decade of work exploring what I call "the robotic moment," the development of a state of mind in which we became willing to grant sociable artifacts more than their due. The machine's prowess was exploiting our vulnerabilities: we wanted to feel recognized, and we didn't want to feel alone.

Significantly, from the mid- to late 1990s, these same two vulnerabilities drove the rise of social media, from online chatrooms to the apps increasingly embedded in mobile technology. Our always-on/always-on-you lives promise that we never have to feel alone, that someone can always hear us. But online-all-the-time social life has built-in limitations. The most important: we are tempted to turn away from the people we are with to the pleasures of our phones. It might be a text or a game. It might an Instagram or Snapchat or Twitter feed. Always distracted, we lose the capacity for solitude. We become accustomed to the constant social stimulation that only connectivity can provide. It isn't just that social media's development led us to ask more of technology than it could realistically or appropriately deliver. We contented ourselves with a text or an e-mail when a conversation would better convey our meaning. We came to ask less of each other. We settled for less empathy, less attention, less care from other human beings.

In education, we said we would skip having in-person classes, because we could design online technology that would deliver a "personalized" class, pitched to each learner's skill level. But in the process, we often stopped asking what we really wanted students to get out of class. Is the class most important as a vector for the transmission of information? What about its role as a place where students learn to listen to each other? To empathize with each other? Those things don't easily happen if students are staring at screens.

Since this book's publication in 2011, the trend line has been clear: we are more distracted in each other's company, and we know it. Recent surveys report that 89 percent of Americans say they interrupted their last social interaction to turn to their phones, and 82 percent say that the conversation suffered for it.[1] Their intuitions are correct. Studies show that if two people are having lunch, a cell phone at the table steers the conversation to lighter subjects, and each party feels less invested in the other.[2] It's not surprising. Each person in the conversation knows that with a phone on the landscape, you can be interrupted at any time.

We pay a price for our distraction. Not surprisingly, one of the most dramatic costs is our empathic capacity. Research has demonstrated a 40 percent decline over a recent thirty-year span in the markers for empathy among college students. And since most of that decline occurred in the last decade of the work, it makes sense to link the empathy gap to the presence of digital communications.[3]

As I worked on *Alone Together*, I came to this formulation: technology makes us forget what we know about life. People will readily say that in face-to-face conversation they learn how to get on with other people and gather important understandings about their children, spouses, parents, and partners. And yet they will also say that they are happy to use technology to flee these conversations. Why? Because face-to-face conversations are difficult. Awkward. Spontaneous. Unscripted. Messy. One young man tells me he will do anything to avoid a conversation. "Conversation? I'll tell you what's wrong with conversation. It takes place in real time, and you can't control what you're going to say."

When offered something that can make things easier, we forget our human purposes. We become fascinated by the idea of a relational world that might be "friction free," kind of like the machine world. In our excitement about how great a technology makes us feel or how amazing it is that a technology can do x or y, however, we forget the more important question: What is the *human* value of x and y? Does it keep its human value if done by a machine?

That is the cautionary message at the heart of this book. We say yes to new technology because we think we are getting something important from it. We give it our time, attention, and money, and increasingly we abandon our privacy to use it. But we need to get into new and more disciplined habits where we examine the assumption that we are getting something important from these new technologies. We must ask whether a technology expands our capacities and possibilities or exploits our vulnerabilities? If we think it does both, what is the balance? Technology offered us sugared soda water, and we embraced it. We took over a hundred years to decide it was no good for us at all. But by the time we declared it toxic, it had become integrated into our most potent images of the American dream. We allowed "Coke and a burger" to became part of our most cherished stories of family cohesiveness and adolescent coming of age. Now, we watch as the myth, slowly and painfully, is deconstructed. We can see, in real time, how much time this takes, how much resistance there is, and how much damage has been done. That should make us ask what new myths we are constructing, right now, about what is good for us. Robots that read to us? Long-distance classes that will save us the trouble of face-to-face encounters with teachers? Computer psychotherapists that will save us the trouble and embarrassment of having to talk to humans about our problems? All of these are on offer. All of these currently make their pitch to become part of our new normal.

And strikingly today, every time a new technology plays to a profound vulnerability, we behave as though we have never heard this story before. Technology, whether food technology or digital technology, can make us forget what we know about life. And these days technology offers ever greater temptations to take the path of forgetting.

So, for example, commerce has dictated that the new normal should involve conversations with a machine that sits permanently on your kitchen table. Amazon's Alexa and Google's Echo are conversational robots at your command. You can ask them to order a pizza or change the song on your sound system. Technology enthusiast Tim O'Reilly describes how Alexa introduced him to the future. No matter what else was going on, even if Alexa was doing multiple tasks, it would always respond to his request to "lower the volume" on his music. This was wonderful, O'Reilly says and asks what enables Alexa to do this very human thing. The answer of course is that, unlike other devices that need to be activated, Alexa is always on. For O'Reilly, this "always listening to me" feature clearly enabled Alexa to surpass the other technologies in the "genie" category. When O'Reilly confronted an executive from Alphabet, Google's new parent company, the man conceded the point, but said that Google couldn't always be listening because of privacy implications. "Can you imagine the blowback if it were Google that was always listening to you?"[4]

To this concern about privacy, O'Reilly responds, "He had a point. But that's how the future happens. Someone breaks the barrier, does the unthinkable, and then it becomes thinkable for everyone."

O'Reilly, too, had a point, but perhaps not the one he was arguing. For me, this exchange makes clear how ready we are to forget the rudiments of why we want to protect our privacy, even that of conversations in our homes, the last place in a surveilled society where we have any privacy at all.

O'Reilly is certainly ready to give it up for the convenience of being able to change the volume on any song from *Hamilton*, without needing to pause from cooking or playing a video game. He is not deterred by the fact that Amazon owns all the information in those kitchen conversations, just as Google owns the information about what we search for and when, about whom we send an e-mail and what we say in it. These, too, are previously unthinkable things that we now take for granted. But are we ready to accept that we cannot rethink our choices?

Even as the threat of authoritarian government becomes greater all over the world, even in democratic countries, our concerns about privacy seem to dissolve if technology offers us a bit of comfort. Ironically, the comfort we seek often entails greater ease in how we use technology. We are harried and too busy. If a mail app reads and sorts our mail, lightening our load even a little, we are content that the company that provides the app may extract data from our mail along the way. We still experience that app as nurturing.

From this point of view, consider a final example, a Furby descendent: Hello Barbie, a robotic doll manufactured by the Mattel Corporation. Hello Barbie comes out of the box and announces that she is a child's friend. If the toy learns that her child owner has a sister, Hello Barbie will announce that she has a sister, too, because this will make the child better able to "relate" to her. Child and doll may go on to discuss feelings about sisters, parents, friends, and teachers. All of this is recorded and accessible to parents. And all of this gets uploaded to the cloud and belongs to Mattel. Yes, the child's confidences are the property of Mattel. The Center for a Commercial-Free Childhood launched a "Hell-No Barbie" campaign against the doll and released seven talking points that summed up why all of this should raise alarms. The reader of *Alone Together* will recognize many of them: A child is being asked to enter a pretend world of as-if feelings, as-if empathy. Hello Barbie pretends empathy but has none to give. It does not allow the space for imaginative play that even the plain-old silent, anorectic Barbie did. It does not allow the child to project and set the agenda, as a traditional doll does. (As I point out in my chapter on sociable robots, a young girl who has just broken her mother's crystal sets her Barbies up in detention. She behaves with the dolls in ways that enable her to express and work through her inner life. A sociable robot has its own agenda.)

And with all of these problems comes a new one, illustrated by Alexa: privacy. O'Reilly welcomes Alexa's always-listening strategy as a transgression that becomes the new normal. But a new normal changes our expectations about the world. In the case of Alexa, we learn that our devices are always on. Over time, a child with Hello Barbie learns that surveillance is the price of play. Kade Crockford of the American Civil Liberties Union of Massachusetts says this about Hello Barbie: "Surveillance camouflaged as play obscures the significant dangers posed by the loss of control of one's personal information. Hello Barbie makes the transfer of sensitive data to unaccountable third parties seem fun and natural. Do we want our children growing up in a world in which they have no consciousness of what it means to have a private life?"[5]

This is our current challenge: combatting the sense that it is normal to lose privacy for a feeling of sociability. In the world of *Alone Together*, I saw the sociable robots as a testing ground, a place where we would allow unjustified relationships with the inanimate, where we would allow relationships that diminished our humanity. And that threat was paired with sociable computing, where we allowed our rich conversations to turn into mere connections. These days, the threat of granting technology too much has exploded far beyond the robotic into the world of apps. There are apps that tend to us, discipline us; on apps we confess, talk about our health, and ask for guidance about how to exercise and how to pray. We live a life of apps, and we see them as nurturing even as they exploit the data we give them. And yet we feel that this bargain is worth it because we refuse to see it as a bargain. Most people, I have learned in my research, see their apps as "free."

In 2011, I looked at sociable robots and social media as two silos. These days, by looking at the robotic moment as it transforms into the world of apps, we can see what talking to machines and talking to each other through machines have in common. Machines want you to keep using them. That is how they can get more data from you in order to sell more to you and others. Machines keep us in a world of machines.

Writing this book inspired me to approach every day with technology with these questions: What is democracy without privacy? What is intimacy without privacy? If your answer is, Let's wait and see; technology always brings change, and people always adjust, then my note to you is this: you are playing with fire.

And now, new questions feel equally pressing: How will intimacy and empathy change in a world where we give toddlers baby bouncers and potty trainers with a slot for a tablet or smart phone? Werner Herzog recently made a film about the Internet in which cosmologist Laurence Krauss asks, "Will our children's children's children need the companionship of humans or will they have evolved in a world where that's not important?" When I began studying computers and people many decades ago, that idea would have sounded far out, literally like something that only a cosmologist would say. But Herzog's film illustrates the thought with the simple picture of a child stabbing at the keys of a smart phone. That mundane image triggers associations that support a world without human companionship. Indeed, as I continue to research the emotional effects of social media, I regularly meet parents who think that their children need people less than they did. Once we accept that there is an app for human confidences and human confessions, whether embodied in a robot or embedded in a small rectangular device, the idea that people won't want or need people becomes thinkable, carried by the technology itself.

When the researcher who discovered that 40 percent decline in empathy among college students over the past thirty years finished her studies, she was depressed by her findings. She told me that her first response was to set out to make "empathy apps" for the iPhone. I understood her reaction. We think that technology has gotten us into trouble, and we would like technology to get us out of it. As I reconsider *Alone Together*, here is how I see its greatest virtues: It provides materials to fight against this tempting idea. It encourages readers to see themselves as the empathy app. For this is my most fundamental belief: We are the empathy app.

Alone Together lays out the terrain for the world we inhabit. I'm hoping that it will encourage today's readers to pick up where it leaves off and write their own call to arms.

NOTES

1. Lee Rainie and Kathryn Zickuhr, "Americans' Views on Mobile Etiquette," Pew Research Center, August 26, 2015, http://www.pewinternet.org/files/2015/08/2015-08-26_mobile-etiquette_FINAL.pdf (accessed March 23, 2017).

2. Andrew Przybyliski and Netta Weinstein, "Can You Connect with Me Now? How the Presence of Mobile Communication Technology Influences Face-to-Face Conversation Quality," *Journal of Social and Personal Relationships* 30, no. 3 (2013), doi: 10.1177/0265407512453827.

3. Sara Konrath, Edward H. O'Brien, and Courtney Hsing, "Changes in Dispositional Empathy in American College Students over Time: A Meta-analysis," *Personality and Social Psychology Review* 15, no. 2 (May 1, 2011), doi:10.1177/1088868310377395.

4. Tim O'Reilly, "What Would Alexa Do?," LinkedIn, August 18, 2016, https://www.linkedin.com/pulse/what-would-alexa-do-tim-o-reilly (accessed March 23, 2017).

5. "Hell No Barbie: 8 Reasons to Leave Hello Barbie on the Shelf," Campaign for a Commercial-Free Childhood, http://www.commercialfreechildhood.org/action/hell-no-barbie-8-reasons-leave-hello-barbie-shelf (accessed March 23, 2017).

alone together

technology proposes itself as the architect of our intimacies. These days, it suggests substitutions that put the real on the run. The advertising for Second Life, a virtual world where you get to build an avatar, a house, a family, and a social life, basically says, "Finally, a place to love your body, love your friends, and love your life."[1] On Second Life, a lot of people, as represented by their avatars, are richer than they are in first life and a lot younger, thinner, and better dressed. And we are smitten with the idea of sociable robots, which most people first meet in the guise of artificial pets. Zhu Zhu pet hamsters, the "it" toy of the 2009–2010 holiday season, are presented as "better" than any real pet could be. We are told they are lovable and responsive, don't require cleanup, and will never die.

Technology is seductive when what it offers meets our human vulnerabilities. And as it turns out, we are very vulnerable indeed. We are lonely but fearful of intimacy. Digital connections and the sociable robot may offer the illusion of companionship without the demands of friendship. Our networked life allows us to hide from each other, even as we are tethered to each other. We'd rather text than talk. A simple story makes this last point, told in her own words by a harried mother in her late forties:

I needed to find a new nanny. When I interview nannies, I like to go to where they live, so that I can see them in their environment, not just in mine. So, I made an appointment to interview Ronnie, who had applied for the job. I show up at her apartment and her housemate answers the door. She is a young woman, around twenty-one, texting on her Black-Berry. Her thumbs are bandaged. I look at them, pained at the tiny thumb splints, and I try to be sympathetic. "That must hurt." But she just shrugs. She explains that she is still able to text. I tell her I am here to speak with Ronnie; this is her job interview. Could she please knock on Ronnie's bed-room door? The girl with the bandaged thumbs looks surprised. "Oh no," she says, "I would never do that. That would be intrusive. I'll text her." And so she sent a text message to Ronnie, no more than fifteen feet away.

This book, which completes a trilogy on computers and people, asks how we got to this place and whether we are content to be here.

In *The Second Self*, I traced the subjective side of personal computers—not what computers do for us but what they do to us, to our ways of thinking about ourselves, our relationships, our sense of being human. From the start, people used interactive and reactive computers to reflect on the self and think about the difference between machines and people. Were intelligent machines alive? If not, why not? In my studies I found that children were most likely to see this new category of object, the computational object, as "sort of" alive—a story that has continued to evolve. In *Life on the Screen*, my focus shifted from how people see computers to how they forge new identities in online spaces. In *Alone To-gether*, I show how technology has taken both of these stories to a new level.

Computers no longer wait for humans to project meaning onto them. Now, sociable robots meet our gaze, speak to us, and learn to recognize us. They ask us to take care of them; in response, we imagine that they might care for us in return. Indeed, among the most talked about robotic designs are in the area of care and companionship. In summer 2010, there are enthusiastic reports in the *New York Times* and the *Wall Street Journal* on robotic teachers, com-panions, and therapists. And Microsoft demonstrates a virtual human, Milo, that recognizes the people it interacts with and whose personality is sculpted by them. Tellingly, in the video that introduces Milo to the public, a young man begins by playing games with Milo in a virtual garden; by the end of the demonstration, things have heated up—he confides in Milo after being told off by his parents.[2]

We are challenged to ask what such things augur. Some people are looking for robots to clean rugs and help with the laundry. Others hope for a mechanical bride. As sociable robots propose themselves as substitutes for people, new networked devices offer us machine-mediated relationships with each other, another kind of substitution. We romance the robot and become inseparable from our smartphones. As this happens, we remake ourselves and our relationships with each other through our new intimacy with machines. People talk about Web access on their BlackBerries as "the place for hope" in life, the place where loneliness can be defeated. A woman in her late sixties describes her new iPhone: "It's like having a little Times Square in my pocketbook. All lights. All the people I could meet." People are lonely. The network is seductive. But if we are always on, we may deny ourselves the rewards of solitude.

THE ROBOTIC MOMENT

In late November 2005, I took my daughter Rebecca, then fourteen, to the Darwin exhibition at the American Museum of Natural History in New York. From the moment you step into the museum and come face-to-face with a full-size dinosaur, you become part of a celebration of life on Earth, what Darwin called "endless forms most beautiful." Millions upon millions of now lifeless specimens represent nature's invention in every corner of the globe. There could be no better venue for documenting Darwin's life and thought and his theory of evolution by natural selection, the central truth that underpins contemporary biology. The exhibition aimed to please and, a bit defensively in these days of attacks on the theory of evolution, wanted to convince.

At the exhibit's entrance were two giant tortoises from the Galápagos Islands, the best-known inhabitants of the archipelago where Darwin did his most famous investigations. The museum had been advertising these tortoises as wonders, curiosities, and marvels. Here, among the plastic models at the museum, was the life that Darwin saw more than a century and a half ago. One tortoise was hidden from view; the other rested in its cage, utterly still. Rebecca inspected the visible tortoise thoughtfully for a while and then said matter-of-factly, "They could have used a robot." I was taken aback and asked what she meant. She said she thought it was a shame to bring the turtle all this way from its island home in the Pacific, when it was just going to sit there in the museum, motionless, doing nothing. Rebecca was both concerned for the imprisoned turtle and unmoved by its authenticity.

It was Thanksgiving weekend. The line was long, the crowd frozen in place. I began to talk with some of the other parents and children. My question— "Do you care that the turtle is alive?"—was a welcome diversion from the boredom of the wait. A ten-year-old girl told me that she would prefer a robot turtle because aliveness comes with aesthetic inconvenience: "Its water looks dirty. Gross." More usually, votes for the robots echoed my daughter's sentiment that in this setting, aliveness didn't seem worth the trouble. A twelve-year-old girl was adamant: "For what the turtles do, you didn't have to have the live ones." Her father looked at her, mystified: "But the point is that they are real. That's the whole point."

The Darwin exhibition put authenticity front and center: on display were the actual magnifying glass that Darwin used in his travels, the very notebook in which he wrote the famous sentences that first described his theory of evolution. Yet, in the children's reactions to the inert but alive Galápagos tortoise, the idea of the original had no place. What I heard in the museum reminded me of Rebecca's reaction as a seven-year-old during a boat ride in the postcard-blue Mediterranean. Already an expert in the world of simulated fish tanks, she saw something in the water, pointed to it excitedly, and said, "Look, Mommy, a jellyfish! It looks so realistic!" When I told this story to a vice president at the Disney Corporation, he said he was not surprised. When Animal Kingdom opened in Orlando, populated by "real"—that is, biological—animals, its first visitors complained that they were not as "realistic" as the animatronic creatures in other parts of Disneyworld. The robotic crocodiles slapped their tails and rolled their eyes—in sum, they displayed archetypal "crocodile" behavior. The biological crocodiles, like the Galápagos tortoises, pretty much kept to themselves.

I believe that in our culture of simulation, the notion of authenticity is for us what sex was for the Victorians—threat and obsession, taboo and fascination. I have lived with this idea for many years; yet, at the museum, I found the children's position strangely unsettling. For them, in this context, aliveness seemed to have no intrinsic value. Rather, it is useful only if needed for a specific purpose. Darwin's endless forms so beautiful were no longer sufficient unto themselves. I asked the children a further question: "If you put a robot instead of a living turtle in the exhibit, do you think people should be told that the turtle is not alive?" Not really, said many children. Data on aliveness can be shared on a "need-to-know basis"—for a purpose. But what are the purposes of living things?

Only a year later, I was shocked to be confronted with the idea that these purposes were more up for grabs than I had ever dreamed. I received a call from a

Scientific American reporter to talk about robots and our future. During that conversation, he accused me of harboring sentiments that would put me squarely in the camp of those who have for so long stood in the way of marriage for homosexual couples. I was stunned, first because I harbor no such sentiments, but also because his accusation was prompted not by any objection I had made to the mating or marriage of people. The reporter was bothered because I had objected to the mating and marriage of people to robots.

The call had been prompted by a new book about robots by David Levy, a British-born entrepreneur and computer scientist. In 1968 Levy, an international chess master, famously wagered four artificial intelligence (AI) experts that no computer program would defeat him at the game in the subsequent decade. Levy won his bet. The sum was modest, 1,250 British pounds, but the AI community was chastened. They had overreached in their predictions for their young science. It would be another decade before Levy was bested in chess by a computer program, Deep Thought, an early version of the program that beat Gary Kasparov, the reigning chess champion in the 1990s.[3] These days, Levy is the chief executive officer at a company that develops "smart" toys for children. In 2009, Levy and his team won—and this for the second time—the prestigious Loebner Prize, widely regarded as the world championship for conversational software. In this contest, Levy's "chat bot" program was best at convincing people that they were talking to another person and not to a machine.

Always impressed with Levy's inventiveness, I found myself underwhelmed by the message of this latest book, *Love and Sex with Robots*.[4] No tongue-in-cheek science fiction fantasy, it was reviewed without irony in the *New York Times* by a reporter who had just spent two weeks at the Massachusetts Institute of Technology (MIT) and wrote glowingly about its robotics culture as creating "new forms of life."[5] *Love and Sex* is earnest in its predictions about where people and robots will find themselves by mid-century: "Love with robots will be as normal as love with other humans, while the number of sexual acts and lovemaking positions commonly practiced between humans will be extended, as robots will teach more than is in all of the world's published sex manuals combined."[6] Levy argues that robots will teach us to be better friends and lovers because we will be able to practice on them. Beyond this, they will substitute where people fail. Levy proposes, among other things, the virtues of marriage to robots. He argues that robots are, of course, "other" but, in many ways, better. No cheating. No heartbreak. In Levy's argument, there is one simple criterion for judging the worth of robots in even the most intimate domains: Does being with a robot make you feel better? The master of today's computerspeak judges future robots by the impact of their behavior. And his next bet is that in a very few years, this is all we will care about as well.

I am a psychoanalytically trained psychologist. Both by temperament and profession, I place high value on relationships of intimacy and authenticity. Granting that an AI might develop its own origami of lovemaking positions, I am troubled by the idea of seeking intimacy with a machine that has no feelings, can have no feelings, and is really just a clever collection of "as if" performances, behaving as if it cared, as if it understood us. Authenticity, for me, follows from the ability to put oneself in the place of another, to relate to the other because of a shared store of human experiences: we are born, have families, and know loss and the reality of death.[7] A robot, however sophisticated, is patently out of this loop.

So, I turned the pages of Levy's book with a cool eye. What if a robot is not a "form of life" but a kind of performance art? What if "relating" to robots makes us feel "good" or "better" simply because we feel more in control? Feeling good is no golden rule. One can feel good for bad reasons. What if a robot companion makes us feel good but leaves us somehow diminished? The virtue of Levy's bold position is that it forces reflection: What kinds of relationships with machines are possible, desirable, or ethical? What does it mean to love a robot? As I read *Love and Sex*, my feelings on these matters were clear. A love relationship involves coming to savor the surprises and the rough patches of looking at the world from another's point of view, shaped by history, biology, trauma, and joy. Computers and robots do not have these experiences to share. We look at mass media and worry about our culture being intellectually "dumbed down." *Love and Sex* seems to celebrate an emotional dumbing down, a willful turning away from the complexities of human partnerships—the inauthentic as a new aesthetic.

I was further discomforted as I read *Love and Sex* because Levy had interpreted my findings about the "holding power" of computers to argue his case. Indeed, Levy dedicated his book to Anthony,* an MIT computer hacker I interviewed in the early 1980s. Anthony was nineteen when I met him, a shy young man who found computers reassuring. He felt insecure in the world of people

---

* This name and the names of others I observed and interviewed for this book are pseudonyms. To protect the anonymity of my subjects, I also change identifying details such as location and profession. When I cite the opinions of scientists or public figures, I use their words with permission. And, of course, I cite material on the public record.

with its emotional risks and shades of gray. The activity and interactivity of computer programming gave Anthony—lonely, yet afraid of intimacy—the feeling that he was not alone.[8] In *Love and Sex*, Levy idealizes Anthony's accommodation and suggests that loving a robot would be a reasonable next step for people like him. I was sent an advance copy of the book, and Levy asked if I could get a copy to Anthony, thinking he would be flattered. I was less sure. I didn't remember Anthony as being at peace with his retreat to what he called "the machine world." I remembered him as wistful, feeling himself a spectator of the human world, like a kid with his nose to the window of a candy store. When we imagine robots as our future companions, we all put our noses to that same window.

I was deep in the irony of my unhappy Anthony as a role model for intimacy with robots when the *Scientific American* reporter called. I was not shy about my lack of enthusiasm for Levy's ideas and suggested that the very fact we were discussing marriage to robots at all was a comment on human disappointments—that in matters of love and sex, we must be failing each other. I did not see marriage to a machine as a welcome evolution in human relationships. And so I was taken aback when the reporter suggested that I was no better than bigots who deny gays and lesbians the right to marry. I tried to explain that just because I didn't think people should marry machines didn't mean that any mix of adult people wasn't fair territory. He accused me of species chauvinism: Wasn't I withholding from robots their right to "realness"? Why was I presuming that a relationship with a robot lacked authenticity? For me, the story of computers and the evocation of life had come to a new place.

At that point, I told the reporter that I, too, was taking notes on our conversation. The reporter's point of view was now data for my own work on our shifting cultural expectations of technology—data, that is, for the book you are reading. His analogizing of robots to gay men and women demonstrated that, for him, future intimacy with machines would not be a second-best substitute for finding a person to love. More than this, the reporter was insisting that machines would bring their own special qualities to an intimate partnership that needed to be honored in its own right. In his eyes, the love, sex, and marriage robot was not merely "better than nothing," a substitute. Rather, a robot had become "better than something." The machine could be preferable—for any number of reasons—to what we currently experience in the sometimes messy, often frustrating, and always complex world of people.

This episode with the *Scientific American* reporter shook me—perhaps in part because the magazine had been for me, since childhood, a gold standard in scientific publication. But the extravagance of the reporter's hopes for robots fell into a pattern I had been observing for nearly a decade. The encounter over *Love and Sex* most reminded me of another time, two years before, when I met a female graduate student at a large psychology conference in New Orleans; she had taken me aside to ask about the current state of research on robots designed to serve as human companions. At the conference, I had given a presentation on *anthropomorphism*—on how we see robots as close to human if they do such things as make eye contact, track our motion, and gesture in a show of friendship. These appear to be "Darwinian buttons" that cause people to imagine that the robot is an "other," that there is, colloquially speaking, "somebody home."

During a session break, the graduate student, Anne, a lovely, raven-haired woman in her mid-twenties, wanted specifics. She confided that she would trade in her boyfriend "for a sophisticated Japanese robot" if the robot would produce what she called "caring behavior." She told me that she relied on a "feeling of civility in the house." She did not want to be alone. She said, "If the robot could provide the environment, I would be happy to help produce the illusion that there is somebody really with me." She was looking for a "no-risk relationship" that would stave off loneliness. A responsive robot, even one just exhibiting scripted behavior, seemed better to her than a demanding boyfriend. I asked her, gently, if she was joking. She told me she was not. An even more poignant encounter was with Miriam, a seventy-two-year-old woman living in a suburban Boston nursing home, a participant in one of my studies of robots and the elderly.

I meet Miriam in an office that has been set aside for my interviews. She is a slight figure in a teal blue silk blouse and slim black pants, her long gray hair parted down the middle and tied behind her head in a low bun. Although elegant and composed, she is sad. In part, this is because of her circumstances. For someone who was once among Boston's best-known interior designers, the nursing home is a stark and lonely place. But there is also something immediate: Miriam's son has recently broken off his relationship with her. He has a job and family on the West Coast, and when he visits, he and his mother quarrel—he feels she wants more from him than he can give. Now Miriam sits quietly, stroking Paro, a sociable robot in the shape of a baby harp seal. Paro, developed in Japan, has been advertised as the first "therapeutic robot" for its ostensibly positive effects on the ill, elderly, and emotionally troubled. Paro can make eye contact by sensing the direction of a human voice, is sensitive to touch, and has a small working English vocabulary for "understanding" its users (the robot's Japanese vocabulary is larger); most importantly, it has "states of mind" affected by how it is treated. For example, it can sense whether it is being stroked gently or with aggression. Now, with Paro, Miriam is lost in her reverie, patting down the robot's soft fur with care. On this day, she is particularly depressed and believes that the robot is depressed as well. She turns to Paro, strokes him again, and says, "Yes, you're sad, aren't you? It's tough out there. Yes, it's hard." Miriam's tender touch triggers a warm response in Paro: it turns its head toward her and purrs approvingly. Encouraged, Miriam shows yet more affection for the little robot. In attempting to provide the comfort she believes it needs, she comforts herself.

Because of my training as a clinician, I believe that this kind of moment, if it happens between people, has profound therapeutic potential. We can heal ourselves by giving others what we most need. But what are we to make of this transaction between a depressed woman and a robot? When I talk to colleagues and friends about such encounters—for Miriam's story is not unusual—their first associations are usually to their pets and the solace they provide. I hear stories of how pets "know" when their owners are unhappy and need comfort. The comparison with pets sharpens the question of what it means to have a relationship with a robot. I do not know whether a pet could sense Miriam's unhappiness, her feelings of loss. I do know that in the moment of apparent connection between Miriam and her Paro, a moment that comforted her, the robot understood nothing. Miriam experienced an intimacy with another, but she was in fact alone. Her son had left her, and as she looked to the robot, I felt that we had abandoned her as well.

Experiences such as these—with the idea of aliveness on a "need-to-know" basis, with the proposal and defense of marriage to robots, with a young woman dreaming of a robot lover, and with Miriam and her Paro—have caused me to think of our time as the "robotic moment." This does not mean that companionate robots are common among us; it refers to our state of emotional—and I would say philosophical—readiness. I find people willing to seriously consider robots not only as pets but as potential friends, confidants, and even romantic partners. We don't seem to care what these artificial intelligences "know" or "understand" of the human moments we might "share" with them. At the robotic moment, the performance of connection seems connection enough. We are poised to attach to the inanimate without prejudice. The phrase "technological promiscuity" comes to mind.

As I listen for what stands behind this moment, I hear a certain fatigue with the difficulties of life with people. We insert robots into every narrative of human frailty. People make too many demands; robot demands would be of a more manageable sort. People disappoint; robots will not. When people talk about relationships with robots, they talk about cheating husbands, wives who fake orgasms, and children who take drugs. They talk about how hard it is to understand family and friends. I am at first surprised by these comments. Their clear intent is to bring people down a notch. A forty-four-year-old woman says, "After all, we never know how another person really feels. People put on a good face. Robots would be safer." A thirty-year-old man remarks, "I'd rather talk to a robot. Friends can be exhausting. The robot will always be there for me. And whenever I'm done, I can walk away."

The idea of sociable robots suggests that we might navigate intimacy by skirting it. People seem comforted by the belief that if we alienate or fail each other, robots will be there, programmed to provide simulations of love.[9] Our population is aging; there will be robots to take care of us. Our children are neglected; robots will tend to them. We are too exhausted to deal with each other in adversity; robots will have the energy. Robots won't be judgmental. We will be accommodated. An older woman says of her robot dog, "It is better than a real dog. . . . It won't do dangerous things, and it won't betray you. . . . Also, it won't die suddenly and abandon you and make you very sad."[10]

The elderly are the first to have companionate robots aggressively marketed to them, but young people also see the merits of robotic companionship. These days, teenagers have sexual adulthood thrust upon them before they are ready to deal with the complexities of relationships. They are drawn to the comfort of connection without the demands of intimacy. This may lead them to a hookup—sex without commitment or even caring. Or it may lead to an online romance—companionship that can always be interrupted. Not surprisingly, teenagers are drawn to love stories in which full intimacy cannot occur—here I think of current passions for films and novels about high school vampires who cannot sexually consummate relationships for fear of hurting those they love. And teenagers are drawn to the idea of technological communion. They talk easily of robots that would be safe and predictable companions.[11]

These young people have grown up with sociable robot pets, the companions of their playrooms, which portrayed emotion, said they cared, and asked to be cared for.[12] We are psychologically programmed not only to nurture what we love but to love what we nurture. So even simple artificial creatures can provoke heartfelt attachment. Many teenagers anticipate that the robot toys of their childhood will give way to full-fledged machine companions. In the psychoanalytic tradition, a symptom addresses a conflict but distracts us from understanding or resolving it; a dream expresses a wish.[13] Sociable robots serve as both symptom and dream: as a symptom, they promise a way to sidestep conflicts about intimacy; as a dream, they express a wish for relationships with limits, a way to be both together and alone.[14]

Some people even talk about robots as providing respite from feeling overwhelmed by technology. In Japan, companionate robots are specifically marketed as a way to seduce people out of cyberspace; robots plant a new flag in the physical real. If the problem is that too much technology has made us busy and anxious, the solution will be another technology that will organize, amuse, and relax us. So, although historically robots provoked anxieties about technology out of control, these days they are more likely to represent the reassuring idea that in a world of problems, science will offer solutions.[15] Robots have become a twenty-first-century deus ex machina. Putting hope in robots expresses an enduring technological optimism, a belief that as other things go wrong, science will go right. In a complicated world, robots seem a simple salvation. It is like calling in the cavalry.

But this is not a book about robots. Rather, it is about how we are changed as technology offers us substitutes for connecting with each other face-to-face. We are offered robots and a whole world of machine-mediated relationships on networked devices. As we instant-message, e-mail, text, and Twitter, technology redraws the boundaries between intimacy and solitude. We talk of getting "rid" of our e-mails, as though these notes are so much excess baggage. Teenagers avoid making telephone calls, fearful that they "reveal too much." They would rather text than talk. Adults, too, choose keyboards over the human voice. It is more efficient, they say. Things that happen in "real time" take too much time. Tethered to technology, we are shaken when that world "unplugged" does not signify, does not satisfy. After an evening of avatar-to-avatar talk in a networked game, we feel, at one moment, in possession of a full social life and, in the next, curiously isolated, in tenuous complicity with strangers. We build a following on Facebook or MySpace and wonder to what degree our followers are friends. We recreate ourselves as online personae and give ourselves new bodies, homes, jobs, and romances. Yet, suddenly, in the half-light of virtual community, we may feel utterly alone. As we distribute ourselves, we may abandon ourselves. Sometimes people experience no sense of having communicated after hours of connection. And they report feelings of closeness when they are paying little attention. In all of this, there is a nagging question: Does virtual intimacy degrade our experience of the other kind and, indeed, of all encounters, of any kind?

The blurring of intimacy and solitude may reach its starkest expression when a robot is proposed as a romantic partner. But for most people it begins when one creates a profile on a social-networking site or builds a persona or avatar for a game or virtual world.[16] Over time, such performances of identity may feel like identity itself. And this is where robotics and the networked life first intersect. For the performance of caring is all that robots, no matter how sociable, know how to do.

I was enthusiastic about online worlds as "identity workshops" when they first appeared, and all of their possibilities remain.[17] Creating an avatar—perhaps of a different age, a different gender, a different temperament—is a way to explore the self. But if you're spending three, four, or five hours a day in an online game or virtual world (a time commitment that is not unusual), there's got to be someplace you're not. And that someplace you're not is often with your family and friends—sitting around, playing Scrabble face-to-face, taking a walk, watching a movie together in the old-fashioned way. And with performance can come disorientation. You might have begun your online life in a spirit of compensation. If you were lonely and isolated, it seemed better than nothing. But online, you're slim, rich, and buffed up, and you feel you have more opportunities than in the real world. So, here, too, better than nothing can become better than something—or better than anything. Not surprisingly, people report feeling let down when they move from the virtual to the real world. It is not uncommon to see people fidget with their smartphones, looking for virtual places where they might once again be more.

Sociable robots and online life both suggest the possibility of relationships the way we want them. Just as we can program a made-to-measure robot, we can reinvent ourselves as comely avatars. We can write the Facebook profile that pleases us. We can edit our messages until they project the self we want to be. And we can keep things short and sweet. Our new media are well suited for accomplishing the rudimentary. And because this is what technology serves up, we reduce our expectations of each other. An impatient high school senior says, "If you really need to reach me, just shoot me a text." He sounds just like my colleagues on a consulting job, who tell me they would prefer to communicate with "real-time texts."

Our first embrace of sociable robotics (both the idea of it and its first exemplars) is a window onto what we want from technology and what we are willing to do to accommodate it. From the perspective of our robotic dreams, networked life takes on a new cast. We imagine it as expansive. But we are just as fond of its constraints. We celebrate its "weak ties," the bonds of acquaintance with people we may never meet. But that does not mean we prosper in them.[18] We often find ourselves standing depleted in the hype. When people talk about the pleasures of these weak-tie relationships as "friction free," they are usually referring to the kind of relationships you can have without leaving your desk. Technology ties us up as it promises to free us up. Connectivity technologies once promised to give us more time. But as the cell phone and smartphone eroded the boundaries between work and leisure, all the time in the world was not enough. Even when we are not "at work," we experience ourselves as "on call"; pressed, we want to edit out complexity and "cut to the chase."

CONNECTIVITY AND ITS DISCONTENTS

Online connections were first conceived as a substitute for face-to-face contact, when the latter was for some reason impractical: Don't have time to make a phone call? Shoot off a text message. But very quickly, the text message became the connection of choice. We discovered the network—the world of connectivity—to be uniquely suited to the overworked and overscheduled life it makes possible. And now we look to the network to defend us against loneliness even as we use it to control the intensity of our connections. Technology makes it easy to communicate when we wish and to disengage at will.

A few years ago at a dinner party in Paris, I met Ellen, an ambitious, elegant young woman in her early thirties, thrilled to be working at her dream job in advertising. Once a week, she would call her grandmother in Philadelphia using Skype, an Internet service that functions as a telephone with a Web camera. Before Skype, Ellen's calls to her grandmother were costly and brief. With Skype, the calls are free and give the compelling sense that the other person is present—Skype is an almost real-time video link. Ellen could now call more frequently: "Twice a week and I stay on the call for an hour," she told me. It should have been rewarding; instead, when I met her, Ellen was unhappy. She knew that her grandmother was unaware that Skype allows surreptitious multitasking.

Her grandmother could see Ellen's face on the screen but not her hands. Ellen admitted to me, "I do my e-mail during the calls. I'm not really paying attention to our conversation."

Ellen's multitasking removed her to another place. She felt her grandmother was talking to someone who was not really there. During their Skype conversations, Ellen and her grandmother were more connected than they had ever been before, but at the same time, each was alone. Ellen felt guilty and confused: she knew that her grandmother was happy, even if their intimacy was now, for Ellen, another task among multitasks.

I have often observed this distinctive confusion: these days, whether you are online or not, it is easy for people to end up unsure if they are closer together or further apart. I remember my own sense of disorientation the first time I realized that I was "alone together." I had traveled an exhausting thirty-six hours to attend a conference on advanced robotic technology held in central Japan. The packed grand ballroom was Wi-Fi enabled: the speaker was using the Web for his presentation, laptops were open throughout the audience, fingers were flying, and there was a sense of great concentration and intensity. But not many in the audience were attending to the speaker. Most people seemed to be doing their e-mail, downloading files, and surfing the Net. The man next to me was searching for a *New Yorker* cartoon to illustrate his upcoming presentation. Every once in a while, audience members gave the speaker some attention, lowering their laptop screens in a kind of curtsy, a gesture of courtesy.

Outside, in the hallways, the people milling around me were looking past me to virtual others. They were on their laptops and their phones, connecting to colleagues at the conference going on around them and to others around the globe. There but not there. Of course, clusters of people chatted with each other, making dinner plans, "networking" in that old sense of the word, the one that implies having a coffee or sharing a meal. But at this conference, it was clear that what people mostly want from public space is to be alone with their personal networks. It is good to come together physically, but it is more important to stay tethered to our devices. I thought of how Sigmund Freud considered the power of communities both to shape and to subvert us, and a psychoanalytic pun came to mind: "connectivity and its discontents."

The phrase comes back to me months later as I interview management consultants who seem to have lost touch with their best instincts for what makes them competitive. They complain about the BlackBerry revolution, yet accept it as inevitable while decrying it as corrosive. They say they used to talk to each other as they waited to give presentations or took taxis to the airport; now they spend that time doing e-mail. Some tell me they are making better use of their "downtime," but they argue without conviction. The time that they once used to talk as they waited for appointments or drove to the airport was never downtime. It was the time when far-flung global teams solidified relationships and refined ideas.

In corporations, among friends, and within academic departments, people readily admit that they would rather leave a voicemail or send an e-mail than talk face-to-face. Some who say "I live my life on my BlackBerry" are forthright about avoiding the "real-time" commitment of a phone call. The new technologies allow us to "dial down" human contact, to titrate its nature and extent. I recently overheard a conversation in a restaurant between two women. "No one answers the phone in our house anymore," the first woman proclaimed with some consternation. "It used to be that the kids would race to pick up the phone. Now they are up in their rooms, knowing no one is going to call them, and texting and going on Facebook or whatever instead." Parents with teenage children will be nodding at this very familiar story in recognition and perhaps a sense of wonderment that this has happened, and so quickly. And teenagers will simply be saying, "Well, what's your point?"

A thirteen-year-old tells me she "hates the phone and never listens to voice-mail." Texting offers just the right amount of access, just the right amount of control. She is a modern Goldilocks: for her, texting puts people not too close, not too far, but at just the right distance. The world is now full of modern Goldilockses, people who take comfort in being in touch with a lot of people whom they also keep at bay. A twenty-one-year-old college student reflects on the new balance: "I don't use my phone for calls any more. I don't have the time to just go on and on. I like texting, Twitter, looking at someone's Facebook wall. I learn what I need to know."

Randy, twenty-seven, has a younger sister—a Goldilocks who got her distances wrong. Randy is an American lawyer now working in California. His family lives in New York, and he flies to the East Coast to see them three or four times a year. When I meet Randy, his sister Nora, twenty-four, had just announced her engagement and wedding date via e-mail to a list of friends and family. "That," Randy says to me bitterly, "is how I got the news." He doesn't know if he is more angry or hurt. "It doesn't feel right that she didn't call," he says. "I was getting ready for a trip home. Couldn't she have told me then? She's my sister, but I didn't have a private moment when she told me in person. Or at least a call, just the two of us. When I told her I was upset, she sort of understood, but laughed and said that she and her fiancé just wanted to do things simply, as simply as possible. I feel very far away from her."

Nora did not mean to offend her brother. She saw e-mail as efficient and did not see beyond. We have long turned to technology to make us more efficient in work; now Nora illustrates how we want it to make us more efficient in our private lives. But when technology engineers intimacy, relationships can be reduced to mere connections. And then, easy connection becomes redefined as intimacy. Put otherwise, cyberintimacies slide into cybersolitudes.

And with constant connection comes new anxieties of disconnection, a kind of panic. Even Randy, who longs for a phone call from Nora on such an important matter as her wedding, is never without his BlackBerry. He holds it in his hands during our entire conversation. Once, he puts it in his pocket. A few moments later, it comes out, fingered like a talisman. In interviews with young and old, I find people genuinely terrified of being cut off from the "grid." People say that the loss of a cell phone can "feel like a death." One television producer in her mid-forties tells me that without her smartphone, "I felt like I had lost my mind." Whether or not our devices are in use, without them we feel disconnected, adrift. A danger even to ourselves, we insist on our right to send text messages while driving our cars and object to rules that would limit the practice.[19]

Only a decade ago, I would have been mystified that fifteen-year-olds in my urban neighborhood, a neighborhood of parks and shopping malls, of front stoops and coffee shops, would feel the need to send and receive close to six thousand messages a month via portable digital devices or that best friends would assume that when they visited, it would usually be on the virtual real estate of Facebook.[20] It might have seemed intrusive, if not illegal, that my mobile phone would tell me the location of all my acquaintances within a ten-mile radius.[21] But these days we are accustomed to all this. Life in a media bubble has come to seem natural. So has the end of a certain public etiquette: on the street, we speak into the invisible microphones on our mobile phones and appear to be talking to ourselves. We share intimacies with the air as though unconcerned about who can hear us or the details of our physical surroundings.

I once described the computer as a second self, a mirror of mind. Now the metaphor no longer goes far enough. Our new devices provide space for the emergence of a new state of the self, itself, split between the screen and the physical real, wired into existence through technology.

Teenagers tell me they sleep with their cell phone, and even when it isn't on their person, when it has been banished to the school locker, for instance, they know when their phone is vibrating. The technology has become like a phantom limb, it is so much a part of them. These young people are among the first to grow up with an expectation of continuous connection: always on, and always on them. And they are among the first to grow up not necessarily thinking of simulation as second best. All of this makes them fluent with technology but brings a set of new insecurities. They nurture friendships on social-networking sites and then wonder if they are among friends. They are connected all day but are not sure if they have communicated. They become confused about companionship. Can they find it in their lives on the screen? Could they find it with a robot? Their digitized friendships—played out with emoticon emotions, so often predicated on rapid response rather than reflection—may prepare them, at times through nothing more than their superficiality, for relationships that could bring superficiality to a higher power, that is, for relationships with the inanimate. They come to accept lower expectations for connection and, finally, the idea that robot friendships could be sufficient unto the day.

Overwhelmed by the volume and velocity of our lives, we turn to technology to help us find time. But technology makes us busier than ever and ever more in search of retreat. Gradually, we come to see our online life as life itself. We come to see what robots offer as relationship. The simplification of relationship is no longer a source of complaint. It becomes what we want. These seem the gathering clouds of a perfect storm.

Technology reshapes the landscape of our emotional lives, but is it offering us the lives we want to lead? Many roboticists are enthusiastic about having robots tend to our children and our aging parents, for instance. Are these psychologically, socially, and ethically acceptable propositions? What are our responsibilities here? And are we comfortable with virtual environments that propose themselves not as places for recreation but as new worlds to live in? What do we have, now that we have what we say we want—now that we have what technology makes easy?[22] This is the time to begin these conversations, together. It is too late to leave the future to the futurists.

ROMANCING THE MACHINE: TWO STORIES

I tell two stories in *Alone Together*: today's story of the network, with its promise to give us more control over human relationships, and tomorrow's story of sociable robots, which promise relationships where we will be in control, even if that means not being in relationships at all. I do not tell tomorrow's story to predict an exotic future. Rather, as a dream in development, sociable robots cast new light on our current circumstances. Our willingness to consider their company says a lot about the dissatisfactions we feel in our networked lives today.

Part One, "The Robotic Moment," moves from the sociable robots in children's playrooms to the more advanced ones in the laboratory and those being developed and deployed for elder care. As the robots become more complex, the intensity of our relationships to them ramps up. I begin my story with a kind of prehistory, going back to the late 1970s and early 1980s and the introduction of the first animated, interactive computer toys into children's lives. It was a time of curiosity about the nature of these new machines. These first computational objects of the playroom provoked a change in children's way of sorting out the question of aliveness. Decisions about whether something was alive would no longer turn on how something moved but on what it knew: physics gave way to psychology. This set the stage for how in the late 1990s, the ground would shift again when children met sociable robots that asked for care. Unlike traditional dolls, the robots wouldn't thrive without attention, and they let you know how you were doing. But even the most primitive of these objects—Tamagotchis and Furbies—made children's evaluation of aliveness less about cognition than about an object's seeming potential for mutual affection. If something asks for your care, you don't want to analyze it but take it "at interface value." It becomes "alive enough" for relationship.

And with this, the heightened expectations begin. Now—for adults and children—robots are not seen as machines but as "creatures," and then, for most people, the quotation marks are dropped. Curiosity gives way to a desire to care, to nurture. From there, we look toward companionship and more. So, for example, when sociable robots are given to the elderly, it is with the suggestion that robots will cure the troubles of their time of life. We go from curiosity to a search for communion. In the company of the robotic, people are alone, yet feel connected: *in solitude, new intimacies.*

Part Two, "Networked," turns to the online life as it reshapes the self. I acknowledge the many positive things that the network has to offer—enhancing friendship, family connections, education, commerce, and recreation. The triumphalist narrative of the Web is the reassuring story that people want to hear and that technologists want to tell. But the heroic story is not the whole story. In virtual worlds and computer games, people are flattened into personae. On social networks, people are reduced to their profiles. On our mobile devices, we often talk to each other on the move and with little disposable time—so little, in fact, that we communicate in a new language of abbreviation in which letters stand for words and emoticons for feelings. We don't ask the open ended "How are you?" Instead, we ask the more limited "Where are you?" and "What's up?" These are good questions for getting someone's location and making a simple plan. They are not so good for opening a dialogue about complexity of feeling. We are increasingly connected to each other but oddly more alone: *in intimacy, new solitudes.*

In the conclusion, I bring my stories together. Relationships with robots are ramping up; relationships with people are ramping down. What road are we travelling? Technology presents itself as a one-way street; we are likely to dismiss discontents about its direction because we read them as growing out of nostalgia or a Luddite impulse or as simply in vain. But when we ask what we "miss," we may discover what we care about, what we believe to be worth protecting. We prepare ourselves not necessarily to reject technology but to shape it in ways that honor what we hold dear. Winston Churchill said, "We shape our buildings and then they shape us."[23] We make our technologies, and they, in turn, shape us. So, of every technology we must ask, Does it serve our human purposes?—a question that causes us to reconsider what these purposes are. Technologies, in every generation, present opportunities to reflect on our values and direction. I intend *Alone Together* to mark a time of opportunity.

I turn now to the story of the robotic moment. It must begin with objects of the playroom because it is there that a generation was introduced to the idea that machines might be partners in mutual affection. But my story is not about child's play. We are on the verge of seeking the company and counsel of sociable robots as a natural part of life. Before we cross this threshold, we should ask why we are doing so. It is one thing to design a robot for an instrumental purpose: to search for explosives in a war zone or, in a more homely register, to vacuum floors and wash dishes. But the robots in this book are designed to *be* with us. As some of the children ask, we must ask, Why do people no longer suffice?

What are we thinking about when we are thinking about robots? We are thinking about the meaning of being alive, about the nature of attachment, about what makes a person. And then, more generally, we are rethinking, What is a relationship? We reconsider intimacy and authenticity. What are we willing to give up when we turn to robots rather than humans? To ask these questions is not to put robots down or deny that they are engineering marvels; it is only to put them in their place.

In the 1960s through the 1980s, debates about artificial intelligence centered on the question of whether machines could "really" be intelligent. These discussions were about the objects themselves, what they could and could not do. Our new encounters with sociable robots—encounters that began in the past decade with the introduction of simple robot toys into children's playrooms—provoke responses that are not about these machines' capabilities but our vulnerabilities. As we will see, when we are asked to care for an object, when an object thrives under our care, we experience that object as intelligent, but, more importantly, we feel ourselves to be in a relationship with it. The attachments I describe do not follow from whether computational objects really have emotion or intelligence, because they do not. The attachments follow from what they evoke in their users. Our new objects don't so much "fool us" into thinking they are communicating with us; roboticists have learned those few triggers that help us fool ourselves. We don't need much. We are ready to enter the romance.

PART ONE

The Robotic Moment

In Solitude, New Intimacies

nearest neighbors

My first brush with a computer program that offered companionship was in the mid-1970s. I was among MIT students using Joseph Weizenbaum's ELIZA, a program that engaged in dialogue in the style of a psychotherapist. So, a user typed in a thought, and ELIZA reflected it back in language that offered support or asked for clarification.[1] To "My mother is making me angry," the program might respond, "Tell me more about your mother," or perhaps, "Why do you feel so negatively about your mother?" ELIZA had no model of what a mother might be or any way to represent the feeling of anger. What it could do was take strings of words and turn them into questions or restate them as interpretations.

Weizenbaum's students knew that the program did not know or understand; nevertheless they wanted to chat with it. More than this, they wanted to be alone with it. They wanted to tell it their secrets.[2] Faced with a program that makes the smallest gesture suggesting it can empathize, people want to say something true. I have watched hundreds of people type a first sentence into the primitive ELIZA program. Most commonly they begin with "How are you today?" or "Hello." But four or five interchanges later, many are on to "My girlfriend left me," "I am worried that I might fail organic chemistry," or "My sister died."

Soon after, Weizenbaum and I were coteaching a course on computers and society at MIT. Our class sessions were lively. During class meetings he would rail against his program's capacity to deceive; I did not share his concern. I saw ELIZA as a kind of Rorschach, the psychologist's inkblot test. People used the program as a projective screen on which to express themselves. Yes, I thought, they engaged in personal conversations with ELIZA, but in a spirit of "as if." They spoke as if someone were listening but knew they were their own audience. They became caught up in the exercise. They thought, I will talk to this program as if it were a person. I will vent; I will rage; I will get things off my chest. More than this, while some learned enough about the program to trip it up, many more used this same inside knowledge to feed ELIZA responses that would make it seem more lifelike. They were active in keeping the program in play.

Weizenbaum was disturbed that his students were in some way duped by the program into believing—against everything they knew to be true—that they were dealing with an intelligent machine. He felt almost guilty about the deception machine he had created. But his worldly students were not deceived. They knew all about ELIZA's limitations, but they were eager to "fill in the blanks." I came to think of this human complicity in a digital fantasy as the "ELIZA effect." Through the 1970s, I saw this complicity with the machine as no more threatening than wanting to improve the working of an interactive diary. As it turned out, I underestimated what these connections augured. At the robotic moment, more than ever, our willingness to engage with the inanimate does not depend on being deceived but on wanting to fill in the blanks.

Now, over four decades after Weizenbaum wrote the first version of ELIZA, artificial intelligences known as "bots" present themselves as companions to the millions who play computer games on the Internet. Within these game worlds, it has come to seem natural to "converse" with bots about a variety of matters, from routine to romantic. And, as it turns out, it's a small step from having your "life" saved by a bot you meet in a virtual world to feeling a certain affection toward it—and not the kind of affection you might feel toward a stereo or car, no matter how beloved. Meantime, in the physical real, things proceed apace. The popular Zhu Zhu robot pet hamsters come out of the box in "nurturing mode." The official biography of the Zhu Zhu named Chuck says, "He lives to feel the love." For the elderly, the huggable baby seal robot Paro is now on sale. A hit in Japan, it now targets the American nursing home market. Roboticists make the case that the elderly need a companion robot because of a lack of human resources. Almost by definition, they say, robots will make things better.

While some roboticists dream of reverse engineering love, others are content to reverse engineer sex.[3] In February 2010, I googled the exact phrase "sex robots" and came up with 313,000 hits, the first of which was linked to an article titled "Inventor Unveils $7,000 Talking Sex Robot." Roxxxy, I learned, "may be the world's most sophisticated, talking sex robot."[4] The shock troops of the robotic moment, dressed in lingerie, may be closer than most of us have ever imagined. And true to the ELIZA effect, this is not so much because the robots are ready but because we are.

In a television news story about a Japanese robot designed in the form of a sexy woman, a reporter explains that although this robot currently performs only as a receptionist, its designers hope it will someday serve as a teacher and companion. Far from skeptical, the reporter bridges the gap between the awkward robot before him and the idea of something akin to a robot wife by referring to the "singularity." He asks the robot's inventor, "When the singularity comes, no one can imagine where she [the robot] could go. Isn't that right? . . . What about these robots after the singularity? Isn't it the singularity that will bring us the robots that will surpass us?"

The singularity? This notion has migrated from science fiction to engineering. The *singularity* is the moment—it is mythic; you have to believe in it—when machine intelligence crosses a tipping point.[5] Past this point, say those who believe, artificial intelligence will go beyond anything we can currently conceive. No matter if today's robots are not ready for prime time as receptionists. At the singularity, everything will become technically possible, including robots that love. Indeed, at the singularity, we may merge with the robotic and achieve immortality. The singularity is technological rapture.

As for Weizenbaum's concerns that people were open to computer psychotherapy, he correctly sensed that something was going on. In the late 1970s, there was considerable reticence about computer psychotherapy, but soon after, opinions shifted.[6] The arc of this story does not reflect new abilities of machines to understand people, but people's changing ideas about psychotherapy and the workings of their own minds, both seen in more mechanistic terms.[7] Thirty years ago, with psychoanalysis more central to the cultural conversation, most people saw the experience of therapy as a context for coming to see the story of your life in new terms. This happened through gaining insight and developing a relationship with a therapist who provided a safe place to address knotty problems. Today, many see psychotherapy less as an investigation of the meaning of our lives and more as an exercise to achieve behavioral change or work on brain chemistry. In this model, the computer becomes relevant in several ways. Computers can help with diagnosis, be set up with programs for cognitive behavioral therapy, and provide information on alternative medications.

Previous hostility to the idea of the computer as psychotherapist was part of a "romantic reaction" to the computer presence, a sense that there were some places a computer could not and should not go. In shorthand, the romantic reaction said, "Simulated thinking might be thinking, but simulated feeling is not feeling; simulated love is never love." Today, that romantic reaction has largely given way to a new pragmatism. Computers "understand" as little as ever about human experience—for example, what it means to envy a sibling or miss a deceased parent. They do, however, perform understanding better than ever, and we are content to play our part. After all, our online lives are all about performance. We perform on social networks and direct the performances of our avatars in virtual worlds. A premium on performance is the cornerstone of the robotic moment. We live the robotic moment not because we have companionate robots in our lives but because the way we contemplate them on the horizon says much about who we are and who we are willing to become.

How did we get to this place? The answer to that question is hidden in plain sight, in the rough-and-tumble of the playroom, in children's reactions to robot toys. As adults, we can develop and change our opinions. In childhood, we establish the truth of our hearts.

I have watched three decades of children with increasingly sophisticated computer toys. I have seen these toys move from being described as "sort of alive" to "alive enough," the language of the generation whose childhood play was with sociable robots (in the form of digital pets and dolls). Getting to "alive enough" marks a watershed. In the late 1970s and early 1980s, children tried to make philosophical distinctions about aliveness in order to categorize computers. These days, when children talk about robots as alive enough for *specific purposes*, they are not trying to settle abstract questions. They are being pragmatic: different robots can be considered on a case-by-case and context-by-context basis. (Is it alive enough to be a friend, a babysitter, or a companion for your grandparents?) Sometimes the question becomes more delicate: If a robot makes you love it, is it alive?

LIFE RECONSIDERED

In the late 1970s and early 1980s, children met their first computational objects: games like Merlin, Simon, and Speak & Spell. This first generation of computers in the playroom challenged children in memory and spelling games, routinely beating them at tic-tac-toe and hangman.[8] The toys, reactive and interactive, turned children into philosophers. Above all else, children asked themselves whether something programmed could be alive.

Children's starting point here is their animation of the world. Children begin by understanding the world in terms of what they know best: themselves. Why does the stone roll down the slope? "To get to the bottom," says the young child, as though the ball had its own desires. But in time, animism gives way to physics. The child learns that a stone falls because of gravity; intentions have nothing to do with it. And so a dichotomy is constructed: physical and psychological properties stand opposed to one another in two great systems. But the computer is a new kind of object: it is psychological and yet a thing. Marginal objects such as the computer, on the lines between categories, draw attention to how we have drawn the lines.[9]

Swiss psychologist Jean Piaget, interviewing children in the 1920s, found that they took up the question of an object's life status by considering its physical movement.[10] For the youngest children, everything that could move was alive, then only things that could move without an outside push or pull. People and animals were easily classified. But clouds that seemed to move on their own accord were classified as alive until children realized that wind, an external but invisible force, was pushing them along. Cars were reclassified as not alive when children understood that motors counted as an "outside" push. Finally, the idea of autonomous movement became focused on breathing and metabolism, the motions most particular to life.

In the 1980s, faced with computational objects, children began to think through the question of aliveness in a new way, shifting from physics to psychology.[11] When they considered a toy that could beat them at spelling games, they were interested not in whether such an object could move on its own but in whether it could think on its own. Children asked if this game could "know." Did it cheat? Was knowing part of cheating? They were fascinated by how electronic games and toys showed a certain autonomy. When an early version of Speak & Spell—a toy that played language and spelling games—had a programming bug and could not be turned off during its "say it" routine, children shrieked with excitement, finally taking out the game's batteries to "kill it" and then (with the reinsertion of the batteries) bring it back to life.

In their animated conversations about computer life and death, children of the 1980s imposed a new conceptual order on a new world of objects.[12] In the 1990s, that order was strained to the breaking point. Simulation worlds—for example the Sim games—pulsed with evolving life forms. And child culture was awash in images of computational objects (from Terminators to digital viruses) all shape-shifting and morphing in films, cartoons, and action figures. Children were encouraged to see the stuff of computers as the same stuff of which life is made. One eight-year-old girl referred to mechanical life and human life as "all the same stuff, just yucky computer 'cy-dough-plasm.'" All of this led to a new kind of conversation about aliveness. Now, when considering computation, children talked about evolution as well as cognition. And they talked about a special kind of mobility. In 1993, a ten-year-old considered whether the creatures on the game SimLife were alive. She decided they were "if they could get out of your computer and go to America Online."[13]

Here, Piaget's narrative about motion resurfaced in a new guise. Children often imbued the creatures in simulation games with a desire to escape their confines and enter a wider digital world. And then, starting in the late 1990s, digital "creatures" came along that tried to dazzle children not with their smarts but with their sociability. I began a long study of children's interactions with these new machines. Of course, children said that a sociable robot's movement and intelligence were signs of its life. But even in conversations specifically about aliveness, children were more concerned about what these new robots might feel. As criteria for life, everything pales in comparison to a robot's capacity to care.

Consider how often thoughts turn to feelings as three elementary school children discuss the aliveness of a Furby, an owl-like creature that plays games and seems to learn English under a child's tutelage. A first, a five-year-old girl, can only compare it to a Tamagotchi, a tiny digital creature on an LED screen that also asks to be loved, cared for, and amused. She asks herself, "Is it [the Furby] alive?" and answers, "Well, I love it. It's more alive than a Tamagotchi because it sleeps with me. It likes to sleep with me." A six-year-old boy believes that something "as alive as a Furby" needs arms: "It might want to pick up something or to hug me." A nine-year-old girl thinks through the question of a Furby's aliveness by commenting, "I really like to take care of it. . . . It's as alive as you can be if you don't eat. . . . It's not like an animal kind of alive."

From the beginning of my studies of children and computers in the late 1970s, children spoke about an "animal kind of alive" and a "computer kind of alive." Now I hear them talk about a "people kind of love" and a "robot kind of love." Sociable robots bring children to the locution that the machines are alive enough to care and be cared for. In speaking about sociable robots, children use the phrase "alive enough" as a measure not of biological readiness but of relational readiness. Children describe robots as alive enough to love and mourn. And robots, as we saw at the American Museum of Natural History, may be alive enough to substitute for the biological, depending on the context. One reason the children at the museum were so relaxed about a robot substituting for a living tortoise is that children were comfortable with the idea of a robot as both machine and creature. I see this flexibility in seven-year-old Wilson, a bright, engaged student at a Boston public elementary school where I bring robot toys for after-school play. Wilson reflects on a Furby I gave him to take home for several weeks: "The Furby can talk, and it looks like an owl," yet "I always hear the machine in it." He knows, too, that the Furby, "alive enough to be a friend," would be rejected in the company of animals: "A real owl would snap its head off." Wilson does not have to deny the Furby's machine nature to feel it would be a good friend or to look to it for advice. His Furby has become his confidant. Wilson's way of keeping in mind the dual aspects of the Furby's nature seems to me a philosophical version of multitasking, so central to our twentieth-century attentional ecology. His attitude is pragmatic. If something that seems to have a self is before him, he deals with the aspect of self he finds most relevant to the context.

This kind of pragmatism has become a hallmark of our psychological culture. In the mid-1990s, I described how it was commonplace for people to "cycle through" different ideas of the human mind as (to name only a few images) mechanism, spirit, chemistry, and vessel for the soul.[14] These days, the cycling through intensifies. We are in much more direct contact with the machine side of mind. People are fitted with a computer chip to help with Parkinson's. They learn to see their minds as program and hardware. They take antidepressants prescribed by their psychotherapists, confident that the biochemical and oedipal self can be treated in one room. They look for signs of emotion in a brain scan. Old jokes about couples needing "chemistry" turn out not to be jokes at all. The compounds that trigger romantic love are forthcoming from the laboratory. And yet, even with biochemical explanations for attraction, nothing seems different about the thrill of falling in love. And seeing that an abused child has a normal brain scan does not mean one feels any less rage about the abuse. Pluralistic in our attitudes toward the self, we turn this pragmatic sensibility toward other things in our path—for example, sociable robots. We approach them like Wilson: they can be machines, and they can be more.

Writing in his diary in 1832, Ralph Waldo Emerson described "dreams and beasts" as "two keys by which we are to find out the secrets of our nature. . . . They are our test objects."[15] If Emerson had lived today, he would have seen the sociable robot as our new test object. Poised in our perception between inanimate program and living creature, this new breed of robot provokes us to reflect on the difference between connection and relationship, involvement with an object and engagement with a subject. These robots are evocative: understanding how people think about them provides a view onto how we think about ourselves. When children talk about these robots, they move away from an earlier cohort's perception of computers as provocative curiosities to the idea that robots might be something to grow old with. It all began when children met the seductive Tamagotchis and Furbies, the first computers that asked for love.[16]

THE TAMAGOTCHI PRIMER

When active and interactive computer toys were first introduced in the late 1970s, children recognized that they were neither dolls nor people nor animals. Nor did they seem like machines. Computers, first in the guise of electronic toys and games, turned children into philosophers, caught up in spontaneous debates about what these objects might be. In some cases, their discussions brought them to the idea that the talking, clever computational objects were close to kin. Children consider the question of what is special about being a person by contrasting themselves with their "nearest neighbors." Traditionally, children took their nearest neighbors to be their dogs, cats, and horses. Animals had feelings; people were special because of their ability to think. So, the Aristotelian definition of man as a rational animal had meaning for even the youngest children. But by the mid-1980s, as thinking computers became nearest neighbors, children considered people special because only they could "feel." Computers were intelligent machines; in contrast, people were emotional machines.[17]

But in the late 1990s, as if on cue, children met objects that presented themselves as having feelings and needs. As emotional machines, people were no longer alone. Tamagotchis and Furbies (both of which sold in the tens of millions) did not want to play tic-tac-toe, but they would tell you if they were hungry or unhappy. A Furby held upside down says, "Me scared," and whimpers as though it means it. And these new objects found ways to express their love.

Furbies, put on the market in 1998, had proper robotic "bodies"; they were small, fur-covered "creatures" with big eyes and ears. Yet, the Tamagotchi, released in 1997, a virtual creature housed in a plastic egg, serves as a reliable primer in the psychology of sociable robotics—and a useful one because crucial elements are simplified, thus stark. The child imagines Tamagotchis as embodied because, like living creatures and unlike machines, they need constant care and are always on. A Tamagotchi has "body enough" for a child to imagine its death.[18] To live, a Tamagotchi must be fed, amused, and cleaned up after. If cared for, it will grow from baby to healthy adult. Tamagotchis, in their limited ways, develop different personalities depending on how they are treated. As Tamagotchis turn children into caretakers, they teach that digital life can be emotionally roiling, a place of obligations and regrets.[19] The earliest electronic toys and games of thirty years ago—such as Merlin, Simon, and Speak & Spell—encouraged children to consider the proposition that something smart might be "sort of alive." With Tamagotchis, needy objects asked for care, and children took further steps.

As they did with earlier generations of hard-to-classify computational objects, curious children go through a period of trying to sort out the new sociable objects. But soon children take them at interface value, not as puzzles but as playmates. The philosophical churning associated with early computer toys (are they alive? do they know?) quickly gives way to new practices. Children don't want to comprehend these objects as much as take care of them. Their basic stance: "I'm living with this new creature. It and many more like it are here to stay." When a virtual "creature" or robot asks for help, children provide it. When its behavior dazzles, children are pleased just to hang out with it.

In the classic children's story *The Velveteen Rabbit*, a stuffed animal becomes "real" because of a child's love. Tamagotchis do not wait passively but demand attention and claim that without it they will not survive. With this aggressive demand for care, the question of biological aliveness almost falls away. We love what we nurture; if a Tamagotchi makes you love it, and you feel it loves you in return, it is alive enough to be a creature. It is alive enough to share a bit of your life. Children approach sociable machines in a spirit similar to the way they approach sociable pets or people—with the hope of befriending them. Meeting a person (or a pet) is not about meeting his or her biochemistry; becoming acquainted with a sociable machine is not about deciphering its programming. While in an earlier day, children might have asked, "What is a Tamagotchi?" they now ask, "What does a Tamagotchi want?"

When a digital "creature" asks children for nurturing or teaching, it seems alive enough to care for, just as caring for it makes it seem more alive. Neil, seven, says that his Tamagotchi is "like a baby. You can't just change the baby's diaper. You have to, like, rub cream on the baby. That is how the baby knows you love it." His eight-year-old sister adds, "I hate it when my Tamagotchi has the poop all around. I am like its mother. That is my job. I don't like it really, but it gets sick if you just leave it messy." Three nine-year-olds consider their Tamagotchis. One is excited that his pet requires him to build a castle as its home. "I can do it. I don't want him to get cold and sick and to die." Another looks forward to her digital pet's demands: "I like it when it says, 'I'm hungry' or 'Play with me.'" The third boils down her relationship to a "deceased" Tamagotchi to its most essential elements: "She was loved; she loved back."[20]

Where is digital fancy bred? Most of all, in the demand for care. Nurturance is the "killer app." In the presence of a needy Tamagotchi, children become responsible parents: demands translate into care and care into the feeling of caring. Parents are enlisted to watch over Tamagotchis during school hours. In the late 1990s, an army of compliant mothers cleaned, fed, and amused their children's Tamagotchis; the beeping of digital pets became a familiar background noise during business meetings.

This parental involvement is imperative because a Tamagotchi is always on. Mechanical objects are supposed to turn off. Children understand that bodies need to be always on, that they become "off" when people or animals die. So, the inability to turn off a Tamagotchi becomes evidence of its life. Seven-year-old Catherine explains, "When a body is 'off,' it is dead." Some Tamagotchis can be asked to "sleep," but nine-year-old Parvati makes it clear that asking her Tamagotchi to sleep is not the same as hitting the pause button in a game. Life goes on: "When they sleep, it is not that they are turned off. They can still get sick and unhappy, even while they are sleeping. They could have a nightmare."

In the late 1970s, computers, objects on the boundary between animate and inanimate, began to lead children to gleeful experiments in which they crashed machines as they talked about "killing" them. And then, there would be elaborate rituals of resuscitation as children talked about bringing machines back to life. After these dramatic rebirths, the machines were, in the eyes of children, what they had been before. Twenty years later, when Tamagotchis die and are reset for a new life, children do not feel that they come back as they were before. Children looked forward to the rebirth of the computers they had crashed, but they dread the demise and rebirth of Tamagotchis. These provoke genuine remorse because, as one nine-year-old puts it, "It didn't have to happen. I could have taken better care."[21]

UNFORGETTABLE

I took care of my first Tamagotchi at the same time that my seven-year-old daughter was nurturing her own. Since I sometimes took a shift attending to her Tamagotchi, I could compare their respective behaviors, and I convinced myself that mine had idiosyncrasies that made it different from hers. My Tamagotchi liked to eat at particular intervals. I thought it prospered best with only small doses of amusement. I worked hard at keeping it happy. I did not anticipate how bad I would feel when it died. I immediately hit the reset button. Somewhat to my surprise, I had no desire to take care of the new infant Tamagotchi that appeared on my screen.

Many children are not so eager to hit reset. They don't like having a new creature in the same egg where their virtual pet has died. For them, the death of a virtual pet is not so unlike the death of what they call a "regular pet." Eight-year-olds talk about what happens when you hit a Tamagotchi's reset button. For one, "It comes back, but it doesn't come back as exactly your same Tamagotchi. . . . You haven't had the same experiences with it. It has a different personality." For another, "It's cheating. Your Tamagotchi is really dead. Your one is really dead. They say you get it back, but it's not the same one. It hasn't had the same things happen to it. It's like they give you a new one. It doesn't remember the life it had." For another, "When my Tamagotchi dies, I don't want to play with the new one who can pop up. It makes me remember the real one [the first one]. I like to get another [a new egg]. . . . If you made it die, you should start fresh." Parents try to convince their children to hit reset. Their arguments are logical: the Tamagotchi is not "used up"; a reset Tamagotchi means one less visit to the toy store. Children are unmoved.

Sally, eight, has had three Tamagotchis. Each died and was "buried" with ceremony in her top dresser drawer. Three times Sally has refused to hit the reset button and convinced her mother to buy replacements. Sally sets the scene: "My mom says mine still works, but I tell her that a Tamagotchi is cheap, and she won't have to buy me anything else, so she gets one for me. I am not going to start up my old one. It died. It needs its rest."

In Sally's "It died. It needs its rest," we see the expansiveness of the robotic moment. Things that never could go together—a program and pity for a weary body—now do go together. The reset button produces objects that are between categories: a creature that seems new but is not really new, a stand-in for something now gone. The new creature, a kind of imposter, is a classic case of Sigmund

Freud's uncanny—it's familiar, yet somehow not.[22] The uncanny is always compelling. Children ask, "What does it mean for a virtual creature to die?" Yet, while earlier generations debated questions about a computer's life in philosophical terms, when faced with Tamagotchis, children quickly move on to day-to-day practicalities. They temper philosophy with tearful experience. They know that Tamagotchis are alive enough to mourn.

Freud teaches us that the experience of loss is part of how we build a self.[23] Metaphorically, at least, mourning keeps a lost person present. Child culture is rich in narratives that take young people through the steps of this fitful process. So, in *Peter Pan*, Wendy loses Peter in order to move past adolescence and become a grown woman, able to love and parent. But Peter remains present in her playful and tolerant way of mothering. Louisa May Alcott's Jo loses her gentle sister Beth. In mourning Beth, Jo develops as a serious writer and finds a new capacity to love. More recently, the young wizard Harry Potter loses his mentor Dumbledore, whose continuing presence within Harry enables him to find his identity and achieve his life's purpose. With the Tamagotchi, we see the beginning of mourning for artificial life. It is not mourned as one would mourn a doll. The Tamagotchi has crossed a threshold. Children breathe life into their dolls. With the Tamagotchi, we are in a realm of objects that children see as having their own agendas, needs, and desires. Children mourn the life the Tamagotchi has led.

A child's mourning for a Tamagotchi is not always a solitary matter. When a Tamagotchi dies, it can be buried in an online Tamagotchi graveyard. The tombstones are intricate. On them, children try to capture what made each Tamagotchi special.[24] A Tamagotchi named Saturn lived to twelve "Tamagotchi years." Its owner writes a poem in its memory: "My baby died in his sleep. I will forever weep. Then his batteries went dead. Now he lives in my head." Another child mourns Pumpkin, dead at sixteen: "Pumpkin, Everyone said you were fat, so I made you lose weight. From losing weight you died. Sorry." Children take responsibility for virtual deaths.[25] These online places of mourning do more than give children a way to express their feelings. They sanction the idea that it is appropriate to mourn the digital—indeed, that there is something "there" to mourn.

alive enough

In the 1990s, children spoke about making their virtual creatures more alive by having them escape the computer. Furbies, the sensation of the 1998 holiday season, embody this documented dream. If a child wished a Tamagotchi to leap off its screen, it might look a lot like the furry and owl-like Furby. The two digital pets have other things in common. As with a Tamagotchi, how a Furby is treated shapes its personality. And both present themselves as visitors from other worlds. But Furbies are more explicit about their purpose in coming to Earth. They are here to learn about humans. So, each Furby is an anthropologist of sorts and wants to relate to people. They ask children to take care of them and to teach them English. Furbies are not ungrateful: they make demands, but they say, "I love you."

Furbies, like Tamagotchis, are "always on," but unlike Tamagotchis, Furbies manifest this with an often annoying, constant chatter.[1] To reliably quiet a Furby, you need a Phillips screwdriver to remove its batteries, an operation that causes it to lose all memory of its life and experiences—what it has learned and how it has been treated. For children who have spent many hours "bringing up" their Furbies, this is not a viable option. On a sunny spring afternoon in 1999, I bring eight Furbies to an afternoon playgroup at an elementary school in western Massachusetts. There are fifteen children in the room, from five to eight years old, from the kindergarten through the third grade. I turn on a tape recorder as I hand the Furbies around. The children start to talk excitedly, greeting the Furbies by imitating their voices. In the cacophony of the classroom, this is what the robotic moment sounds like:

> He's a baby! He said, "Yum." Mine's a baby? Is this a baby? Is he sleeping now? He burped! What is "be-pah?" He said, "Be-pah." Let them play together. What does "a lee koo wah" mean? Furby, you're talking to me. Talk! C'mon boy. Good boy! Furby, talk! Be quiet everybody! Oh, look it, he's in love with another one! Let them play together! It's tired. It's asleep. I'm going to try to feed him. How come they don't have arms? Look, he's in love! He called you "Mama." He said, "Me love you." I have to feed him. I have to feed mine too. We love you, Furby. How do you make him fall asleep? His eyes are closed. He's talking with his eyes closed. He's sleeptalking. He's dreaming. He's snoring. I'm giving him shade.
>
> C'mon, Furby, c'mon—let's go to sleep, Furby. Furby, shh, shh. Don't touch him. I can make him be quiet. This is a robot. Is this a robot? What has this kind of fur? He's allergic to me. It's kind of like it's alive. And it has a body. It has a motor. It's a monster. And it's kind of like it's real because it has a body. It was alive. It is alive. It's not alive. It's a robot.

From the very first, the children make it clear that the Furby is a machine but alive enough to need care. They try to connect with it using everything they have: the bad dreams and scary movies that make one child see the Furby "as a monster" and their understanding of loneliness, which encourages another to exhort, "Let them play together!" They use logic and skepticism: Do biological animals have "this kind of fur?" Do real animals have motors? Perhaps, although this requires a new and more expansive notion of what a motor can be. They use the ambiguity of this new object to challenge their understanding of what they think they already know. They become more open to the idea of the biological as mechanical and the mechanical as biological. Eight-year-old Pearl thinks that removing the batteries from a Furby causes it to die and that people's death is akin to "taking the batteries out of a Furby."

Furbies reinforce the idea that they have a biology: each is physically distinct, with particular markings on its fur, and each has some of the needs of living things. For example, a Furby requires regular feeding, accomplished by depressing its tongue with one's finger. If a Furby is not fed, it becomes ill. Nursing a Furby back to health always requires more food. Children give disease names to Furby malfunctions. So, there is Furby cancer, Furby flu, and Furby headache.

Jessica, eight, plays with the idea that she and her Furby have "body things" in common, for example, that headache. She has a Furby at home; when her sisters pull its hair, Jessica worries about its pain: "When I pull my hair it really hurts, like when my mother brushes the tangles. So, I think [the Furby's hair pulls] hurt too." Then, she ponders her stomach. "There's a screw in my belly button," she says. "[The screw] comes out, and then blood comes out." Jessica thinks that people, like Furbies, have batteries. "There are hearts, lungs, and a big battery inside." People differ from robots in that our batteries "work forever like the sun." When children talk about the Furby as kin, they experiment with the idea that they themselves might be almost machine. Ideas about the human as machine or as joined to a machine are played out in classroom games.[2] In their own way, toy robots prepare a bionic sensibility. There are people who do, after all, have screws and pins and chips and plates in their flesh. A recent recipient of a cochlear implant describes his experience of his body as "rebuilt."[3]

We have met Wilson, seven, comfortable with his Furby as both machine and creature. Just as he always "hears the machine" in the Furby, he finds the machine in himself. As the boy sings improvised love songs about the robot as a best friend, he pretends to use a screwdriver on his own body, saying, "I'm a Furby." Involved in a second-grade class project of repairing a broken Furby by dismantling it, screw by screw, Wilson plays with the idea of the Furby's biological nature: "I'm going to get [its] baby out." And then he plays with the idea of his own machine nature: he applies the screwdriver to his own ankle, saying, "I'm unscrewing my ankle."

Wilson enjoys cataloguing what he and the Furby have in common. Most important for Wilson is that they "both like to burp." In this, he says, the Furby "is just like me—I love burping." Wilson holds his Furby out in front of him, his hands lightly touching the Furby's stomach, staring intently into its eyes. He burps just after or just before his Furby burps, much as in the classic bonding scene in *E.T.: The Extraterrestrial* between the boy Elliott and the visitor from afar. When Wilson describes his burping game, he begins by saying that he makes his Furby burp, but he ends up saying that his Furby makes him burp. Wilson likes the sense that he and his Furby are in sync, that he can happily lose track of where he leaves off and the Furby begins.[4]

WHAT DOES A FURBY WANT?

When Wilson catalogues what he shares with his Furby, there are things of the body (the burping) and there are things of the mind. Like many children, he thinks that because Furbies have language, they are more "peoplelike" than a "regular" pet. They arrive speaking Furbish, a language with its own dictionary, which many children try to commit to memory because they would like to meet their Furbies more than half way. The Furby manual instructs children, "I can learn to speak English by listening to you talk. The more you play with me, the more I will use your language." Actually, Furby English emerges over time, whether or not a child talks to the robot. (Furbies have no hearing or language-learning ability.[5]) But until age eight, children are convinced by the illusion and believe they are teaching their Furbies to speak. The Furbies are alive enough to need them.

Children enjoy the teaching task. From the first encounter, it gives them something in common with their Furbies and it implies that the Furbies can grow to better understand them. "I once didn't know English," says one six-year-old. "And now I do. So I know what my Furby is going through." In the classroom with Furbies, children shout to each other in competitive delight: "My Furby speaks more English than yours! My Furby speaks English."

I have done several studies in which I send Furbies home with schoolchildren, often with the request that they (and their parents) keep a "Furby diary." In my first study of kindergarten to third graders, I loan the Furbies out for two weeks at a time. It is not a good decision. I do not count on how great will be children's sense of loss when I ask them to return the Furbies. I extend the length of the loans, often encouraged by parental requests. Their children have grown too attached to give up the robots. Nor are they mollified by parents' offers to buy them new Furbies. Even more so than with Tamagotchis, children attach to a particular Furby, the one they have taught English, the one they have raised.

For three decades, in describing people's relationships with computers, I have often used the metaphor of the Rorschach, the inkblot test that psychologists use as a screen onto which people can project their feelings and styles of thought. But as children interact with sociable robots like Furbies, they move beyond a psychology of projection to a new psychology of engagement. They try to deal with the robot as they would deal with a pet or a person. Nine-year-old Leah, in an after-school playgroup, admits, "It's hard to turn it [the Furby] off when it is talking to me." Children quickly understand that to get the most out of your Furby, you have to pay attention to what it is telling you. When you are with a Furby, you can't play a simple game of projective make-believe. You have to continually assess your Furby's "emotional" and "physical" state. And children fervently believe that the child who loves his or her Furby best will be most loved in return.

This mutuality is at the heart of what makes the Furby, a primitive exemplar of sociable robotics, different from traditional dolls. As we've seen, such relational artifacts do not wait for children to "animate" them in the spirit of a Raggedy Ann doll or a teddy bear. They present themselves as already animated and ready for relationship. They promise reciprocity because, unlike traditional dolls, they are not passive. They make demands. They present as having their own needs and inner lives. They teach us the rituals of love that will make them thrive. For decades computers have asked us to *think* with them; these days, computers and robots, deemed sociable, affective, and relational, ask us to *feel* for and with them.

Children see traditional dolls as they want them or need them to be. For example, an eight-year-old girl who feels guilty about breaking her mother's best crystal pitcher might punish a row of Barbie dolls. She might take them away from their tea party and put them in detention, doing unto the dolls what she imagines should be done unto her. In contrast, since relational artifacts present themselves as having minds and intentions of their own, they cannot be so easily punished for one's own misdeeds. Two eight-year-old girls comment on how their "regular dolls" differ from the robotic Furbies. The first says, "A regular doll, like my Madeleine doll . . . you can make it go to sleep, but its eyes are painted open, so, um, you cannot get them to close their eyes. . . . Like a Madeleine doll cannot go, 'Hello, good morning.'" But this is precisely the sort of thing a Furby can do. The second offers, "The Furby tells you what it wants."

Indeed, Furbies come with manuals that provide detailed marching orders. They want language practice, food, rest, and protestations of love. So, for example, the manual instructs, "Make sure you say 'HEY FURBY! I love you!' frequently so that I feel happy and know I'm loved." There is general agreement among children that a penchant for giving instructions distinguishes Furbies from traditional dolls. A seven-year-old girl puts it this way: "Dolls let you tell them what they want. The Furbies have their own ideas." A nine-year-old boy sums up the difference between Furbies and his action figures: "You don't play with the Furby, you sort of hang out with it. You do try to get power over it, but it has power over you too."

Children say that traditional dolls can be "hard work" because you have to do all the work of giving them ideas; Furbies are hard work for the opposite reason.

They have plenty of ideas, but you have to give them what they want and when they want it. When children attach to a doll through the psychology of projection, they attribute to the doll what is most on their mind. But they need to accommodate a Furby. This give-and-take prepares children for the expectation of relationship with machines that is at the heart of the robotic moment.

Daisy, six, with a Furby at home, believes that each Furby's owner must help his or her Furby fulfill its mission to learn about people. "You have to teach it; when you buy it, that is your job." Daisy tells me that she taught her Furby about Brownie Girl Scouts, kindergarten, and whales. "It's alive; I teach it about whales; it loves me." Padma, eight, says that she likes meeting what she calls "Furby requests" and thinks that her Furby is "kind of like a person" because "it talks." She goes on: "It's kind of like me because I'm a chatterbox." After two weeks, it is time for Padma to return her Furby, and afterward she feels regret: "I miss how it talked, and now it's so quiet at my house. . . . I didn't get a chance to make him a bed."

After a month with her Furby, Bianca, seven, speaks with growing confidence about their mutual affection: "I love my Furby because it loves me. . . . It was like he really knew me."[6] She knows her Furby well enough to believe that "it doesn't want to miss fun . . . at a party." In order to make sure that her social butterfly Furby gets some rest when her parents entertain late into the evening, Bianca clips its ears back with clothespins to fool the robot into thinking that "nothing is going on . . . so he can fall asleep." This move is ineffective, and all of this activity is exhausting, but Bianca calmly sums up her commitment: "It takes lots of work to take care of these."

When Wilson, who so enjoys burping in synchrony with his Furby, faces up to the hard work of getting his Furby to sleep, he knows that if he forces sleep by removing his Furby's batteries, the robot will "forget" whatever has passed between them—this is unacceptable. So Furby sleep has to come naturally. Wilson tries to exhaust his Furby by keeping it up late at night watching television. He experiments with Furby "sleep houses" made of blankets piled high over towers of blocks. When Wilson considers Furby sleep, his thoughts turn to Furby dreams. He is sure his Furby dreams "when his eyes are closed." What do Furbies dream of? Second and third graders think they dream "of life on their flying saucers."[7] And they dream about learning languages and playing with the children they love.

David and Zach, both eight, are studying Hebrew. "My Furby dreams about Hebrew," says David. "It knows how to say *Eloheinu*. . . . I didn't even try to teach it; it was just from listening to me doing Hebrew homework." Zach agrees: "Mine said *Dayeinu* in its sleep." Zach, like Wilson, is proud of how well he can make his Furby sleep by creating silence and covering it with blankets. He is devoted to teaching his Furby English and has been studying Furbish as well; he has mastered the English/Furbish dictionary that comes with the robot. A week after Zach receives his Furby, however, his mother calls my office in agitation. Zach's Furby is broken. It has been making a "terrible" noise. It sounds as though it might be suffering, and Zach is distraught. Things reached their worst during a car trip from Philadelphia to Boston, with the broken Furby wailing as though in pain. On the long trip home, there was no Phillips screwdriver for the ultimate silencing, so Zach and his parents tried to put the Furby to sleep by nestling it under a blanket. But every time the car hit a bump, the Furby woke up and made the "terrible" noise. I take away the broken Furby, and give Zach a new one, but he wants little to do with it. He doesn't talk to it or try to teach it. His interest is in "his" Furby, the Furby he nurtured, the Furby he taught. He says, "The Furby that I had before could say 'again'; it could say 'hungry.'" Zach believes he was making progress teaching the first Furby a bit of Spanish and French. The first Furby was never "annoying," but the second Furby is. *His* Furby is irreplaceable.

After a few weeks, Zach's mother calls to ask if their family has my permission to give the replacement Furby to one of Zach's friends. When I say yes, Zach calmly contemplates the loss of Furby #2. He has loved; he has lost; he is not willing to reinvest. Neither is eight-year-old Holly, who becomes upset and withdrawn when her mother takes the batteries out of her Furby. The family was about to leave on an extended vacation, and the Furby manual suggests taking out a Furby's batteries if it will go unused for a long time. Holly's mother did not understand the implications of what she saw as commonsense advice from the manual. She insists, with increasing defensiveness, that she was only "following the instructions." Wide-eyed, Holly tries to make her mother understand what she has done: when the batteries are removed, Holly says, "the Furby forgets its life."

Designed to give users a sense of progress in teaching it, when the Furby evolves over time, it becomes the irreplaceable repository and proof of its owner's care. The robot and child have traveled a bit of road together. When a Furby forgets, it is as if a friend has become amnesic. A new Furby is a stranger. Zach and Holly cannot bear beginning again with a new Furby that could never be the Furby into which each has poured time and attention.

OPERATING PROCEDURES

In the 1980s, the computer toy Merlin made happy and sad noises depending on whether it was winning or losing the sound-and-light game it played with children. Children saw Merlin as "sort of alive" because of how well it played memory games, but they did not fully believe in Merlin's shows of emotion. When a Merlin broke down, children were sorry to lose a playmate. When a Furby doesn't work, however, children see a creature that might be in pain.

Lily, ten, worries that her broken Furby is hurting. But she doesn't want to turn it off, because "that means you aren't taking care of it." She fears that if she shuts off a Furby in pain, she might make things worse. Two eight-year-olds fret about how much their Furbies sneeze. The first worries that his sneezing Furby is allergic to him. The other fears his Furby got its cold because "I didn't do a good enough job taking care of him." Several children become tense when Furbies make unfamiliar sounds that might be signals of distress. I observe children with their other toys: dolls, toy soldiers, action figures. If these toys make strange sounds, they are usually put aside; broken toys lead easily to boredom. But when a Furby is in trouble, children ask, "Is it tired?" "Is it sad?" "Have I hurt it?" "Is it sick?" "What shall I do?"

Taking care of a robot is a high-stakes game. Things can—and do—go wrong. In one kindergarten, when a Furby breaks down, the children decide they want to heal it. Ten children volunteer, seeing themselves as doctors in an emergency room. They decide they'll begin by taking it apart.

The proceedings begin in a state of relative calm. When talking about their sick Furby, the children insist that this breakdown does not mean the end: people get sick and get better. But as soon as scissors and pliers appear, they become anxious. At this point, Alicia screams, "The Furby is going to die!" Sven, to his classmates' horror, pinpoints the moment when Furbies die: it happens when a Furby's skin is ripped off. Sven considers the Furby as an animal. You can shave an animal's fur, and it will live. But you cannot take its skin off. As the operation continues, Sven reconsiders. Perhaps the Furby can live without its skin, "but it will be cold." He doesn't back completely away from the biological (the Furby is sensitive to the cold) but reconstructs it. For Sven, the biological now includes creatures such as Furbies, whose "insides" stay "all in the same place" when their skin is removed. This accommodation calms him down. If a Furby is simultaneously biological and mechanical, the operation in process, which is certainly removing the Furby's skin, is not necessarily destructive. Chil-

dren make theories when they are confused or anxious. A good theory can reduce anxiety.

But some children become more anxious as the operation continues. One
suggests that if the Furby dies, it might haunt them. It is alive enough to turn
into a ghost. Indeed, a group of children start to call the empty Furby skin "the
ghost of Furby" and the Furby's naked body "the goblin." They are not happy
that this operation might leave a Furby goblin and ghost at large. One girl comes
up with the idea that the ghost of the Furby will be less fearful if distributed.
She asks if it would be okay "if every child took home a piece of Furby skin."
She is told this would be fine, but, unappeased, she asks the same question two
more times. In the end, most children leave with a bit of Furby fur.[8] Some talk
about burying it when they get home. They leave room for a private ritual to
placate the goblin and say good-bye.

Inside the classroom, most of the children feel they are doing the best they
can with a sick pet. But from outside the classroom, the Furby surgery looks
alarming. Children passing by call out, "You killed him." "How dare you kill
Furby?" "You'll go to Furby jail." Denise, eight, watches some of the goings-on
from the safety of the hall. She has a Furby at home and says that she does not
like to talk about its problems as diseases because "Furbies are not animals." She
uses the word "fake" to mean nonbiological and says, "Furbies are fake, and they
don't get diseases." But later, she reconsiders her position when her own Furby's
batteries run out and the robot, so chatty only moments before, becomes inert.
Denise panics: "It's dead. It's dead right now. . . . Its eyes are closed." She then declares her Furby "both fake and dead." Denise concludes that worn-out batteries
and water can kill a Furby. It is a mechanism, but alive enough to die.

Linda, six, is one of the children whose family has volunteered to keep a
Furby for a two-week home study. She looked forward to speaking to her Furby,
sure that unlike her other dolls, this robot would be worth talking to. But on its
very first night at her home, her Furby stops working: "Yeah, I got used to it,
and then it broke that night—the night that I got it. I felt like I was broken or
something. . . . I cried a lot. . . . I was really sad that it broke, 'cause Furbies talk,
they're like real, they're like real people." Linda is so upset about not protecting
her Furby that when it breaks she feels herself broken.

Things get more complicated when I give Linda a new Furby. Unlike children
like Zach who have invested time and love in a "first Furby" and want no replacements, Linda had her original Furby in working condition for only a few
hours. She likes having Furby #2: "It plays hide-and-seek with me. I play red

light, green light, just like in the manual." Linda feeds it and makes sure it gets enough rest, and she reports that her new Furby is grateful and affectionate. She makes this compatible with her assessment of a Furby as "just a toy" because she has come to see gratitude, conversation, and affection as something that toys can manage. But now she will not name her Furby or say it is alive. There would be risk in that: Linda might feel guilty if the new Furby were alive enough to die and she had a replay of her painful first experience.

Like the child surgeons, Linda ends up making a compromise: the Furby is both biological and mechanical. She tells her friends, "The Furby is kind of real but just a toy." She elaborates that "[the Furby] is real because it is talking and moving and going to sleep. It's kind of like a human and a pet." It is a toy because "you had to put in batteries and stuff, and it could stop talking."

So hybridity can offer comfort. If you focus on the Furby's mechanical side, you can enjoy some of the pleasures of companionship without the risks of attachment to a pet or a person. With practice, says nine-year-old Lara, reflecting on her Furby, "you can get it to like you. But it won't die or run away. That is good." But hybridity also brings new anxieties. If you grant the Furby a bit of life, how do you treat it so that it doesn't get hurt or killed? An object on the boundaries of life, as we've seen, suggests the possibility of real pain.

AN ETHICAL LANDSCAPE

When a mechanism breaks, we may feel regretful, inconvenienced, or angry. We debate whether it is worth getting it fixed. When a doll cries, children know that they are themselves creating the tears. But a robot with a body can get "hurt," as we saw in the improvised Furby surgical theater. Sociable robotics exploits the idea of a robotic body to move people to relate to machines as subjects, as creatures in pain rather than broken objects. That even the most primitive Tamagotchi can inspire these feelings demonstrates that objects cross that line not because of their sophistication but because of the feelings of attachment they evoke. The Furby, even more than the Tamagotchi, is alive enough to suggest a body in pain as well as a troubled mind. Furbies whine and moan, leaving it to their users to discover what might help. And what to make of the moment when an upside down Furby says, "Me scared!"?

Freedom Baird takes this question very seriously.[9] A recent graduate of the MIT Media Lab, she finds herself engaged with her Furby as a creature and a machine. But how seriously does she take the idea of the Furby as a creature? To determine this, she proposes an exercise in the spirit of the Turing test.

In the original Turing test, published in 1950, mathematician Alan Turing, inventor of the first general-purpose computer, asked under what conditions people would consider a computer intelligent. In the end, he settled on a test in which the computer would be declared intelligent if it could convince people it was not a machine. Turing was working with computers made up of vacuum tubes and Teletype terminals. He suggested that if participants couldn't tell, as they worked at their Teletypes, if they were talking to a person or a computer, that computer would be deemed "intelligent."[10]

A half century later, Baird asks under what conditions a creature is deemed alive enough for people to experience an ethical dilemma if it is distressed. She designs a Turing test not for the head but for the heart and calls it the "upside-down test." A person is asked to invert three creatures: a Barbie doll, a Furby, and a biological gerbil. Baird's question is simple: "How long can you hold the object upside down before your emotions make you turn it back?" Baird's experiment assumes that a sociable robot makes new ethical demands. Why? The robot performs a psychology; many experience this as evidence of an inner life, no matter how primitive. Even those who do not think a Furby has a mind— and this, on a conscious level, includes most people—find themselves in a new place with an upside-down Furby that is whining and telling them it is scared. They feel themselves, often despite themselves, in a situation that calls for an ethical response. This usually happens at the moment when they identify with the "creature" before them, all the while knowing that it is "only a machine."

This simultaneity of vision gives Baird the predictable results of the upside-down test. As Baird puts it, "People are willing to be carrying the Barbie around by the feet, slinging it by the hair . . . no problem. . . . People are not going to mess around with their gerbil." But in the case of the Furby, people will "hold the Furby upside down for thirty seconds or so, but when it starts crying and saying it's scared, most people feel guilty and turn it over."

The work of neuroscientist Antonio Damasio offers insight into the origins of this guilt. Damasio describes two levels of experiencing pain. The first is a physical response to a painful stimulus. The second, a far more complex reaction, is an emotion associated with pain. This is an internal representation of the physical.[11] When the Furby says, "Me scared," it signals that it has crossed the line between a physical response and an emotion, the internal representation. When people hold a Furby upside down, they do something that would be painful if done to an animal. The Furby cries out—as if it were an animal. But then it says, "Me scared"—as if it were a person.

People are surprised by how upset they get in this theater of distress. And then they get upset that they are upset. They often try to reassure themselves, saying things like, "Chill, chill, it's only a toy!" They are experiencing something new: you can feel bad about yourself for how you behave with a computer program. Adults come to the upside-down test knowing two things: the Furby is a machine and they are not torturers. By the end, with a whimpering Furby in tow, they are on new ethical terrain.[12]

We are at the point of seeing digital objects as both creatures and machines. A series of fractured surfaces—pet, voice, machine, friend—come together to create an experience in which knowing that a Furby is a machine does not alter the feeling that you can cause it pain. Kara, a woman in her fifties, reflects on holding a moaning Furby that says it is scared. She finds it distasteful, "not because I believe that the Furby is really scared, but because I'm not willing to hear anything talk like that and respond by continuing my behavior. It feels to me that I could be hurt if I keep doing this." For Kara, "That is not what I do. . . . In that moment, the Furby comes to represent how I treat creatures."

When the toy manufacturer Hasbro introduced its My Real Baby robot doll in 2000, it tried to step away from these complex matters. My Real Baby shut down in situations where a real baby might feel pain. This was in contrast to its prototype, a robot called "IT," developed by a team led by MIT roboticist Rodney Brooks. "IT" evolved into "BIT" (for Baby IT), a doll with "states of mind" and facial musculature under its synthetic skin to give it expression.[13] When touched in a way that would induce pain in a child, BIT cried out. Brooks describes BIT in terms of its inner states:

> If the baby were upset, it would stay upset until someone soothed it or it finally fell asleep after minutes of heartrending crying and fussing. If BIT . . . was abused in any way—for instance, by being swung upside down—it got very upset. If it was upset and someone bounced it on their knee, it got more upset, but if the same thing happened when it was happy, it got more and more excited, giggling and laughing, until eventually it got overtired and started to get upset. If it were hungry, it would stay hungry until it was fed. It acted a lot like a real baby.[14]

BIT, with its reactions to abuse, became the center of an ethical world that people constructed around its responses to pleasure and pain. But when Hasbro put BIT into mass production as My Real Baby, the company decided not to present children with a toy that responded to pain. The theory was that a robot's response to pain could "enable" sadistic behavior. If My Real Baby were touched, held, or bounced in a way that would hurt a real baby, the robot shut down.

In its promotional literature, Hasbro marketed My Real Baby as "the most real, dynamic baby doll available for young girls to take care of and nurture." They presented it as a companion that would teach and encourage reciprocal social behavior as children were trained to respond to its needs for amusement as well as bottles, sleep, and diaper changes. Indeed, it was marketed as realistic in all things—except that if you "hurt" it, it shut down. When children play with My Real Baby, they do explore aggressive possibilities. They spank it. It shuts down. They shake it, turn it upside down, and box its ears. It shuts down.

Hasbro's choice—maximum realism, but with no feedback for abuse—inspires strong feelings, especially among parents. For one group of parents, what is most important is to avoid a child's aggressive response. Some believe that if you market realism but show no response to "pain," children are encouraged to inflict it because doing so seems to have no cost. Others think that if a robot simulates pain, it enables mistreatment.

Another group of parents wish that My Real Baby would respond to pain for the same reason that they justify letting their children play violent video games: they see such experiences as "cathartic." They say that children (and adults too) should express aggression (or sadism or curiosity) in situations that seem "realistic" but where nothing "alive" is being hurt. But even these parents are sometimes grateful for My Real Baby's unrealistic show of "denial." They do not want to see their children tormenting a screaming baby.

No matter what position one takes, sociable robots have taught us that we do not shirk from harming realistic simulations of life. This is, of course, how we now train people for war. First, we learn to kill the virtual. Then, desensitized, we are sent to kill the real. The prospect of studying these matters raises awful questions. Freedom Baird had people hold a whining, complaining Furby upside down, much to their discomfort. Do we want to encourage the abuse of increasingly realistic robot dolls?

When I observe children with My Real Baby in an after-school playgroup for eight-year-olds, I see a range of responses. Alana, to the delight of a small band of her friends, flings My Real Baby into the air and then shakes it violently while holding it by one leg. Alana says the robot has "no feelings." Watching her, one wonders why it is necessary then to "torment" something without feelings. She does not behave this way with the many other dolls in the playroom. Scott, upset, steals the robot and brings it to a private space. He says, "My Real Baby is like a baby and like a doll. . . . I don't think she wants to get hurt."

As Scott tries to put the robot's diaper back on, some of the other children stand beside him and put their fingers in its eyes and mouth. One asks, "Do you think that hurts?" Scott warns, "The baby's going to cry!" At this point, one girl tries to pull My Real Baby away from Scott because she sees him as an inadequate protector: "Let go of her!" Scott resists. "I was in the middle of changing her!" It seems a good time to end the play session. As the research team, exhausted, packs up to go, Scott sneaks behind a table with the robot, gives it a kiss, and says good-bye, out of the sight of the other children.

In the pandemonium of Scott and Alana's playgroup, My Real Baby is alive enough to torment and alive enough to protect. The adults watching this—a group of teachers and my research team—feel themselves in an unaccustomed quandary. If the children had been tossing around a rag doll, neither we, nor presumably Scott, would have been as upset. But it is hard to see My Real Baby treated this way. All of this—the Furbies that complain of pain, the My Real Babies that do not—creates a new ethical landscape. The computer toys of the 1980s only suggested ethical issues, as when children played with the idea of life and death when they "killed" their Speak & Spells by taking out the toys' batteries. Now, relational artifacts pose these questions directly.

One can see the new ethics at work in my students' reactions to Nexi, a humanoid robot at MIT. Nexi has a female torso, an emotionally expressive face, and the ability to speak. In 2009, one of my students, researching a paper, made an appointment to talk with the robot's development team. Due to a misunderstanding about scheduling, my student waited alone, near the robot. She was upset by her time there: when not interacting with people, Nexi was put behind a curtain and blindfolded.

At the next meeting of my graduate seminar, my student shared her experience of sitting alongside the robot. "It was very upsetting," she said. "The curtain—and why was she blindfolded? I was upset because she was blindfolded." The story of the shrouded and blindfolded Nexi ignited the seminar. In the conversation, all the students talked about the robot as a "she." The designers had done everything they could to give the robot gender. And now, the act of blindfolding signaled sight and consciousness. In class, questions tumbled forth: Was the blindfold there because it would be too upsetting to see Nexi's eyes? Perhaps when Nexi was turned off, "her" eyes remained open, like the eyes of a dead person? Perhaps the robot makers didn't want Nexi to see "out"? Perhaps they didn't want Nexi to know that when not in use, "she" is left in a corner behind a curtain? This line of reasoning led the seminar to an even more unsettling question: If Nexi is smart enough to need a blindfold to protect "her" from fully grasping "her" situation, does that mean that "she" is enough of a subject to make "her" situation abusive? The students agreed on one thing: blindfolding the robot sends a signal that "this robot can see." And seeing implies understanding and an inner life, enough of one to make abuse possible.

I have said that Sigmund Freud saw the uncanny as something long familiar that feels strangely unfamiliar. The uncanny stands between standard categories and challenges the categories themselves. It is familiar to see a doll at rest. But we don't need to cover its eyes, for it is we who animate it. It is familiar to have a person's expressive face beckon to us, but if we blindfold that person and put them behind a curtain, we are inflicting punishment. The Furby with its expressions of fear and the gendered Nexi with her blindfold are the new uncanny in the culture of computing.

I feel even more uncomfortable when I learn about a beautiful "female" robot, Aiko, now on sale, that says, "Please let go . . . you are hurting me," when its artificial skin is pressed too hard. The robot also protests when its breast is touched: "I do not like it when you touch my breasts." I find these programmed assertions of boundaries and modesty disturbing because it is almost impossible to hear them without imagining an erotic body braced for assault.

FROM THE ROMANTIC REACTION TO THE ROBOTIC MOMENT

Soon, it may seem natural to watch a robot "suffer" if you hurt it. It may seem natural to chat with a robot and have it behave as though pleased you stopped by. As the intensity of experiences with robots increases, as we learn to live in new landscapes, both children and adults may stop asking the questions "Why am I talking to a robot?" and "Why do I want this robot to like me?" We may simply be charmed by the pleasure of its company.

The romantic reaction of the 1980s and 1990s put a premium on what only people can contribute to each other: the understanding that grows out of shared human experience. It insisted that there is something essential about the human spirit. In the early 1980s, David, twelve, who had learned computer programming at school, contrasted people and programs this way: "When there are computers who are just as smart as the people, the computers will do a lot of the jobs, but there will still be things for the people to do. They will run the restaurants, taste the food, and they will be the ones who will love each other, have families and love each other. I guess they'll still be the only ones who go to church."[15] Adults, too, spoke of life in families. To me, the romantic reaction was captured by how one man rebuffed the idea that he might confide in a computer psychotherapist: "How can I talk about sibling rivalry to something that never had a mother?"

Of course, elements of this romantic reaction are still around us. But a new sensibility emphasizes what we share with our technologies. With psychopharmacology, we approach the mind as a bioengineerable machine.[16] Brain imaging trains us to believe that things—even things like feelings—are reducible to what they look like. Our current therapeutic culture turns from the inner life to focus on the mechanics of behavior, something that people and robots might share.

A quarter of a century stands between two conversations I had about the possibilities of a robot confidant, the first in 1983, the second in 2008. For me, the differences between them mark the movement from the romantic reaction to the pragmatism of the robotic moment. Both conversations were with teenage boys from the same Boston neighborhood; they are both Red Sox fans and have close relationships with their fathers. In 1983, thirteen-year-old Bruce talked about robots and argued for the unique "emotionality" of people. Bruce rested his case on the idea that computers and robots are "perfect," while people are "imperfect," flawed and frail. Robots, he said, "do everything right"; people "do the best they know how." But for Bruce it was human imperfection that makes for the ties that bind. Specifically, his own limitations made him feel close to his father ("I have a lot in common with my father. . . . We both have chaos"). Perfect robots could never understand this very important relationship. If you ever have a problem, you go to a person.

Twenty-five years later, a conversation on the same theme goes in a very different direction. Howard, fifteen, compares his father to the idea of a robot confidant, and his father does not fare well in the comparison. Howard thinks the robot would be better able to grasp the intricacies of high school life: "Its database would be larger than Dad's. Dad has knowledge of basic things, but not enough of high school." In contrast to Bruce's sense that robots are not qualified to have an opinion about the goings-on in families, Howard hopes that robots might be specially trained to take care of "the elderly and children"—something he doesn't see the people around him as much interested in.

Howard has no illusions about the uniqueness of people. In his view, "they don't have a monopoly" on the ability to understand or care for each other. Each human being is limited by his or her own life experience, says Howard, but "computers and robots can be programmed with an infinite amount of information." Howard tells a story to illustrate how a robot could provide him with better advice than his father. Earlier that year, Howard had a crush on a girl at school who already had a boyfriend. He talked to his father about asking her out. His father, operating on an experience he had in high school and what Howard considers an outdated ideal of "macho," suggested that he ask the girl out even though she was dating someone else. Howard ignored his father's advice, fearing it would lead to disaster. He was certain that in this case, a robot would have been more astute. The robot "could be uploaded with many experiences" that would have led to the right answer, while his father was working with a limited data set. "Robots can be made to understand things like jealousy from observing how people behave. . . . A robot can be fully understanding and open-minded." Howard thinks that as a confidant, the robot comes out way ahead. "People," he says, are "risky." Robots are "safe."

> There are things, which you cannot tell your friends or your parents, which . . . you could tell an AI. Then it would give you advice you could be more sure of. . . . I'm assuming it would be programmed with prior knowledge of situations and how they worked out. Knowledge of you, probably knowledge of your friends, so it could make a reasonable decision for your course of action. I know a lot of teenagers, in particular, tend to be caught up in emotional things and make some really bad mistakes because of that.

I ask Howard to imagine what his first few conversations with a robot might be like. He says that the first would be "about happiness and exactly what that is, how do you gain it." The second conversation would be "about human fallibility," understood as something that causes "mistakes." From Bruce to Howard, human fallibility has gone from being an endearment to a liability.

No generation of parents has ever seemed like experts to their children. But those in Howard's generation are primed to see the possibilities for relationships their elders never envisaged. They assume that an artificial intelligence could monitor all of their e-mails, calls, Web searches, and messages. This machine could supplement its knowledge with its own searches and retain a nearly infinite amount of data. So, many of them imagine that via such search and storage an artificial intelligence or robot might tune itself to their exact needs. As they see it, nothing technical stands in the way of this robot's understanding, as Howard puts it, "how different social choices [have] worked out." Having knowledge and your best interests at heart, "it would be good to talk to . . . about life. About romantic matters. And problems of friendship."

Life? Romantic matters? Problems of friendship? These were the sacred spaces of the romantic reaction. Only people were allowed there. Howard thinks that all of these can be boiled down to information so that a robot can be both expert resource and companion. We are at the robotic moment.

As I have said, my story of this moment is not so much about advances in technology, impressive though these have been. Rather, I call attention to our strong response to the relatively little that sociable robots offer—fueled it would seem by our fond hope that they will offer more. With each new robot, there is a ramp-up in our expectations. I find us vulnerable—a vulnerability, I believe, not without risk.

true companions

In April 1999, a month before AIBO's commercial release, Sony demonstrated the little robot dog at a conference on new media in San Jose, California. I watched it walk jerkily onto an empty stage, followed by its inventor, Toshitado Doi. At his bidding, AIBO fetched a ball and begged for a treat. Then, with seeming autonomy, AIBO raised its back leg to some suggestion of a hydrant. Then, it hesitated, a stroke of invention in itself, and lowered its head as though in shame. The audience gasped. The gesture, designed to play to the crowd, was wildly successful. I imagined how audiences responded to Jacques de Vaucanson's eighteenth-century digesting (and defecating) mechanical duck and to the chess-playing automata that mesmerized Edgar Alan Poe. AIBO, like these, was applauded as a marvel, a wonder.[1]

Depending on how it is treated, an individual AIBO develops a distinct personality as it matures from a fall-down puppy to a grown-up dog. Along the way, AIBO learns new tricks and expresses feelings: flashing red and green eyes direct our emotional traffic; each of its moods comes with its own soundtrack. A later version of AIBO recognizes its primary caregiver and can return to its charging station, smart enough to know when it needs a break. Unlike a Furby, whose English is "destined" to improve as long as you keep it turned on, AIBO stakes a claim to intelligence and impresses with its ability to show what's on its mind.

If AIBO is in some sense a toy, it is a toy that changes minds. It does this in several ways. It heightens our sense of being close to developing a postbiological life and not just in theory or in the laboratory. And it suggests how this passage will take place. It will begin with our seeing the new life as "as if" life and then deciding that "as if" may be life enough. Even now, as we contemplate "creatures" with artificial feelings and intelligence, we come to reflect differently on our own. The question here is not whether machines can be made to think like people but whether people have always thought like machines.

The reconsiderations begin with children. Zane, six, knows that AIBO doesn't have a "real brain and heart," but they are "real enough." AIBO is "kind of alive" because it can function "*as if* it had a brain and heart." Paree, eight, says that AIBO's brain is made of "machine parts," but that doesn't keep it from being "like a dog's brain. . . . Sometimes, the way [AIBO] acted, like he will get really frustrated if he can't kick the ball. That seemed like a real emotion . . . so that made me treat him like he was alive, I guess." She says that when AIBO needs its batteries charged, "it is like a dog's nap." And unlike a teddy bear, "an AIBO needs its naps."

As Paree compares her AIBO's brain to that of a dog, she clears the way for other possibilities. She considers whether AIBO might have feelings like a person, wondering if AIBO "knows its own feelings"—or "if the controls inside know them." Paree says that people use both methods. Sometimes people have spontaneous feelings and "just become aware" of them (this is "knowing your own feelings"). But other times, people have to program themselves to have the feelings they want. "If I was sad and wanted to be happy"—here Paree brings her fists up close to her ears to demonstrate concentration and intent—"I would have to *make* my brain say that I am set on being happy." The robot, she thinks, probably has the second kind of feelings, but she points out that both ways of getting to a feeling get you to the same place: a smile or a frown if you are a person, a happy or sad sound if you are an AIBO. Different inner states lead to the same outward states, and so inner states cease to matter. AIBO carries a behaviorist sensibility.

SPARE PARTS

Keith, seventeen, is going off to college next year and taking his AIBO with him. He treats the robot as a pet, all the while knowing that it is not a pet at all. He says, "Well, it's not a pet like others, but it *is* a damn good pet. . . . I've taught it everything. I've programmed it to have a personality that matches mine. I've never let it reset to its original personality. I keep it on a program that lets it develop to show the care I've put into it. But of course, it's a robot, so you have to keep it dry, you have to take special care with it." His classmate Logan also has an AIBO. The two have raised the robots together. If anything, Logan's feelings are even stronger than Keith's. Logan says that talking to AIBO "makes you better, like, if you're bored or tired or down . . . because you're actually, like, interacting with something. It's nice to get thoughts out."

The founders of artificial intelligence were much taken with the ethical and theological implications of their enterprise. They discussed the mythic resonance of their new science: Were they people putting themselves in the place of gods?[2] The impulse to create an object in one's own image is not new—think Galatea, Pygmalion, Frankenstein. These days, what is new is that an off-the-shelf technology as simple as an AIBO provides an experience of shaping one's own companion. But the robots are shaping us as well, teaching us how to behave so that they can flourish.[3] Again, there is psychological risk in the robotic moment. Logan's comment about talking with the AIBO to "get thoughts out" suggests using technology to know oneself better. But it also suggests a fantasy in which we cheapen the notion of companionship to a baseline of "interacting with something." We reduce relationship and come to see this reduction as the norm.

As infants, we see the world in parts. There is the good—the things that feed and nourish us. There is the bad—the things that frustrate or deny us. As children mature, they come to see the world in more complex ways, realizing, for example, that beyond black and white, there are shades of gray. The same mother who feeds us may sometimes have no milk. Over time, we transform a collection of parts into a comprehension of wholes.[4] With this integration, we learn to tolerate disappointment and ambiguity. And we learn that to sustain realistic relationships, one must accept others in their complexity. When we imagine a robot as a true companion, there is no need to do any of this work.

The first thing missing if you take a robot as a companion is *alterity*, the ability to see the world through the eyes of another.[5] Without alterity, there can be no empathy. Writing before robot companions were on the cultural radar, the psychoanalyst Heinz Kohut described barriers to alterity, writing about fragile people—he calls them narcissistic personalities—who are characterized not by love of self but by a damaged sense of self. They try to shore themselves up by turning other people into what Kohut calls *selfobjects*. In the role of selfobject, another person is experienced as part of one's self, thus in perfect tune with a fragile inner state. The selfobject is cast in the role of what one needs, but in these relationships, disappointments inevitably follow. Relational artifacts (not only as they exist now but as their designers promise they will soon be) clearly present themselves as candidates for the role of selfobject.

If they can give the appearance of aliveness and yet not disappoint, relational artifacts such as sociable robots open new possibilities for narcissistic experience. One might even say that when people turn other people into selfobjects, they are trying to turn a person into a kind of spare part. A robot is already a spare part. From this point of view, relational artifacts make a certain amount of "sense" as successors to the always-resistant human material. I insist on underscoring the "scare quotes" around the word "sense." For, from a point of view that values the richness of human relationships, they don't make any sense at all. Selfobjects are "part" objects. When we fall back on them, we are not taking in a whole person. Those who can only deal with others as part objects are highly vulnerable to the seductions of a robot companion. Those who succumb will be stranded in relationships that are only about one person.

This discussion of robots and psychological risks brings us to an important distinction. Growing up with robots in roles traditionally reserved for people is different from coming to robots as an already socialized adult. Children need to be with other people to develop mutuality and empathy; interacting with a robot cannot teach these. Adults who have already learned to deal fluidly and easily with others and who choose to "relax" with less demanding forms of social "life" are at less risk. But whether child or adult, we are vulnerable to simplicities that may diminish us.

GROWING UP AIBO

With a price tag of $1,300 to $2,000, AIBO is meant for grown-ups. But the robot dog is a harbinger of the digital pets of the future, and so I present it to children from age four to thirteen as well as to adults. I bring it to schools, to after-school play centers, and, as we shall see in later chapters, to senior centers and nursing homes. I offer AIBOs for home studies, where families get to keep them for two or three weeks. Sometimes, I study families who have bought an AIBO of their own. In these home studies, just as in the home studies of Furbies, families are asked to keep a "robot diary." What is it like living with an AIBO?

The youngest children I work with—the four- to six-year-olds—are initially preoccupied with trying to figure out what the AIBO is, for it is not a dog and not a doll. The desire to get such things squared away is characteristic of their age. In the early days of digital culture, when they met their first electronic toys and games, children of this age would remain preoccupied with such questions of categories. But now, faced with this sociable machine, children address them and let them drop, taken up with the business of a new relationship.

Maya, four, has an AIBO at home. She first asks questions about its origins ("How do they make it?") and comes up with her own answer: "I think they start with foil, then soil, and then you get some red flashlights and then put them in the eyes." Then she pivots to sharing the details of her daily life with AIBO: "I love to play with AIBO every day, until the robot gets tired and needs to take a nap." Henry, four, follows the same pattern. He begins with an effort to categorize AIBO: AIBO is closest to a person, but different from a person because it is missing a special "inner power," an image borrowed from his world of Pokémon.[6] But when I see Henry a week later, he has bonded with AIBO and is stressing the positive, all the things they share. The most important of these are "remembering and talking powers, the strongest powers of all." Henry is now focused on the question of AIBO's affection: How much does this robot like him? Things seem to be going well: he says that AIBO favors him "over all his friends."

By eight, children move even more quickly from any concern over AIBO's "nature" to the pleasures of everyday routines. In a knowing tone, Brenda claims that "people make robots and . . . people come from God or from eggs, but this doesn't matter when you are playing with the robot." In this dismissal of origins we see the new pragmatism. Brenda embraces AIBO as a pet. In her robot diary, she reminds herself of the many ways that this pet should not be treated as a dog. One early entry reminds her not to feed it, and another says, "Do *not* take AIBO on walks so it can poop." Brenda feels guilty if she doesn't keep AIBO entertained. She thinks that "if you don't play with it," its lights get red to show its discontent at "playing by itself and getting all bored." Brenda thinks that when bored, AIBO tries to "entertain itself." If this doesn't work, she says, "it tries to get my attention." Children believe that AIBO asks for attention when it needs it. So, for example, a sick AIBO will want to get better and know it needs human help. An eight-year-old says, "It would want more attention than anything in the whole world."

AIBO also "wants" attention in order to learn. And here children become invested. Children don't just grow up with AIBO around; they grow AIBO up. Oliver is a lively, engaged nine-year-old who lives in a suburban house with many pets. His mother smilingly describes their home life as "controlled chaos," and for two weeks an AIBO has been part of this scene. Oliver has been very active in raising his AIBO. First came simple things: "I trained it to run to certain things and wave its tail." And then came more complicated things, like teaching AIBO soccer. Oliver also spends time just "keeping AIBO company" because he says, "AIBO prefers to be with people." Oliver says, "I went home with a puppy, but now it knows me. . . . It recognizes so many things. . . . It can feel when you pet him. . . . The electricity in AIBO is like blood in people. . . . People and robots both have feelings, but people have more feelings. Animals and robots both have feelings, but robots have more feelings that they can say."

But when Oliver has a problem, he doesn't talk to AIBO but to his hamster. He says that although AIBO can "*say* more of his feelings, my hamster *has* more feelings." Oliver does not see AIBO's current lack of emotionality as a fixed thing. On the contrary. "Give him six months," Oliver says. "That's how long it took Peanut [the hamster] to really love. . . . If it advanced more, if it had more technology, it could certainly love you in the future." In the meantime, taking care of AIBO involves more than simply keeping it busy. "You also have to watch out for his feelings. AIBO is very moody." This does not bother Oliver because it makes AIBO more like the pets he already knows. The bottom line for Oliver: "AIBO loves me. I love AIBO." As far as Oliver is concerned, AIBO is alive enough for them to be true companions.

The fact that AIBO can develop new skills is very important to children; it means that their time and teaching make a difference. Zara, eight, says of her time with AIBO, "The more you play with it, the more actful [Zara's word!] it gets, the more playful. And I think the less you play with it, the lazier it gets." Zara and her eleven-year-old cousin Yolanda compare their AIBO puppies to their teddy bears. Both girls make it clear that AIBO is no doll. Yolanda says that turning a teddy bear into a companion requires "work" because her teddy's feelings "come from my brain." The AIBO, on the other hand, "has feelings all by itself."[7] Zara agrees. You can tell a teddy bear what it should feel, but AIBO "can't feel something else than what it is expressing." AIBO has its "own feelings." She says, "If AIBO's eyes are flashing red, you can't say that the puppy is happy just because you want it to be."

A teddy bear may be irreplaceable because it has gone
child. It calls up memories of one's younger self. And, of cour
teddy calls up the experiences a child had in its company.
don't want to replace an AIBO, something else is in play. A
irreplaceable because it calls back memories not only of one's younger self but
of the robot's younger self as well, something we already saw as children connected to their Tamagotchis and Furbies. In comparing her AIBO to her teddy
bear, Yolanda stresses that AIBO is "more real" because as it grows up, "it goes
through all the stages."

FROM BETTER THAN NOTHING TO BETTER THAN ANYTHING

Yolanda's feelings about AIBO also go through all the stages. She first sees AIBO
as a substitute: "AIBO might be good practice for all children whose parents
aren't ready to take care of a real dog." But then she takes another step: in some
ways AIBO might be better than a real dog. "The AIBO," says Yolanda, "doesn't
shed, doesn't bite, doesn't die." More than this, a robotic companion can be made
as you like it. Yolanda muses about how nice it would be to "keep AIBO at a
puppy stage for people who like to have puppies." Children imagine that they
can create a customized AIBO close to their heart's desire.[8] Sometimes their
heart's desire is to have affection when that pleases them and license to walk
away, something not possible with a biological pet.

Two nine-year-olds—Lydia and Paige—talk through the steps that take a
robot from better than nothing to better than anything. Lydia begins by thinking
of AIBO as a substitute for a real pet if you can't have one: "An AIBO, since you
can't be allergic to a robot, that would be very nice to have." But as she gets to
know AIBO better, she sees a more enticing possibility. "Sometimes," she says,
"I might like [AIBO] more than a real living animal, like a real cat or a real dog,
because, like if you had a bad day . . . then you could just turn this thing off and
it wouldn't bug you." Paige has five pets—three dogs, two cats—and when she is
sad, she says, "I cuddle with them." This is a good thing, but she complains that
pets can be trouble: "All of them want your attention. If you give one attention
you have to give them all attention, so it's kinda hard. . . . When I go somewhere,
my kitten misses me. He'll go into my room and start looking for me." AIBO
makes things easy: "AIBO won't look at you like 'play with me'; it will just go to
sleep if there is nothing else to do. It won't mind."

Paige explains that the worst thing that ever happened to her was when her family "had to put their dog to sleep." She hasn't wanted a new one since. "But the thing about AIBO," she says, "is that you don't have to put him to sleep.... I think you could fix [AIBO] with batteries . . . but when your dog actually dies, you can't fix it." For now, the idea that AIBO, as she puts it, "will last forever" makes it better than a dog or cat. Here, AIBO is not practice for the real. It offers an alternative, one that sidesteps the necessity of death.[9] For Paige, simulation is not necessarily second best.

Pets have long been thought good for children because they teach responsibility and commitment. AIBO permits something different: attachment without responsibility. Children love their pets, but at times, like their overextended parents, they feel burdened by their pets' demands. This has always been true. But now children see a future where something different may be available. With robot pets, children can give enough to feel attached, but then they can turn away. They are learning a way of feeling connected in which they have permission to think only of themselves. And yet, since these new pets seem betwixt and between what is alive and what is not, this turning away is not always easy. It is not that some children feel responsible for AIBO and other do not. The same children often have strong feelings on both sides of the matter.

So for example, Zara likes the idea that AIBO won't get sick if she forgets to walk or feed it. She likes the idea that she can "get credit" for training AIBO even without the burden of being consistent. Yet, Zara also says that "AIBO makes you feel responsible for it." Her cousin Yolanda also likes it that AIBO does not make her feel guilty if she doesn't give it attention, but she feels an even greater moral commitment: "I would feel just as bad if my puppy's or my AIBO's arms broke. I love my AIBO."

Zara and Yolanda are tender with their AIBO. But other children, equally attached to the robot, are very rough. AIBO is alive enough to provoke children to act out their hostility, something we have seen with Furbies and My Real Babies and something we will see again with more advanced robots. Of course, this hostility causes us to look at what else is going on in a child's life, but in the case of AIBO, we see how it can be provoked by anxiety about the robot itself. Uncanny objects are disquieting as well as compelling.

Recall four-year-old Henry who categorized robots by their degree of Pokémon powers. He believes that his AIBO recognizes him and that they have a special relationship. Nevertheless, Henry takes to increasingly aggressive play with AIBO. Over and over, he knocks it down, slapping its side, as he makes two contradictory claims about the robot. First he says that "AIBO doesn't really have feelings," which would make his aggression permissible. But he also says that AIBO prefers him to his friends, something that indicates feelings: "AIBO doesn't really like my friend Ramon," he says with a smile. The more Henry talks about how AIBO dislikes other children, the more he worries that his aggression toward AIBO might have consequences. AIBO, after all, could come to dislike him. To get out of his discomfort, Henry demotes AIBO to "just pretend." But then he is unhappy because his belief in AIBO's affection increases his self-esteem. Henry is caught in a complicated, circular love test. In our passage to post-biological relationships, we give ourselves new troubles.

As soon as children met computers and computer toys in the late 1970s and early 1980s, they used aggression as a way to animate them and to play with ideas about life and death. Children crashed and revived computer programs; they "killed" Merlin, Simon, and Speak & Spell by pulling out their batteries and then made them come back to life. Aggression toward sociable robots is more complex because children are trying to manage more significant attachments. To take only one example, robots disappoint when they do not display the affection children lead themselves to expect. To avoid hurt, children want to dial things down. Turning robots into objects that can be hurt with impunity is a way to put them in their place. Whether we have permission to hurt or kill an object influences how we think about its life.[10] To children, being able to kill spiders without punishment makes spiders seem less alive, and hurting a robot can make it seem less alive as well. But as in the discussion about whether My Real Baby should cry in "pain," things are complicated. For the idea that you can hurt a robot can also make it seem *more* alive.

Like Henry, twelve-year-old Tamara is aggressive toward AIBO and troubled by what this implies. She wants to play with AIBO in the same way that she plays with her much-loved cat. But she worries that AIBO's responses to her are generic. She says, "AIBO acts the same to everyone. It doesn't attach herself to one person like most animals do." Tamara says that sometimes she stops herself from petting AIBO: "I start to pet it, and then, like, I would start to be, like, 'Oh wait. You're not a cat. You're not alive.'" And sometimes she gives in to an urge to "knock it over because it was just so cute when it was getting up and then it would, like, shake its head, because then it seemed really alive because that's what dogs do." She tries to reassure me: "I'm not like this with my animals."

From their earliest experiences with the electronic toys and games of the late 1970s, children split the notion of consciousness and life. You didn't have to be biologically alive to have awareness. And so, Tamara who knows AIBO is not alive, imagines that it still might feel pain. In the end, her aggression puts her in a tough spot; AIBO is too much like a companion to be a punching bag. For Tamara, the idea that AIBO might "see" well enough to recognize her is frightening because it might know she is hitting it. But the idea of AIBO as aware and thus more lifelike is exciting as well.

Tamara projects her fear that AIBO knows she is hurting it and gives herself something to be afraid of.[11] She says of her AIBO, "I was afraid it would turn evil or something." She worries that another AIBO, a frightening AIBO with bad intentions and a will of its own, lives within the one she complains of as being too generic in its responses. This is a complicated relationship, far away from dreaming of adventures with your teddy bear.

The strong feelings that robots elicit may help children to a better understanding of what is on their minds, but a robot cannot help children find the meaning behind the anger it provokes. In the best case, behavior with an AIBO could be discussed in a relationship with a therapist. One wonders, for example, if in her actions with AIBO, Tamara shows her fears of something within herself that is only partially mastered. Henry and Tamara are in conflicted play with a robot that provokes them to anger that they show no signs of working through.

AIBO excites children to reach out to it as a companion, but it cannot be a friend. Yet, both children and adults talk as though it can. Such yearnings can be poignant. As Yolanda's time with AIBO is ending, she becomes more open about how it provides companionship when she is "down" and suggests that AIBO might help if someone close to you died. "For the person to be happy, they would have to focus on someone that is special to them, someone that is alive. . . . That could be an AIBO."

SIMULTANEOUS VISIONS AND COLD COMFORTS

Ashley, seventeen, is a bright and active young woman who describes herself as a cat lover. I have given her an AIBO to take home for two weeks, and now she is at my office at MIT to talk about the experience. During the conversation, Ashley's AIBO plays on the floor. We do not attend to it; it does tricks on its own—and very noisily. After a while, it seems as though the most natural thing would be to turn AIBO off, in the same spirit that one might turn off a radio whose volume interferes with a conversation. Ashley moves toward the AIBO, hesitates, reaches for its off switch, and hesitates again. Finally, with a small grimace, she hits the switch. AIBO sinks to the ground, inert. Ashley comments, "I know it's not alive, but I would be, like, talking to it and stuff, and then it's just a weird experience to press a[n off] button. It made me nervous. . . . [I talk to it] how I would talk to my cat, like he could actually hear me and understand praise and stuff like that." I am reminded of Leah, nine, who said of her Furby, "It's hard to turn it off when it is talking to me."

Ashley *knows* AIBO is a robot, but she experiences it as a biological pet. It becomes alive for her not only because of its intelligence but because it seems to her to have real emotions. For example, she says that when AIBO's red lights shone in apparent frustration, "it seemed like a real emotion. . . . So that made me treat him like he was alive. . . . And that's another strange thing: he's not really physically acting those emotions out, but then you see the colors and you think, 'Oh, he's upset.'"

Artificial intelligence is often described as the art and science of "getting machines to do things that would be considered intelligent if done by people." We are coming to a parallel definition of artificial emotion as the art of "getting machines to express things that would be considered feelings if expressed by people." Ashley describes the moment of being caught between categories: she realizes that what the robot is "acting out" is not emotion, yet she feels the pull of seeing "the colors" and experiencing AIBO as "upset." Ashley ends up seeing AIBO as both machine and creature.

So does John Lester, a computer scientist coming from a far more sophisticated starting point. From the early 1990s, Lester pioneered the use of online communities for teaching, learning, and collaboration, including recent work developing educational spaces on the virtual world of Second Life. Lester bought one of the first AIBOs on the market. He called it Alpha in deference to its being "one of the first batch."[12] When Lester took Alpha out of its box, he shut the door to his office and spent the entire day "hanging out with [my] new puppy." He describes the experience as "intense," comparing it to the first time he saw a computer or typed into a Web browser. He quickly mastered the technical aspects of AIBO, but this kind of understanding did not interfere with his pleasure in simply being with the puppy. When Sony modified the robot's software, Lester bought a second AIBO and named it Beta. Alpha and Beta are machines, but Lester does not like anyone to treat them as inanimate metal and plastic. "I think about my AIBOs in different ways at the same time," Lester says.

In the early days of cubism, the simultaneous presentation of many perspectives of the human face was subversive. But at a certain point, one becomes accustomed to looking at a face in this new way. A face, after all, does have multiple aspects; only representational conventions keep us from appreciating them together. But once convention is challenged, the new view of the face suggests depth and new complexities. Lester has a cubist view of AIBO; he is aware of it as machine, bodily creature, and mind. An AIBO's sentience, he says, is "awesome." The creature is endearing. He appreciates the programming behind the exact swing of the "floppy puppy ears." To Lester, that programming gives AIBO a mind.

Lester understands the mechanisms that AIBO's designers have used to draw him in: AIBO's gaze, its expressions of emotion, and the fact that it "grows up" under his care. But this understanding does not interfere with his attachment, just as knowing that infants draw him in with their big, wide eyes does not threaten his connection with babies. Lester says that when he is with AIBO, he does not feel alone. He says that "from time to time" he "catches himself" in engineer mode, remarking on a technical detail of AIBO that he admires, but these moments do not pull him away from enjoying the companionship of his AIBO puppies. This is not a connection he plays at.

It is a big step from accepting AIBO as a companion, and even a solace, to the proposals of David Levy, the computer scientist who imagines robots as intimate partners. But today's fantasies and Levy's dreams share something important: the idea that after a robot serves as a better-than-nothing substitute, it might become equal, or even preferable, to a pet or person. In Yolanda's terms, if your pet is a robot, it might always stay a cute puppy. By extension, if your lover were a robot, you would always be the center of its universe. A robot would not just be better than nothing or better than something, but better than anything. From watching children play with objects designed as "amusements," we come to a new place, a place of cold comforts. Child and adult, we imagine made to measure companions. Or, at least we imagine companions who are always interested in us.

Harry, a forty-two-year-old architect, enjoys AIBO's company and teaching it new tricks. He knows that AIBO is not aware of him as a person but says, "I don't feel bad about this. A pet isn't as aware of me as a person might be. . . . Dogs don't measure up to people. . . . Each level of creature simply does their best. I like it that he [AIBO] recognizes me as his master." Jane, thirty-six, a grade school teacher, is similarly invested in her AIBO. She says she has "adopted my husband's AIBO . . . because it is so cute. I named it and love to spend time with it." Early in our conversation, Jane claims that she turns to AIBO for "amusement," but she ends up saying that she also turns to it when she is lonely.

Jane looks forward to its company after a long workday. Jane talks to her AIBO. "Spend[ing] time" with AIBO means sharing the events of her day, "like who I'm having lunch with at school, which students give me trouble." Her husband, says Jane, is not interested in these topics. It is more comfortable to talk to AIBO than to force him to listen to stories that bore him. In the company of their robots, Jane and Harry are alone in a way that encourages them to give voice to their feelings. Is there harm here?

In the case of children, I am concerned about their getting comfortable with the idea that a robot's companionship is even close to a replacement for a person. Later, we will hear teenagers talk about their dread of conversation as they explain why "texting is always better than talking." Some comment that "sometime, but not now," it would be good to learn how to have a conversation. The fantasy of robotic companionship suggests that sometime might not have to come. But what of an adult who says he prefers a robot for a reason?

Wesley, sixty-four, provides us with such a case. He has come to see his own self-centeredness as an intractable problem. He imagines a robot helpmate as a way to satisfy himself without hurting others. Divorced three times, Wesley hopes a robot would "learn my psychology. How I get depressed, how I get over it. A robot that could anticipate my cycles, never criticize me over them, learn how to just let me get over them." Wesley says, "I'd want from the robot a lot of what I want from a woman, but I think the robot would give me more in some ways. With a woman, there are her needs to consider. . . . That's the trouble I get into. If someone loves me, they care about my ups and downs. And that's so much pressure."

Wesley knows he is difficult to live with. He once saw a psychiatrist who told him that his "cycles" were out of the normal range. Ex-wives, certainly, have told him he is "too moody." He sees himself as "pressure" on a woman, and he feels pressure as well because he has not been able to protect women he cared for from his "ups and downs." He likes the idea of a robot because he could act naturally—it could not be hurt by his dark moods. Wesley considers the possibility of two "women," one real and the other artificial: "Maybe I would want a robot that would be the perfect mate—less needs—*and* a real woman. The robot could take some of the pressure off the real woman. She wouldn't have to perform emotionally at such a high level, really an unrealistic level. . . . I could stay in my comfort zone."

Rudimentary versions of Wesley's fantasy are in development. I have spoken briefly of the Internet buzz over Roxxxy, put on the market in January 2010, advertised as "the world's first sex robot." Roxxxy cannot move, although it has electronically warmed skin and internal organs that pulse. It does, however, make conversation. The robot's creator, Douglas Hines, helpfully offers, "Sex only goes so far—then you want to be able to talk to the person."[13] So, for example, when Roxxxy senses that its hand is being held, the robot says, "I love holding hands with you," and moves into more erotic conversation when the physical caresses become more intimate. One can choose different personalities for Roxxxy, ranging from wild to frigid. The robot will be updated over the Internet to expand its capabilities and vocabulary. It can already discuss soccer.

Hines, an engineer, says that he got into the robot business after a friend died in the September 11 attacks on the Twin Towers. Hines wanted to preserve his friend's personality so that his children could interact with him as they grew up. Like AI scientist and inventor Raymond Kurzweil, who dreams of a robotic incarnation of his father who died tragically young, Hines committed himself to the project of building an artificial personality. At first, he considered building a home health aid for the elderly but decided to begin with sex robots, a decision that he calls "only marketing." His long-term goal is to take artificial personalities into the mainstream. He still wants to recreate his lost friend.

The well-publicized launch of Roxxxy elicits a great deal of online discussion. Some postings talk about how "sad" it is that a man would want such a doll. Others argue that having a robot companion is better than being lonely. For example, "There are men for who attaining a real woman is impossible. . . . This isn't simply a matter of preference. . . . In the real world, sometimes second best is all they can get."

I return to the question of harm. Dependence on a robot presents itself as risk free. But when one becomes accustomed to "companionship" without demands, life with people may seem overwhelming. Dependence on a person is risky—it makes us subject to rejection—but it also opens us to deeply knowing another. Robotic companionship may seem a sweet deal, but it consigns us to a closed world—the loveable as safe and made to measure.[14]

Roboticists insist that the artificial can be made unpredictable so that relating to robots will never feel rote or mechanical. Robots, they say, will be surprising, helpful, and meaningful in their own right. Yet, in my interviews, fantasies about robot companions do not dwell on robots full of delightful surprises. Rather, they return, again and again, to how robots might, as Yolanda suggested, be made to order, a safe haven in an unsafe world.

CHAPTER 4

enchantment

a little over a year after AIBO's release, My Real Baby became available in stores. In November 2000, I attended a party at MIT to celebrate its launch. The air was festive: My Real Babies were being handed around liberally to journalists, designers, toy-industry executives, and members of the MIT faculty and their guests.

An editor from *Wired* magazine made a speech at the party, admiring how much advanced technology was now available off the shelf. The robot was impressive, certainly. But it was also surprisingly clunky; its motors whirred as its limited range of facial expressions changed. Engineering students around me expressed disappointment, having hoped for more. As I chatted with one of them, my eyes wandered to a smiling faculty wife who had picked up a My Real Baby and was holding it to her as she would a real child. She had the robot resting over her shoulder, and I noticed her moment of shocked pleasure when the robot burped and then settled down. The woman instinctively kissed the top of My Real Baby's head and gently massaged its back as she talked with a friend— all of these the timeless gestures of maternal multitasking. Later, as she was leaving, I asked her about the experience. "I loved it," she said. "I can't wait to get one." I asked why. "No reason. It just gives me a good feeling."

My Real Baby tells you when it is happy and when it wants to play. But it adds a lot more to the mix: it blinks and sucks its thumb; with facial musculature under its skin, it can smile, laugh, frown, and cry. As with all sociable robots, getting along with the robot baby requires learning to read its states of mind. It gets tired and wants to sleep; it gets overexcited and wants to be left alone. It wants to be touched, fed, and have its diaper changed. Over time, My Real Baby develops from infant into two-year-old; baby cries and moans give way to coherent sentences. As it matures, the robot becomes more independent, more likely to assert its needs and preferences. The Tamagotchi primer is followed in the essential: My Real Baby demands care, and its personality is shaped by the care it receives.

Both AIBO and My Real Baby encourage people to imagine robots in everyday life. That is not surprising. After all, these are not extraterrestrials: one is a dog and one is a baby. What is surprising is that time spent with these robots provokes not just fantasies about mutual affection, as we've already seen, but the notion that robots will be there to care for us in the sense of *taking* care of us. To put it too simply, conversations about My Real Baby easily lead to musing about a future in which My Real Baby becomes My Real Babysitter. In this, My Real Baby and AIBO are evocative objects—they give people a way to talk about their disappointments with the people around them—parents and babysitters and nursing home attendants—and imagine being served more effectively by robots. When one fifth-grade boy objects that the AIBO before him wouldn't be useful to an elderly person, he is corrected. His classmates make it clear that they are not talking about AIBO specifically. "AIBO is one, but there will be more."

The first time I heard this fantasy—children suggesting that the descendants of such primitive robots might someday care for them—I was stunned. But in fact, the idea of robot caretaking is now widespread in the culture. Traditional science fiction, from *Frankenstein* to the Chucky movies, portrays the inanimate coming to life as terrifying. Recently, however, it has also been portrayed as gratifying, nearly redemptive. In *Star Wars*, R2D2 is every child's dream of a helpmate. In Steven Spielberg's *A.I.: Artificial Intelligence*, a robot's love brings hope to a grieving mother. In Disney's *WALL-E*, a robot saves the planet, but more than this, it saves the people: it reminds them how to love. In *9*, the humans are gone, but the robots that endure are committed to salvaging human values. An emerging mythology depicts benevolent robots.

I study My Real Baby among children five through fourteen. Some play with the robot in my office. Some meet it in classrooms and after-school settings. Others take it home for two or three weeks. Because this is a robot that represents a baby, it gets children talking about family things, care and attention, how much they have and how much more they want. Children talk about working mothers, absent fathers, and isolated grandparents. There is much talk of divorce. Some children wonder whether one of this robot's future cousins might be a reasonable babysitter; something mechanical might be more reliable than the caretaking they have.[1]

Many of the children I study return to empty homes after school and wait for a parent or older family member to come home from work. Often their only babysitter is the television or a computer game, so in comparison a robot looks like pretty good company. Nicole is eleven. Both of her parents are nurses. Sometimes their shifts overlap, and when this happens, neither is home until late. Nicole thinks a robot might be comforting: "If you cut yourself and you want some sympathy. Or you had a bad day at school—even your best friend was mad at you. It would be better to not be alone when you came home." Twelve-year-old Kevin is not so sure: "If robots don't feel pain, how could they comfort you?" But the philosophical conversations of the late 1970s and 1980s are cut short: these children are trying to figure out if a robot might be good for them in the most practical terms.

The twenty children in Miss Grant's fifth-grade class, in a public school on Boston's North Shore, are nine and ten. They have all spent time with the AIBOs and My Real Babies that I brought to their school. Now we are about to begin a home study where one group of children after another will take a My Real Baby home for two weeks. Most take the position Wilson staked out with his Furby and Lester settled into with his AIBO. They are content to be with a machine that they treat as a living creature. Noah remarks that My Real Baby is very noisy when it changes position, but he is quick to point out that this is insignificant: "The whirring doesn't bother me," he says. "I forget it right away."

In the robotic moment, what you are made of—silicon, metal, flesh—pales in comparison with how you behave. In any given circumstance, some people and some *robots* are competent and some not. Like people, any particular robot needs to be judged on its own merits. Tia says, "Some robots would be good companions because they are more efficient and reliable," and then she pauses. I ask her to say more, and she tells me a story. She was at home alone with her pregnant mother, who quite suddenly went into labor. On short notice, they needed to find a babysitter for Tia. Luckily, her grandmother was close by and able to take over, but nevertheless, Tia found the incident frightening. "Having a robot babysitter would mean never having to panic about finding someone at the last minute. It is always ready to take care of you." In only a few years, children have moved from taking care of Tamagotchis and Furbies to fantasies of being watched over by benign and competent digital proctors. The Tamagotchis and Furbies were always on. Here, a robot is thought of as "always ready."

These fifth graders know that AIBO and My Real Baby are not up to the job of babysitter, but these robots inspire optimism that scientists are within striking distance. The fifth graders think that a robot could be a babysitter if it could manage babysitter *behavior*. In their comments about how a robot might pass that test, one hears about the limitations of the humans who currently have the job: "They [robots] would be more efficient than a human if they had to call for an emergency and had a phone right inside them. . . . They are more practical because if someone gets hurt they are not going to stress or freak out." "They would be very good if you were sick and your mother worked." "Robots would always be sure that you would have fun. People have their own problems." Rather than a mere understudy, a robot could be better qualified to serve. Hesitations are equally pragmatic. One fifth grader points out how much air conditioners and garbage disposals break. "The robot might shut down" too.

In the 1980s, most children drew a line—marking a kind of sacred space— between the competencies of computers and what was special about being a person. In Miss Grant's class, the sacred space of the romantic reaction is less important than getting the job done. Most of the children are willing to place robots and humans on an almost-level playing field and debate which can perform better in a given situation. To paraphrase, these pragmatic children say that if people are better at fun, let's put them in charge of fun. If a robot will pay more attention to them than a distracted babysitter, let the robot babysit. If the future holds robots that behave lovingly, these children will be pleased to feel loved. And they are not dissuaded if they see significant differences between their way of thinking and how they imagine robots think. They are most likely to say that if these differences don't interfere with how a robot performs its job, the differences are not worth dwelling on.

Children are not afraid to admit that when robots become caretakers, some things will be lost, things they will miss. But they also make it clear that when they say they will "miss" something (like having a mother at home to watch them when they are sick), it is not necessarily something they have or ever hope to. Children talk about parents who work all day and take night shifts. Conversations about families are as much about their elusiveness as about their resources.

On this almost-level playing field, attitudes about robotic companionship are something of a litmus test for how happy children are with those who care for them. So, children who have incompetent or boring babysitters are interested in robots. Those who have good babysitters would rather stick with what they have.

FROM MY REAL BABY TO MY REAL BABYSITTER

Jude is happy with his babysitter. "She is creative. She finds ways for us to have fun together." He worries that a robot in her place might be too literal minded: "If parents say [to a person], 'Take care of the kid,' they [the person] won't just go, 'Okay, I'm just going to make sure you don't get hurt.' They'll play with you; they'll make sure you have fun too." Jean-Baptiste agrees. Robot babysitters are "only in some ways alive. . . . It responds to you, but all it really thinks about is the job. If their job is making sure you don't get hurt, they're not going to be thinking about ice cream." Or it might know that children like ice cream, but wouldn't understand what ice cream was all about. How bad would this be? Despite his concerns, Jean-Baptiste says he "could love a robot if it was very, very nice to me." It wouldn't understand it was being nice, but for Jean-Baptiste, kindness is as kindness does.

Some children are open to a robot companion because people are so often disappointing. Colleen says, "I once had a babysitter just leave and go over to a friend's house. A robot babysitter wouldn't do that." Even when they stayed around, her babysitters were preoccupied. "I would prefer to have a robot babysitter. . . . A robot would give me all its attention." Octavio says that human babysitters are better than robots "if you are bored"—humans are able to make up better games. But they often get meals wrong: "What's with the cereal for dinner? That's boring. I should have pasta or chicken for dinner, not cereal." Because of their "programming," robots would know that cereal at night is not appropriate. Or, at least, says Octavio, robots would be programmed to take interest in his objections. In this way, the machines would know that cereal does not make a good dinner. Programming means that robots can be trusted. Octavio's classmate Owen agrees. It is easier to trust a robot than a person: "You can only trust a person if you know who they are. You would have to know a person more [than a robot]. . . . You wouldn't have to know the robot, or you would get to know it much faster."

Owen is not devaluing the "human kind" of trust, the trust built as people come through for each other. But he is saying that human trust can take a long time to develop, while robot trust is as simple as choosing and testing a program. The meaning of intelligence changed when the field of artificial intelligence declared it was something computers could have. The meaning of memory changed when it was something computers used. Here the word "trust" is under siege, now that it is something of which robots are worthy. But some of the children are concerned that a trustworthy, because consistent, robot might still fall short as babysitter for lack of heart. So Bridget says she could love a robot babysitter if it did a good job, but she is skeptical about the possibility. She describes what might occur if a robot babysitter were taking care of her and she scraped her knee: "It's just going to be like, [in a robot voice] 'Okay, what do I do, get a Band-Aid and put it on, that's it. That's my job, just get a Band-Aid and put it on.' . . . [stops using robot's voice] But to love somebody, you need a body and a heart. These computers don't really have a heart. It's just a brain. . . . A robot can get hurt, but it doesn't really hurt. The robot just shuts down. When hurt, the robot says, 'Right. Okay, I'm hurt, now I'll shut down.'"

As Bridget speaks, I feel a chill. This "shutdown" is, of course, the behavior of My Real Baby, which shuts down when treated roughly. Bridget seizes upon that detail as a reason why a robot cannot have empathy. How easy it would be, how small a technical thing, to give robots "pretend empathy." With some trepidation, I ask Bridget, "So, if the robot showed that it felt pain, would that make a difference?" Without hesitation she answers, "Oh yes, but these robots shut down if they are hurt." From my perspective, the lack of robotic "empathy" depends on their not being part of the human life cycle, of not experiencing what humans experience. But these are not Bridget's concerns. She imagines a robot that could be comforting if it performed pain. This is the behaviorism of the robotic moment.

There is little sentimentality in this classroom. Indeed, one of Miss Grant's students sees people as potential obstacles to relationships with robots: "If you are already attached to your babysitter, you won't be able to bond with a robot." And this might be a shame. For the babysitter is not necessarily better, she just got there first. The children's lack of sentimentality does not mean that the robots always come out ahead. After a long conversation about robot babysitters, Octavio, still dreaming of pasta instead of cereal, imagines how a robot might be programmed both to play with him and feed him "chicken and pasta because that is what you are supposed to have at night." But Bridget dismisses Octavio's plan as "just a waste. You could have just had a person." Jude concurs: "What's the point of buying a robot for thousands and thousands of dollars when you could have just kept the babysitter for twenty dollars an hour?"

DON'T WE HAVE PEOPLE FOR THESE JOBS?

Children speak fondly of their grandparents, whose care is often a source of family tension. Children feel a responsibility, and they want their parents to take responsibility. And yet, children see that their parents struggle with this. Might robots be there to fill in the gaps?

Some children are taken with the idea that machines could help with purely practical matters. They talk about a robot "getting my grandmother water in the middle of the night," "watching over my grandmother when she sleeps," and being outfitted with "emergency supplies." The robots might be more reliable than people—they would not need sleep, for example—and they might make it easier for grandparents to continue living in their own homes.

But other children's thinking goes beyond emergencies to offering grandparents the pleasures of robotic companionship. Oliver, the nine-year-old owner of Peanut the hamster, says that his grandparents are frail and don't get out much. He considers in detail how their days might be made more interesting by an AIBO. But the robots might come with their own problems. Oliver points out that his grandparents are often confused, and it would be easy for them to confuse the robots. "Like, the old people might tell them [the AIBOs] the wrong people to obey or to do the opposite or not listen to the right person." His sister Emma, eleven, sees only the bright side of a robotic companion. "My grandmother had a dog and the dog died before she did. My grandmother said she would die when her dog died. . . . I'm not sure that it is good for old people to have dogs. I think the AIBO would have been better for her." Back in Miss Grant's class, Bonnie thinks a robot might be the ultimate consolation. "If you had two grandparents and one died," she says, "a robot would help the one that was alone."

Jude, also in Miss Grant's class, knows that his grandmother enjoys talking about the past, when she was a young mother, during what she calls "her happiest time." He thinks that My Real Baby can bring her back to that experience. "She can play at that." But it is Jude who first raises a question that will come to preoccupy these children. He thinks that his grandparents might *prefer* a robot to visits from a real baby.

Jude thinks aloud: "Real babies require work and then, well, they stop being babies and are harder for an older person to care for." Jude says that while he and other kids can easily tell the difference between robots and a real baby, his grandparents might be fooled. "It will cry if it's bored; when it gets its bottle, it will be happy."

This association to the idea that robots might "double" for family members brings to mind a story I heard when I first visited Japan in the early 1990s. The problems of the elderly loomed large. Unlike in previous generations, children were mobile, and women were in the workforce. Aging and infirm parents were unlikely to live at home. Visiting them was harder; they were often in different cities from their children. In response, some Japanese children were hiring actors to substitute for them and visit aging parents.[2] The actors would visit and play their parts. Some of the elderly parents had dementia and might not have known the difference. Most fascinating were reports about the parents who knew that they were being visited by actors. They took the actors' visits as a sign of respect, enjoyed the company, and played the game. When I expressed surprise at how satisfying this seemed for all concerned, I was told that in Japan being elderly is a role, just as being a child is a role. Parental visits are, in large part, the acting out of scripts. The Japanese valued the predictable visits and the well-trained and courteous actors. But when I heard of it, I thought, "If you are willing to send in an actor, why not send in a robot?"

Eighteen years later, a room of American fifth graders are actively considering that proposition. The children know that their grandparents value predictability. When the children visit, they try their best to accommodate their elders' desire for order. This is not always easy: "My grandmother," says Dennis, "she really likes it if my glass, like with water, is only placed in a certain place. She doesn't like it if I don't wheel her only in a certain way through the hospital. It's hard." In this arena, children think that robots might have an edge over them. They begin to envision robots as so much a part of the family circle that they provoke a new kind of sibling rivalry.

One girl describes a feeling close to dread: "If my grandmother started loving the robot, she might start thinking it is her family and that her real family might not be as important to her anymore." Children worry that the robots could spark warm—too warm—feelings. They imagine their grandparents as grateful to, dependent on, and fond of their new caretakers. The robot that begins as a "solution" ends up a usurper. Owen worries that "grandparents might love the robot more than you. . . . They would be around the robot so much more." I ask if the robot would love the grandparents back. "Yes," says Owen, "a little bit. I might feel a little jealous at the robot."

Hunter's grandmother lives alone. She has a button to press if she needs help—for example, if she falls or feels ill. Although Hunter knows that My Real Baby and AIBO couldn't help his grandmother, he thinks future robots might. Hunter has mixed feelings: "I worry that if a robot came in that could help her with falls, then she might really want it. . . . She might like it more than me. It would be more helpful than I am." Hunter wants to be the one to help his grandmother, but he doesn't live with her. He realizes the practicality of the robot but is "really upset that the robot might be the hero for her."

This is the sentiment of fourteen-year-old Chelsea, an eighth grader in Hartford. Her grandmother, eighty-four, lives in a nursing home. Chelsea and her mother visit once a week. Her grandmother's forgetfulness frightens her. "I don't want her forgetting about me." When I introduce her to My Real Baby, Chelsea talks about her grandmother: "She would like this. She really would. I kind of hate that. But this does a lot of what she wants. . . . Actually, I think she would like that it would remember her and it wouldn't ask her too many questions. I worry that when I go with my mom, we ask her so many questions. I wonder if she is relieved when we leave sometimes. My Real Baby would just love her, and there wouldn't be any stress."

I ask Chelsea if she would like to bring a My Real Baby to her grandmother. Her response is emphatic: "No! I know this sounds freaky, but I'm a little jealous. I don't like it that I could be replaced by a robot, but I see how I could be." I ask Chelsea about the things that only she can offer her grandmother, such as memories of their time together. Chelsea nods but says little. For the time being she can only think of the calm presence of the robot stand-in. The next time I see Chelsea, she is with her mother. They have discussed the idea of the robot companion. From Chelsea's point of view, the conversation did not go well; she is upset that her mother seems taken by the idea.[3] Chelsea is sharp with her mother: "It is better that grandma be lonely than forget us because she is playing with her robot. This whole thing makes me jealous of a robot."

In Miss Grant's class, the conversation about robots and grandparents ends up on a skeptical note. Some children become jealous, while others come to see the substitution as wrong. One says, "I wouldn't let that thing [a robot] touch my grandmother." For another, "That would be too weird." A third worries that a robot might "blow up . . . stop working . . . put the house on fire." A conversation that began as matter-of-fact becomes more animated. An anxious consensus emerges: "Don't we have people for these jobs?"

RORSCHACH TO RELATIONSHIP

My Real Baby was primitive, the first of its kind, and not a commercial success. Nevertheless, it was able to reach the "real baby" in us, the part that needs care and worries it will not come. It made it possible for children to project their hopes of getting what they are missing onto the idea of a robot.

Callie, ten, is serious and soft-spoken. When I first bring My Real Baby to her school, she says that "they were probably confused about who their mommies and daddies were because they were being handled by so many different people." She thinks this must have been stressful and is convinced that things will be easier on the robots when they are placed in homes. Like any adoptive mother, she is concerned about bonding with her baby and wants to be the first in her class to take My Real Baby home. She imagines that future study participants will have a harder time with the robot, which is sure to "cry a lot" because "she doesn't know, doesn't think that this person is its mama." As soon as Callie brought My Real Baby home, she stepped into the role of its mother. Now, after three weeks of the home study, our conversation takes place in her suburban home outside of Providence, Rhode Island.

Callie begins with a diversionary tactic: she notes small differences between My Real Baby and a biological child (the size of their pupils, for example) in a seeming effort to minimize the much larger differences between them. She works hard to sustain her feeling that My Real Baby is alive and has emotions. She wants this to be the case. Taking care of My Real Baby makes her feel more cared for. She explains that her parents are very busy and don't have a lot of time to spend with her. She and her four-year-old brother compete for their attention.

For the most part, Callie is taken care of by nannies and babysitters. She sees her mother only "if she [is] not going out." Callie describes her as "very busy . . . with very important work." But what Callie says she misses most is spending time with her father, of whom she speaks throughout her interviews and play sessions. Sometimes he comes to our sessions, but he is visibly distracted. He usually has his BlackBerry with him and checks his e-mail every few minutes. He seems to have little time to concentrate exclusively on his daughter. Nevertheless Callie is intensely loyal to him. She explains that he works all day and often has to go out to important meetings at night. He needs time to travel. Tellingly, Callie thinks that grown-ups would like My Real Baby as much as children do because, in its presence, adults would be "reminded of being parents."

Callie loves to babysit. Caring for others makes her feel wanted in a way that life at home sometimes does not. Her relationship with My Real Baby during the three-week home study comes to play something of the same role: loving the robot makes her feel more loved. She knows the robot is mechanical but has little concern for its (lack of) biology. It is alive enough to be loved because it has feelings, among them an appreciation of her motherly love. She sees the robot as capable of complex and mixed emotions. "It's got similar-to-human feelings, because she can really tell the differences between things, and she's happy a lot. She gets happy, and she gets sad, and mad, and excited. I think right now she's excited and happy at the same time." When My Real Baby says, "I love you," Callie sees the robot's expressed feelings as genuine. "I think she really does," says Callie, almost tearfully. "I feel really good when it says that. Her expressions change. Sort of like Robbie [her four-year-old brother]." Playing with My Real Baby, she says, "makes me incredibly happy." She worries about leaving the robot at home when she goes to school. She knows what it's like to feel abandoned and worries that My Real Baby is sad during the day because no one is paying attention to it. Callie hopes that during these times, My Real Baby will play with one of Callie's pets, a strategy that Callie uses when she feels lonely.

My Real Baby sleeps near Callie's bed on a silk pillow. She names the robot after her three-year-old cousin Bella. "I named her like my cousin . . . because she [My Real Baby] was sort of demanding and said most of the things that Bella does." But Callie often compares My Real Baby to her brother Robbie. Robbie is four, and Callie thinks My Real Baby is "growing up" to be his age. After feeding the robot, Callie tries several times to burp it, saying, "This is what babies need to do." She holds the robot closer with increasing tenderness. She believes that it is getting to know her better as they spend more time together. With time, she says, "Our relationship, it grows bigger. . . . Maybe when I first started playing with her, she didn't really know me . . . but now that she's . . . played with me a lot more she really knows me and is a lot more outgoing."

When Callie plays with other dolls, she says she is "pretending." Time with My Real Baby is different: "I feel like I'm her real mom. I bet if I really tried, she could learn another word. Maybe 'Da-da.' Hopefully if I said it a lot, she would pick it up. It's sort of like a real baby, where you wouldn't want to set a bad example." In Callie's favorite game with My Real Baby, she imagines that she and the robot live in their own condo. She takes herself out of her own family and creates a new one in which she takes care of the robot and the robot is her constant companion. It is a fantasy in which this child, hungry for attention, finally gets as much attention as she wants.

In my study, Callie takes home both an AIBO and a My Real Baby. But very soon, the AIBO begins to malfunction: it develops a loud mechanical wheeze and its walking becomes wobbly. When this happens, Callie treats the AIBO as ill rather than broken—as a sick animal in need of "veterinary care." Callie thinks it has "a virus, maybe the flu. Poor AIBO. I felt sad for it. It was a good AIBO." Most important to Callie is maintaining her sense of herself as a successful mother. Once AIBO is her baby, she cannot not fail "him." She ministers to AIBO—keeps it warm, shows it love—but when it does not recover, her attitude changes. She cannot not tolerate that the AIBO is sick and she cannot help. So she reinterprets AIBO's problem. It is not ill; it is playing. When AIBO can walk no more, Callie says, "Oh, that's what my dog does when he wants attention. I think it might be sleeping. Or just stretching in a different way than a normal dog would." When she hears the troubling mechanical sounds, Callie considers that AIBO might be "just going to sleep." Once she interprets the inert AIBO as sleeping, she is able to relax. She takes AIBO in her arms, holds it close, and pets it gently. She says, "Aww, man! How playful. AIBO! . . . He is sort of tired and wants to rest." Callie focuses on what is most important to her: that AIBO should feel loved. She says, "He knows that I'm holding him."

As Callie plays out scenarios in the imaginary condo, her parents and some of the researchers are charmed by the ease of her relationship with the robots, the way she accepts them as good company. But Callie's earnestness of connection is compelled; she *needs* to connect with these robots.

Callie is very sad when her three weeks with My Real Baby and AIBO come to an end. She has used the time to demonstrate her ability to be a loving mother, a good caretaker to her pets, her brother, and her robots. Before leaving My Real Baby, Callie opens its box and gives the robot a final, emotional good-bye. She reassures My Real Baby that it will be missed and that "the researchers will take good care of you." Callie has tried to work through a desire to feel loved by becoming indispensable to her robots. She fears that her parents forget her during their time away; now, Callie's concern is that My Real Baby and AIBO will forget her.

With the best of intentions, roboticists hope we can use their inventions to practice our relationship skills. But for someone like Callie, practice may be too perfect. Disappointed by people, she feels safest in the sanctuary of an as-if world. Of course, Callie's story is not over. Her parents love her and may become more present. She may find a caring teacher. But at ten, ministering to her robots, Callie reminds us of our vulnerability to them. More than harmless amusements, they are powerful because they invite our attachment. And such attachments change our way of being in the world.

Seven-year-old Tucker, severely ill, is afraid of his body, afraid of dying, and afraid to talk about it. A relationship with AIBO gives voice to these feelings. Home-administered treatments help Tucker to breathe, but even so, he spends several months a year in hospitals. Enthusiastic play with AIBO sometimes leaves him too tired to speak. His parents are reassuring that when this happens, he just needs to rest, and, indeed, after some time sitting quietly, Tucker is always able to continue.

Tucker's mother explains that safety is always his first concern, something that, she admits, can become trying when he second-guesses her driving. When Tucker plays his favorite computer game, Roller Coaster Tycoon, rather than build the wildest roller coaster possible, he builds the safest one. The game allows you choices for how to spend your money in developing your amusement park. Tucker likes to put his cash into maintenance and staffing. He says that very often the game declares him the winner of the award for the "safest park." So, when he first meets AIBO in my office, Tucker's priority is that it be kept safe. His anxiety about this is so great that he denies any reality in which it is, in fact, endangered. So, when AIBO smashes into a fence of red siding that defines its space, Tucker interprets this as AIBO "scratching a door, wanting to go in . . . because it hasn't been there yet." Defense mechanisms are the responses we use to deal with realities too threatening to face. Like Callie ignoring the reality of her broken AIBO, Tucker sees only what he can handle.

Like Callie, Tucker sees AIBO's feelings as real; he says that the robot recognizes and loves him. Tucker explains that when he goes to school, his dog Reb misses him and sometimes wants to jump into the car with him. He thinks that when he takes AIBO home, it will have the same loving desires. Indeed, Tucker finds few differences between AIBO and Reb, most of them unflattering to the biological pet. When Tucker learns to interpret AIBO's blinking lights, he concludes that the robot and Reb have "the same feelings," although he decides that AIBO seems the angrier of the two.

Tucker wishes he himself were stronger and projects this wish onto AIBO: he likes to talk about the robot as a superhero dog that shows up the limitations of his biological dog. Tucker says, "AIBO is probably as smart as Reb and at least he isn't as scared as my dog." While freely celebrating AIBO's virtues, Tucker avoids answering any questions about what Reb can do that AIBO cannot. I am reminded of Chelsea, who, once having decided that a calm robot might be more comforting to her grandmother than her own anxious and talkative self, could not be engaged on what only she had to offer.

So, it is not uncommon for AIBO to do a trick and for Tucker to comment, "My dog couldn't do that." AIBO is the better dog, and we hear why. AIBO is alive even if his heart is made of batteries and wires. AIBO will never get sick or die. In fact, AIBO is everything that Tucker wishes to be. Tucker identifies with AIBO as a being that can resist death through technology. AIBO gives Tucker the idea that people, like this robot, may someday be recharged and rewired. Just as no blood is needed for AIBO's heart to feel emotion, batteries and wires might someday keep a person alive. Tucker uses care for AIBO to dream himself into a cyborg future.

At one point Tucker says that he "would miss AIBO as much as Reb if either of them died." Tucker seems startled when he realizes that in fantasy he has allowed that AIBO could die. He immediately explains that AIBO *could* die but does not *have* to die. And AIBO will *not* die if Tucker protects him. In this moment of poignant identification, Tucker sees AIBO as both potentially immortal and a creature like him, someone who needs to be kept out of harm's way. In Tucker's case, precautions have often been futile. Despite the best of care, he has often landed in the hospital. In AIBO's case, Tucker believes that precautions will work. They will require vigilance. Tucker tells us his elaborate plans to care for the robot when he takes it home. As he speaks, Tucker's anxiety about AIBO's possible death comes through: "He'll probably be in my room most of the time. And I'm probably going to keep him downstairs so he doesn't fall down the stairs. Because he probably, in a sense he would die if he fell down the stairs. Because he could break."

After the robot goes home with him, Tucker reports on their progress. On AIBO's first day, Tucker says, "AIBO was charging and probably didn't miss me." By the second day, Tucker is sure that AIBO cares. But of course, AIBO is not always at his best, something that helps Tucker identify with the robot, for Tucker, too, has good and bad days. Tucker says that after he returns his AIBO, he will miss the robot and that the robot "will probably miss me."

With AIBO at home, Tucker dreams up duels between the robot and his Bio Bugs. Bio Bugs are robot creatures that can walk and engage in combat with each other, gaining "survival skills" along the way. They can end up very aggressive. With great excitement, Tucker describes their confrontations with AIBO. The battles between AIBO and the Bio Bugs seem to reassure him that, no matter what, AIBO will survive. It reinforces the image of the robot as a life form able to defy death, something Tucker would like to become. The "bugs" are the perfect representation of a bacterium or virus, such as those that Tucker continually fights off. AIBO easily defeats them.

When it is time to return the robot, Tucker seems concerned that his healthy older brother, Connor, twelve, barely played with AIBO during the weeks they had the robot at home. Tucker brings this up with a shaky voice. He explains that his brother didn't play with the robot because "he didn't want to get addicted to him so he would be sad when we had to give him back." Tucker wishes that he had more of his brother's attention; the two are not close. Tucker fears that his brother does not spend time with him because he is so frail. In general, he worries that his illness keeps people away because they don't want to invest in him. AIBO, too, is only passing through their home. Tucker is upset by Connor's hesitancy to bond with something "only passing in his life." Tucker tells us that he is making the most of his time with AIBO.

Callie and Tucker nurture robots that offer a lot more room for relationship than Furbies and Tamagotchis. Yet, both My Real Baby and AIBO are commercially available pastimes. I've studied other children who come to MIT laboratories to visit more advanced robots. These robots are not toys; they have their own toys. Grown-ups don't just play with them; these robots have their own grown-up attendants. Is this a game for grown-ups or a more grown-up game? Is it a game at all? To treat these robots as toys is to miss the point—and even the children know it.

CHAPTER 5

complicities

I first met Cog in July 1994, in Rodney Brooks's Artificial Intelligence Laboratory at MIT. The institute was hosting an artificial-life workshop, a conference that buzzed with optimism about science on its way to synthesizing what contributors called "the living state." Breathtaking though they were in capturing many of the features of living systems, most of the "life forms" this field had developed had no physical presence more substantial than images on a computer screen; these creatures lived in simulation. Not so Cog, a life-size human torso, with mobile arms, neck, and head.

Cog grew out of a long research tradition in Brooks's lab. He and his colleagues work with the assumption that much of what we see as complex behavior is made up of simple responses to a complex environment. Consider how artificial intelligence pioneer Herbert Simon describes an ant walking across a sand dune: the ant is not thinking about getting from point A to point B. Instead, the ant, in its environment, follows a simple set of rules: keep moving and avoid obstacles. After more than fifteen years of using this kind of strategy to build robots that aspired to insect-level intelligence, Brooks said he was ready "to go for the whole iguana."[1] In the early 1990s, Brooks and his team began to build Cog, a robotic two-year-old. The aspiration was to have Cog "learn" from its environment, which included the many researchers who dedicated themselves to its education. For some, Cog was a noble experiment on the possibilities of embodied, "emergent" intelligence. For others, it was a grandiose fantasy. I decided to see for myself.

I went to Brooks's lab with Christopher Langton, one of the founders of the field of artificial life—indeed, the man who had coined the term. In town from New Mexico for the A-Life conference, Langton was as eager as I to see the robot. At the AI lab, robot parts were stacked in compartments and containers; others were strewn about in riots of color. In the midst of it all was Cog, on a pedestal, immobile, almost imperial—a humanoid robot, one of the first, its face rudimentary, but with piercing eyes.

Trained to track the movement of human beings (typically those objects whose movements are not constant), Cog "noticed" me soon after I entered the room. Its head turned to follow me, and I was embarrassed to note that this made me happy—unreasonably happy. In fact, I found myself competing with Langton for the robot's attention. At one point, I felt sure that Cog's eyes had "caught" my own, and I experienced a sense of triumph. It was noticing *me*, not its other guest. My visit left me surprised—not so much by what Cog was able to accomplish but by my own reaction to it. For years, whenever I had heard Brooks speak about his robotic "creatures," I had always been careful to mentally put quotation marks around the word. But now, with Cog, I had an experience in which the quotation marks disappeared. There I stood in the presence of a robot and I wanted it to favor me. My response was involuntary, I am tempted to say visceral. Cog had a face, it made eye contact, and it followed my movements. With these three simple elements in play, although I knew Cog to be a machine, I had to fight my instinct to react to "him" as a person.

MECHANICAL TODDLERS

Cog's builders imagined a physically agile toddler that responds to what it sees, touches, and hears. An adjacent laboratory houses another robot designed to simulate that toddler's emotions. This is the facially and vocally expressive Kismet, with large doll eyes and eyelashes and red rubber tubing lips. It speaks in a soft babble that mimics the inflections of human speech. Kismet has a range of "affective" states and knows how to take its turn in conversation. It can repeat a requested word, most often to say its own name or to learn the name of the person talking to it.[2]

Like Cog, Kismet learns through interaction with people. Brooks and his colleagues hoped that by building learning systems, we would learn about learning.[3]

And robots that learn through social interaction are the precursors to machines that can actively collaborate with people. A sociable robot would, for example, know how to interpret human signaling. So, to warn an astronaut of danger, a robot working alongside could lift the palm of its hand in that universal cue that says "stop." And the person working with the robot could also communicate with simple gestures.[4] But more than marking progress toward such practical applications, Cog and Kismet generate feelings of kinship. We've already seen that when this happens, two ideas become more comfortable. The first is that people are not so different from robots; that is, people are built from information. The second is that robots are not so different from people; that is, robots are more than the sum of their machine parts.

From its very beginnings, artificial intelligence has worked in this space between a mechanical view of people and a psychological, even spiritual, view of machines. Norbert Weiner, the founder of cybernetics, dreamed in the 1960s that it was "conceptually possible for a human being to be sent over a telegraph line," while in the mid-1980s, one MIT student mused that his teacher, AI pioneer Marvin Minsky, really wanted to "create a computer beautiful enough that a soul would want to live in it."[5] Whether or not a soul is ready to inhabit any of our current machines, reactions to Cog and Kismet bring this fantasy to mind. A graduate student, often alone at night in the lab with Kismet, confides, "I say to myself it's just a machine, but then after I leave, I want to check on it at night, just to make sure it's okay." Not surprisingly, for we have seen this as early as the ELIZA program, both adults and children are drawn to do whatever it takes to sustain a view of these robots as sentient and even caring.[6] This complicity enlivens the robots, even as the people in their presence are enlivened, sensing themselves in a relationship.

Over the years, some of my students have even spoken of time with Cog and Kismet by referring to a robotic "I and thou."[7] Theologian Martin Buber coined this phrase to refer to a profound meeting of human minds and hearts. It implies a symmetrical encounter. There is no such symmetry between human beings and even the most advanced robots. But even simple actions by Cog and Kismet inspire this extravagance of description, touching, I think, on our desire to believe that such symmetry is possible. In the case of Cog, we build a "thou" through the body. In the case of Kismet, an expressive face and voice do the work. And both robots engage with the power of the gaze. A robotic face is an enabler; it encourages us to imagine that robots can put themselves in our place and that we can put ourselves in theirs.[8]

When a robot holds our gaze, the hardwiring of evolution makes us think that the robot is interested in us. When that happens, we feel a possibility for deeper connection. We want it to happen. We come to sociable robots with the problems of our lives, with our needs for care and attention. They promise satisfactions, even if only in fantasy. Getting satisfaction means helping the robots, filling in where they are not yet ready, making up for their lapses. We are drawn into necessary complicities.

I join with Brian Scassellati and Cynthia Breazeal, the principal designers for Cog and Kismet respectively, on a study of children's encounters with these robots.[9] We introduce them to sixty children, from ages five to fourteen, from a culturally and economically diverse cross section of local communities. We call it our "first-encounters" study because in most cases, the children meet Cog or Kismet just once and have never previously seen anything like them.

When children meet these robots, they quickly understand that these machines are not toys—indeed, as I have said, these robots have their *own* toys, an array of stuffed animals, a slinky, dolls, and blocks. The laboratory setting in which adults engage with the robots says, "These robots don't belong *to* you, they belong *with* you." It says, "They are not *for* you; in some important way, they are *like* you." Some children wonder, if these robots belong with people, then what *failings* in people require robots? For one thirteen-year-old boy, Cog suggests that "humans aren't good enough so they need something else."

In our first-encounters study children's time with the robots is unstructured. We ask questions, but not many. The children are encouraged to say whatever comes to mind. Our goal is to explore some rather open questions: How do children respond to an encounter with a novel form of social intelligence? What are they looking for?

To this last, the answer is, most simply, that children want to connect with these machines, to teach them and befriend them. And they want the robots to like, even love, them. Children speak of this directly ("Cog loves me"; "Kismet is like my sister; she loves me"; "He [Cog] is my pal; he wants to do things with me, everything with me. Like a best friend."). Even the oldest children are visibly moved when Kismet "learns" their names, something that this robot can do but only rarely accomplishes. Children get unhappy if Kismet says the name of another child, which they often take as evidence of Kismet's disinterest.

Children are willing to work hard, really hard, to win the robots' affection. They dance for the robots and sing favorite childhood songs: "The Farmer in the Dell," "Happy Birthday," "Three Blind Mice." They try to make the robots happy with stuffed animals and improvised games. One ten-year-old boy makes clay treats for Kismet to eat and tells us that he is going "to take care of it and protect it against all evil." But because Cog and Kismet cannot like or dislike, children's complicity is required to give the impression that there is an emerging fondness. Things can get tense. These more sophisticated robots seem to promise more intimacy than their simpler "cousins." So when they do not gratify, they seem more "withholding."

During our study Cog has a broken arm, and Kismet is being modified for research purposes. On many days both robots are "buggy." Children work gamely around these limitations. So, on a day when there are problems with Kismet's microphone, some children try out the idea that Kismet is having trouble talking because it speaks a foreign language. A five-year-old decides that this language is Korean, his own language. A twelve-year-old argues for French, then changes her mind and decides on Spanish. When Kismet finally does speak to her, she is pleased. She says that she was right about the Spanish. "He trusts me," she says happily, and bids the robot good-bye with a wave and an *adios*. Of course, children are sometimes exhausted by a robot's quirky malfunctions or made anxious when attempts to charm a broken machine fail. There are disappointments and even tears. And yet, the children persevere. The robots are alive enough to keep them wanting more.

As we saw with simpler robots, the children's attachments speak not simply to what the robots offer but to what children are missing. Many children in this study seem to lack what they need most: parents who attend to them and a sense of being important. Children imagine sociable machines as substitutes for the people missing in their lives. When the machines fail, it is sometimes a moment to revisit past losses. What we ask of robots shows us what we need.

BUILDING A "THOU" THROUGH THE BODY

When children realize that Cog will not speak, they do not easily give up on a feeling that it should. Some theorize that it is deaf. Several of the children have learned a bit of American Sign Language at school and seize on it as a way to communicate. They do not question the idea that Cog has things it wants to say and that they would be interested to hear.

When Allegra, nine, meets Cog, she reaches out to shake its hand. Cog returns her gesture, and they have a moment when their eyes and hands lock. Allegra then wants to know if it is possible to make a mouth for Cog. The robot has a mouth, but Allegra means a mouth that can speak. Like the five-year-old who thought that a Furby should have arms "because it might want to hug me," Allegra explains that Cog "probably wants to talk to other people . . . and it might want to smile." Allegra also thinks that an "improved" Cog should know how to dance. Scassellati asks, "Should it just dance for you or should it be able to dance with you?" Allegra's answer is immediate: "Dance with me!" Inspired, she begins to dance, first hip-hop, then the slow and graceful turns of ballet. In response, Cog moves its head and its one functional arm. Robot and child are bound together. After a few minutes, Allegra says, "If his [Cog's] other arm could move, I think that I would teach him to hug me." Cog has become alive enough to love her. Later, Allegra makes her dance steps more complex and rapid. Now she dances not with but for Cog. She wants to please it, and she says, "a little bit I want to show off for him."

Brooke, seven, comes to her session with Cog hoping that it has "a heart . . . and tonsils" so that it will be able to talk and sing with her. When this doesn't work out, she moves on to teaching Cog to balance its toys—stuffed animals, a slinky, blocks—on its arms, shoulders, and neck. When things go awry, as they often do (Cog can rarely balance the toys), she gently chides the robot: "Are you paying attention to me, mister?" She says that Cog's failures are perhaps due to her not having identified its favorite toy, and she remains Cog's dedicated tutor. Cog finally succeeds in balancing its slinky and this reanimates the robot in her eyes. When Cog fails in successive attempts, Brooke assumes it has lost interest in her game. She asks it, "What's the matter?" She never questions her pupil's competency, only its desire.

But Brooke yearns to talk to the robot. She tells Cog that at home she feels ignored, in the shadow of her eleven-year-old sister Andrea, who is scheduled to meet Cog later that day: "Nobody talks to me. . . . Nobody listens to me." When Cog responds with silence, she is distressed. "Is he trying to tell me to go away?" she asks. "Cog, Cog, Cog . . . why aren't you listening to me?" Suddenly, she has an idea and declares, "I didn't think of this before. . . . This is what you have to do." She begins to use sign language. "I know how to say 'house'. . . . I can teach him to say 'house' [she taps her head with her right palm, making the sign for house]." Then she signs "eat" and "I love you" as Cog focuses on her hands. She is happy that Cog pays attention: "He loves me, definitely."

Now, feeling both successful and competitive, Brooke boasts that she has a better relationship with Cog than her sister will have: "She's probably just going to talk to Cog. I'm not just talking. I'm teaching." As Brooke leaves, she announces to the research team, "I wanted him to speak to me. I know the robot down the hall [Kismet]is the talking one. But I really wanted *him* to talk."

Scassellati is used to hearing such sentiments. He has worked on Cog for seven years and seen a lot of people behave as though smitten with his robot and frustrated that it will not talk with them. He uses the first-encounters study for an experiment in what he considers "responsible pedagogy." Thirty of the children in our study participate in a special session during which Scassellati demystifies Cog. One by one, Scassellati disables each element of Cog's intelligence and autonomy. A robot that began the session able to make eye contact and imitate human motion ends up a simple puppet—the boy Pinocchio reduced to wood, pins, and string.

So later that day, Scassellati "debriefs" Brooke and Andrea. He shows the sisters what Cog sees on its vision monitors and then covers its "eyes"—two cameras for close vision, two for distance vision—and the girls watch the four monitors go blank, one after another. They are given a computer mouse that controls Cog's movement and they get to "drive" it.

Together, the sisters direct Cog's eyes toward them. When Cog "sees" them, as evidenced by their appearance on its vision monitors, the quiet, didactic tone of the debriefing breaks down. Brooke screams out, "He's looking at us" and the carefully built-up sense of Cog as mechanism is gone in a flash. Even as the girls control the robot as though it were a puppet, they think back to the more independent Cog and are certain that it "likes" looking at them.

As Scassellati proceeds with this debriefing, he tries to demonstrate that Cog's "likes and dislikes" are determined by its programming. He shows the girls that what has Cog's attention appears in a red square on a computer screen. They can control what gets into the square by changing what its program interprets as being of the highest value. So, for example, Cog can be told to look for red things and skin-colored things, a combination that would have Cog looking for a person with a red shirt.

Despite this lesson, the sisters refer to the red square as "the square that says what Cog likes," and Brooke is joyful when Cog turns toward her hand: "Yep, he likes it." They try to get Cog's interest with a multicolored stuffed caterpillar, which, to their delight, makes it into Cog's red square as well. Cog also likes Brooke's leg. But she is troubled that Cog does not like a Mickey Mouse toy. On one hand, she understands that Cog's lack of interest is due to Mickey's coloration, half black and half red. The black is keeping Mickey from being registered as a favorite. "I see," says Brooke, "Mickey is only half red." But she continues to talk as though it is within Cog's power to make Mickey a favorite. "I really want Cog to like Mickey. I like Mickey. Maybe he's *trying* to like Mickey."

The children imbue Cog with life even when being shown, as in the famous scene from the *Wizard of Oz*, the man (or, in this case, the machines) behind the magic. Despite Scassellati's elegant explanations, the children want Cog to be alive enough to have autonomy and personality. They are not going to let anyone take this away. Scassellati's efforts to make the robot "transparent" seem akin to telling someone that his or her best friend's mind is made up of electrical impulses and chemical reactions. Such an explanation is treated as perhaps accurate but certainly irrelevant to an ongoing relationship.

Scassellati is concerned that Cog's lifelike interface is deceptive; most of his colleagues take a different view. They want to build machines that people will relate to as peers. They don't see lifelike behaviors as deceptions but as enablers of relationship. In *The Republic*, Plato says, "Everything that deceives may be said to enchant."[10] The sentiment also works when put the other way around. Once Cog enchants, it is taken as kin. That which enchants, deceives.

Children have met this idea before; it is a fairy tale staple. More recently, in the second volume of the Harry Potter series, a tale of young wizards in training, Harry's friend Ginny Weasley falls under the spell of an interactive diary. She writes in it; it writes back. It is the wizarding version of the ELIZA program. Even in a world animated by living objects (here, people in photographs get to move around and chat), a caution is served. Ginny's father, himself a wizard, asks, "Haven't I taught you anything? What have I always told you? Never trust anything that can think for itself *if you can't see where it keeps its brain*."[11] But, of course, it is too late. When something seems to thinks for itself, we put it in the category of "things we form relationships with." And then we resist having information about mechanisms—or a detail such as where it keeps its brain— derail our connection. Children put Cog in that charmed circle.

When Scassellati turns Cog into a limp puppet, showing where Cog "keeps its brain," children keep the autonomous and responsive Cog in mind. They see Cog's malfunctions as infirmities, reasons to offer support. Part of complicity is "covering" for a robot when it is broken. When Cog breaks its arm, children talk about its "wounds." They are solicitous: "Do you think it needs some sort of, well, bandage?"

BUILDING A THOU THROUGH A FACE AND A VOICE

As with Cog, children will describe a "buggy" Kismet as sick or needing rest. So, on days when Kismet does not speak, children talk to the "deaf" Kismet and discuss how they will chat with it when it "gets better." Robyn, nine, is chatting with an expressive and talkative Kismet that suddenly goes mute and immobile. Robyn's reaction: "He is sleeping."

Sometimes children weave complex narratives around Kismet's limitations. Lauren, ten, gets into a happy rhythm of having Kismet repeat her words. When Kismet begins to fail, Lauren likens the robot's situation to her own. It is not always possible to know what Kismet is learning just from watching "what is happening on the outside" just as we cannot observe what is happening inside of her as she grows up. Despite its silence, Lauren believes that Kismet is growing up "inside." Lauren says that Kismet is "alive enough" to have parents and brothers and sisters, "and I don't see them around here." Lauren wonders if their absence has caused Kismet to fall silent.

Fred, eight, greets Kismet with a smile and says, "You're cool!" He tells us that he is terrorized by two older brothers whose "favorite pastime is to beat me up." A robot might help. He says, "I wish I could build a robot to save me from my brothers. . . . I want a robot to be my friend. . . . I want to tell my secrets." Fred stares intently into Kismet's large blue eyes and seems to have found his someone. In response to Fred's warm greeting, Kismet vocalizes random sounds, but Fred hears something personal. He interprets Kismet as saying, "What are you doing, Rudy [one of Fred's brothers]?" Fred is not happy that Kismet has confused him with one of his roughhousing brothers and corrects Kismet's error. "I'm Fred, not Rudy. I'm here to play with you." Fred is now satisfied that Kismet has his identity squared away as the robot continues its soft babble. Fred is enchanted by their interchange. When Fred presents a dinosaur toy to Kismet, it says something that sounds like "derksherk," which Fred inteprets as Kismet's pronunciation of dinosaur. During one back-and-forth with Kismet about his favorite foods, Fred declares victory: "See! It said cheese! It said potato!"

When Kismet sits in long silence, Fred offers, "Maybe after a while he gets bored." When Kismet shows no interest in its toys, Fred suggests, "These toys probably distract Kismet." At this point, the research team explains Kismet's workings to Fred—the Kismet version of Scassellati's "Cog demystification" protocol. We show Fred the computer monitor that displays what Kismet is "hearing." Fred, fascinated, repeats what he sees on the monitor, hoping this will make it easier for Kismet to understand him. When this strategy doesn't prompt a response, Fred blames Kismet's bad hearing. But in the end, Fred concludes that Kismet has stopped talking to him because it likes his brothers better. Fred would rather feel rejected than see Kismet as a less than adequate relational partner.

Amber, six, also fights to keep Kismet alive enough to be a friend. On the day Amber visits MIT, Kismet's face is expressive but its voice is having technical difficulty. The young girl, unfazed, attends to this problem by taking Kismet's part in conversation. So, Amber engages Kismet with a toy and asks Kismet if she is happy. When Kismet doesn't answer, Amber answers for it with a hearty "Yep!"

When after many minutes, Kismet haltingly begins to speak, Amber's response is immediate: "He likes me!" Now, Kismet babbles, and Amber interprets. The young girl says aloud what Kismet *meant* to say and then engages in a conversation with Kismet based on her interpretation. Before leaving Kismet, Amber tries hard to have the robot say, "I love you." After a half dozen prompts, Kismet says something close enough. Amber thanks Kismet, says, "I love you too," and kisses the robot good-bye.

In some ways, Amber's time with Kismet resembles play with a traditional doll, during which a child must "fill in" both sides of the interaction. But even at its worst, Kismet gives the appearance of *trying* to relate. At its best, Kismet appears to be in continuous, expressive conversation. As with Cog, Kismet's failures can be interpreted as disappointments or rejections—very human behaviors. Your Raggedy Ann doll cannot actively reject you. When children see a sociable robot that does not pay attention to them, they see something alive enough to mean it.

BUILDING A THOU BY CARING

Children try to get close to Cog and Kismet by tending them. Children ask the robots how they are feeling, if they are happy, if they like their toys. Robyn, the nine-year-old who imagines Kismet asleep when it mysteriously stops speaking, thinks that the robot is alive because "it talks and moves like a person." When Kismet develops problems, Robyn wants to take it home to "feed it and give it water to drink so that it wouldn't die; I would give it a Tylenol if it felt sick and I would make Kismet his own room." The room, Robyn explains, would have a television on which Kismet could "see other robots so it wouldn't miss its family and friends."

As children see it, they teach the robots, and the robots appreciate it, even if they are imperfect pupils. Over half the children in the first-encounters study say, unprompted, that they love the robots and the robots love them back. From those who don't speak of love, there is still talk about Cog and Kismet having made a "good effort" during their lessons. When children congratulate the robots one hears something akin to parental pride. When the robots succeed, in even the smallest thing, children take credit and present each success as evidence that their own patience has borne fruit. During our study the robots' performance is subpar. But the children's investment—their desire, connection, and pride—makes the sessions sparkle.

This is clear in the relationship that Neela, eleven, forms with Cog. When Neela first sees Cog, she exclaims, "Oh, it's so cute!" and then explains, "He has such innocent eyes, and a soft-looking face." After teaching the robot to balance a stuffed caterpillar on its arm, she says, "I could never get tired of Cog. . . . It's not like a toy because you can't teach a toy; it's like something that's part of you, you know, something you love, kind of like another person, like a baby." When Cog raises its arm, Neela says, "I wonder what he's thinking?" She asks, "What do you want?" "What do you like?" When Cog hesitates in his play—for example, when he is slow to raise his arm in response to her actions—Neela never uses a mechanical explanation for Cog's trouble. Her reasoning is always psychological. She says that Cog reminds her of the "slow kids" in her class and she is sympathetic. "He's slow—it takes him a while to run through his brain." And she wants to help. "I want to be its friend, and the best part of being his friend would be to help it learn. . . . In some ways Cog would be better than a person-friend because a robot would never try to hurt your feelings." (This is an eleven-year-old's version of the comment made by the graduate student who wanted a robot boyfriend.) For Neela, a silent Cog is simply disabled: "Being with Cog was like being with a deaf or blind person because it was confused, it didn't understand what you were saying." In fact, Neela says that Cog does "see"—just not very well during her visit. To compensate, Neela treats the robot as a person having a bout of temporary blindness. "I was just like, 'Hello!' because a blind person would have to listen." Neela hopes that Cog will get over its problems or that "he might grow out of it. . . . He's very young you know."

Neela has recently arrived from India and is having trouble fitting in at school. She explains that a group of girls seemed to accept her but then made fun of her accent: "Girls are two-faced. They say they like you and then they don't. They can't make up their mind." Cog poses fewer risks. At school the girls who taunted her finally begged her forgiveness, but Neela hasn't been able to accept their apology. In this last regard, "Cog could be a better friend than a person because it is easier to forgive. . . . It's easier to forgive because it doesn't really understand." Recall that Neela speaks of Cog as "part of you . . . something you love." This is love safe from rejection. Like any object of love, the robot becomes "part of you." But for Neela, Cog, unlike a person, does not have enough independence to hurt you. In Neela's feelings for Cog we see how easily a robot can become a part object: it will meet our emotional needs because we can make it give us what we want. Is this an object for our times? If so, it is not an object that teaches us how to be with people.

Some children, particularly with Kismet, explicitly put themselves in the role of sibling or parent. In either of these roles, the relationship with Kismet may become a place to reenact the tensions in a family, something we have already seen with AIBO and My Real Baby. In the pursuit of Kismet, brothers come to blows and sisters bitterly compete. And efforts to parent Kismet can be a critique of what goes on at home. Rain, ten, lives with her mother and is preoccupied by her father's absence. She explains that she would never abandon Kismet: "My father doesn't live at home; he moved away. If Kismet came to live with me, I would never move away, ever. I would leave him juice every morning. I would make him a comfortable bed. And I would teach it to really talk, not just the little bit it knows now." There is much similarity between this kind of talk and what happens in a therapist's office when children act out their conflicts with their dolls. A doll can let you vent feelings, enjoy imaginary companionship, and teach you what is on your mind. But unlike dolls, these robots "push back." Children move beyond using the robot to relive past relationships. They hope for a relationship with the robot in the real.

Madison, nine, works with Kismet on a day when the robot is at its best. Its emotive face is responsive and appropriate. It remembers words and repeats them back in humanlike cadences. The result looks like Madison is speaking earnestly to someone whose inflection and tone make her feel perfectly understood.

Madison asks Kismet questions in a gentle and soft-spoken manner, "What is your name? Do you have parents?" Kismet responds warmly. Encouraged, Madison continues. "Do you have brothers and sisters?" Kismet moves its head in a way that suggests to Madison that the answer is yes. Madison tells us that Kismet is a little girl (she was "born from a stomach"), but a new kind of little girl. And like any baby, "she" doesn't know when "her" birthday is. Madison wants to be "her" good parent. "Do you like ice cream?" Madison asks, and when Kismet quietly responds to this question, the two go on to discuss ice cream flavors, favorite colors, and best toys.

Madison begins to dangle one toy after another in front of Kismet's face, laughing at its changing expressions. Madison tells Kismet that some of the girls in her school are mean; she says that Kismet is nicer than they are. Kismet looks at Madison with interest and sounds encouraging. In this warm atmosphere, Madison tells Kismet that she looks forward to introducing the robot to her baby sister. Playing with her sister, Madison says, is her favorite thing to do, and she expects Kismet will feel the same way. Kismet nods and purrs happily. Again, projection onto an object becomes engagement with a subject; Rorschach gives way to relationship.

Madison believes that Kismet learns from every child who comes to play. But you can't be impatient. "Babies learn slowly," she offers. Like a baby, Kismet, too, will learn over time. "I taught Kismet to smile," Madison says. "[Kismet] is still little, but it grows up." To justify this claim, Madison, like Lauren, distinguishes between what you can see of a child's learning and what is hidden from view: "You can't always tell what babies are learning by looking at them on any day." The same is true for Kismet. Kismet is learning "inside" even if we can't see it. A mother knows her child has secrets.

In the hour she plays with Kismet, Madison becomes increasingly happy and relaxed. Watching girl and robot together, it is easy to see Kismet as increasingly happy and relaxed as well. Child and robot are a happy couple. It is almost impossible not to see Madison as a gratified mother and Kismet as a content child. Certainly, Kismet seems to prefer Madison to the children who have visited with it earlier that day. For me, their conversation is one of the most uncanny moments in the first-encounters study, stunning in its credibility because Kismet does not know about ice cream flavors, baby sisters, or mean girls. Kismet does not like Madison; it is not capable of liking anything or anybody.

BUILDING A THOU IN DISAPPOINTMENT AND ANGER

The children in the study care about having the robots' attention and affection far more than I anticipated. So their interpretation of robot malfunctions as illness is ingenious; they can walk away without feeling dismissed. But the most vulnerable children take disappointments with a robot very personally. The children most upset by a robot's indifference are those who feel least tended to. They seem almost desperate for Kismet and Cog to recognize and respond to them. Since the children in our study come from a wide range of backgrounds, some tell us that the snack they get during their session at MIT is the best meal of their day. Some find ways to make it clear that their time at MIT is the most attention they have received that week. Children from affluent as well as economically disadvantaged homes talk about parents they rarely see. When these children interpret robotic technical limitations as rejection, they become withdrawn, depressed, or angry. Some take foolish chances.

My field notes taken after one session with Kismet describe a conversation with the junior members of my research team, two college seniors and two graduate students: "Emergency meeting with team after session with Estelle. Disappointment with Kismet provokes her binge eating, withdrawal. Team feels responsible. How to handle such children? What did child want? A friend? A future?" My team meets at a local coffee shop to discuss the ethics of exposing a child to a sociable robot whose technical limitations make it seem uninterested in the child.

We have spent the afternoon with twelve-year-old Estelle, who had seen the flyer describing our work on the bulletin board of her after-school center: "Children wanted for study. Meet MIT Robots!" She brought it to her counselor and asked to participate. Estelle tells us that she "stood over my counselor while she called MIT." Estelle has taken special care with her appearance in preparation for her day with us. She is dressed in her best clothes, her hair brushed to a fine polish. As soon as we picked her up, Estelle talks nonstop about "this wonderful day." She has never been to MIT, but she knows it is a "very important place." No one in her family has been to college. "I am the first one to go into a college . . . today."

On the day of Estelle's visit, Kismet engages people with its changing facial expressions but is not at its vocal best. We explain Kismet's technical problems to Estelle, but nonetheless, she makes every effort to get Kismet to speak. When her efforts bear no fruit, Estelle withdraws, sullen. She goes to the room where we interview children before and after they meet the robots. There we have set out some simple snacks. Estelle begins to eat, not stopping until we finally ask her to leave some of the crackers, cookies, and juice boxes for other children. She briefly stops eating but begins again as we wait for the car service that will bring her back to the after-school program. She tells us that the robot does not like her. We explain this is not the case. She is unappeased. From her point of view, she has failed on her most important day. As Estelle leaves, she takes four boxes of cookies from our supply box and puts them into her backpack. We do not stop her. Exhausted, we reconvene to ask ourselves a hard question: Can a broken robot break a child? We would not consider the ethics of having children play with a damaged copy of Microsoft Word or a torn Raggedy Ann doll. But sociable robots provoke enough emotion to make this ethical question feel very real.

The question comes up again with Leon, twelve. Timid and small for his age, Leon usually feels like the odd man out. In Cog, Leon sees another figure who "probably doesn't have a lot of friends," and Leon says they have a good chance to connect. But, like Estelle, Leon has not come to the laboratory on a good day. Cog is buggy and behaves as though bored. The insecure child is quick to believe that the robot is not interested in him. Leon had been shown Cog's inner workings, and Scassellati gently reminds Leon that Cog's "interests" are set by people adjusting its program. Leon sees the monitor that reflects these preset values, but he insists that "Cog doesn't really care about me." He explodes in jealousy when he sees Cog looking at a tall, blond researcher, even as Scassellati points to the researcher's red T-shirt, the true lure that mobilizes Cog's attention. Leon cannot focus. He insists that Cog "likes" the researcher and does not like him. His anxieties drive his animation of the robot.

Now Leon embarks on an experiment to determine whether Cog cares about him. Leon lifts and then lowers his arm and waits for Cog to repeat what he has done. Cog lifts its arm and then, as the robot's arm moves down, Leon puts his head directly in its path. This is a love test: if Cog stops before hitting him, Leon will grant that Cog cares about him. If the falling arm hits Leon, Cog doesn't like him. Leon moves swiftly into position for the test. We reach out to stop him, appalled as the child puts his head in harm's way. Cog's arm stops before touching Leon's head. The researchers exhale. Leon is jubilant. Now he knows that Cog is not indifferent. With great pleasure, he calls out "Cog!" and the robot turns toward him. "He heard me! He heard me!"

After Leon has been with Cog for about an hour, the boy becomes preoccupied with whether he has spent enough time with Cog to make a lasting impression. His thoughts return to the tall blond researcher who "gets to be with Cog all the time." Leon is sure that Cog is in love with her. Leon chides her: "He keeps looking at you. He is in love with you." Leon then settles on a new idea: "Cog is a boy and so obviously likes girls more than boys." This at least is a reason why he doesn't stand a chance here. Leon wonders whether he might have more success with Kismet, which the children usually see as a female because of its doll eyes, red lips, and long eyelashes.

Most children find a way to engage with a faltering robot, imagining themselves as parents or teachers or healers. But both Estelle and Leon became depressed when they were not "recognized." Other frustrated children persevere in anger. Edward, six, is small for his age. What he lacks in size he makes up for in energy. From the start, he announces that he wants to be the "best at everything about the robots." His father tells us that at home and at school, Edward likes to be "in charge." He plays rough and gets into fights. With no prologue, Edward walks up to Kismet and asks, "Can you talk?" When Kismet doesn't answer, Edward repeats his question at greater volume. Kismet stares into space. Again, Edward asks, "Can you talk?" Now, Kismet speaks in the emotionally layered babble that has delighted other children or puzzled them into inventive games. This is not Edward's reaction to this winsome speaker of nonsense. He tries to understand Kismet: "What?" "Say that again?" "What exactly?" "Huh? What are you saying?" After a few minutes, Edward decides that Kismet is making no sense. He tells the robot, "Shut up!" And then, Edward picks up objects in the laboratory and forces them into Kismet's mouth—first a metal pin, then a pencil, then a toy caterpillar. Edward yells, "Chew this! Chew this!" Absorbed by hostility, her remains engaged with the robot.

Shawn, six years older than Edward, has a similar reaction. He visits the lab with his two younger brothers on whom he rains insults as they all wait to visit the robots. When Shawn meets Kismet, he calms down, and his tone is friendly: "What's your name?" But when Kismet is silent, Shawn becomes enraged. He covers the cameras that serve as Kismet's eyes and orders, "Say something!" Kismet remains silent. Shawn sits silently too, staring at Kismet as though sizing up an opponent. Suddenly, he shouts, "Say, 'Shut up!' Say, 'Shut up!'" "Say, 'Hi!' . . . Say, 'Blah!'" The adults in the room are silent; we gave the children no rules about what they could and could not say. Suddenly, Kismet says, "Hi." Shawn smiles and tries to get Kismet to speak again. When Kismet does not respond, Shawn forces his pen into Kismet's mouth. "Here! Eat this pen!" Shawn, like Edward, does not tire of this exercise.

One way to look at Estelle and Leon, Edward and Shawn is to say that these children are particularly desperate for attention, control, and a sense of connection. And so, when the robots disappoint, they are more affected than other children. Of course, this is true. But this explanation puts the full burden on the children. Another way to look at their situation puts more of the burden on us.

What would we have given to these children if the robots had been in top form? In the cases of Edward and Shawn, we have two "class bullies," the kids everyone is afraid of. But these boys are lonely. As bullies, they are isolated, often alone or surrounded by children who are not friends but whom they simply boss around. They see robots as powerful, technological, and probably expensive. It is exciting to think about controlling something like that. For them, a sociable robot is a possible friend—one that would not ask for too much in return and would never reject them, but in whom they might confide. But like the insecure Estelle and Leon, these are the children who most need relationships that will model mutuality, where control is not the main thing on the table. Why do we propose machine companionship to them in the first place? From this perspective, problems aren't limited to when the robots break down. Vulnerable children are not helped even when the robots are doing just fine.

AGAIN, ON AN ETHICAL TERRAIN

In the robot laboratory, children are surrounded by adults talking to and teaching robots. The children quickly understand that Cog needs Brian Scassellati and Kismet needs Cynthia Breazeal. The children imagine Scassellati and Breazeal to be the robots' parents. Both are about to leave the Artificial Intelligence laboratory, where they have been graduate students, and move on to faculty positions.

Breazeal will be staying at MIT but leaving the AI Lab for the Media Lab. The two are down the street from each other, but the tradition of academic property rights demands that Kismet, like Cog, be left behind in the laboratory that paid for its development. The summer of the first-encounters study is the last time Breazeal will have access to Kismet. Breazeal describes a sharp sense of loss. Building a new Kismet will not be the same. This is the Kismet she has "raised" from a "child." She says she would not be able to part with Kismet if she weren't sure it would remain with people who would treat it well.

It comes as no surprise that separation is not easy for Breazeal; more striking is how hard it is for those around Kismet to imagine the robot without her. A ten-year-old who overhears a conversation among graduate students about how Kismet will remain in the lab quietly objects, "But Cynthia is Kismet's mother."[12] Watching Breazeal interact with Kismet, one does sense a maternal connection, one that Breazeal describes as "going beyond its being a mere machine." She knows Kismet's every move, and yet, she doesn't. There are still surprises that delight. Her experience calls to mind a classic science fiction story by Brian Ald-iss, "Supertoys Last All Summer Long," best known through its movie adapta-tion, the Steven Spielberg film *A.I.: Artificial Intelligence*.[13] In *A.I.*, scientists build a humanoid robot, David, who is programmed to love. David expresses his love to a woman, Monica, who has adopted him as her child.

The pressing issue raised by this film is not the potential reality of a robot that "loves"—we are far from building anything like the robot David—but how Monica's feelings come about. Monica is a human being who responds to a ma-chine that asks for nurturance by caring for it. Her response to a robot that reaches out to her is confusion mixed with love and attachment.

It would be facile to make a simple analogy between Breazeal's situation and that of Monica in *A.I.*, but Breazeal is, in fact, one of the first people to have one of the signal experiences in that story—sadness caused by separation from a robot to which one has formed an attachment based on nurturance. At issue here is not Kismet's achieved level of intelligence but Breazeal's journey: in a very limited sense, Breazeal "brought up" Kismet. But even that very limited ex-perience provokes strong emotion. Being asked to nurture a machine constructs us as its parents. This new relationship creates its own loop, drawing us into the complicities that make it possible. We are asked to nurture. We want to help. We become open to playing along, willing to defer to what the robot is able to do.

In fiction and myth, human beings imagine themselves "playing God" and creating new forms of life. Now, in the real, sociable robots suggest a new dy-namic. We have created something that we relate to as an "other," an equal, not something over which we wield godlike power. As these robots get more sophisticated—more refined in their ability to target us—these feelings grow stronger. We are drawn by our humanity to give to these machines something of the consideration we give to each other. Because we reach for mutuality, we want them to care about us as we care for them. They can hurt us.

I noted earlier the chilling credibility of the interaction between Madison and Kismet and the desperation of children who seem to need these robots too much. Cog and Kismet are successful in getting children to relate to them "for real." It is the robots' success that gives me pause, as does the prospect of "con-versations" between the most needy among us—the disadvantaged young, the deprived elderly, the emotionally and physically disabled—and ever more lifelike sociable robots. Roboticists want us to consider a "best-case" scenario in which robotic companions serve as mentors, first steps toward more complex encounters. Even My Real Baby was marketed as a robot that could teach your child "socialization." I am skeptical. I believe that sociable technology will always disappoint because it promises what it cannot deliver. It promises friendship but can only deliver performances. Do we really want to be in the business of manufacturing friends that will never be friends?

Roboticists will argue that there is no harm in people engaging in conversations with robots; the conversations may be interesting, fun, educational, or comforting. But I find no comfort here. A machine taken as a friend demeans what we mean by friendship. Whom we like, who likes us—these things make us who we are. When Madison felt joyful in Kismet's "affection," I could not be glad. I felt in the shadow of an experiment, just beginning, in which humans are the subjects.

Even now, our excitement about the possibilities for robot/human interaction moves us to play fast and loose with our emotions. In one published experiment, two young children are asked to spend time with a man and a robot designed to be his clone.[14] The experiment has a significant backstory. Japanese roboticist Hiroshi Ishiguro built androids that duplicate himself, his wife, and his five-year-old daughter. The daughter's first reaction when she saw her android clone was to flee. She refused to go near it and would no longer visit her father's laboratory. Years later, when the daughter was ten, a group of psychologists designed a study in which this girl and a four-year-old boy (a child of one of the researchers) were asked to interact with both Ishiguro and his android double. Both children begin the study reluctant to interact with the android. Then, both (by measures such as "makes eye contact" and "speaks") become willing to engage almost equally with the man and with the robot. Ishiguro's daughter is finally able to sit in a room alone with her father's android clone. It is hard to know how to comment on this narrative of a frightened child who makes ever-fainter objections to her part in this experiment. It seems to have little in it that is positive. Yet, the authors use this narrative as evidence of success: children will be open to humanlike robots as teachers, babysitters, and companions. But what could it mean to this child to sit with her father's machine double? What could she want from it? Why does it matter that she is finally willing to make eye contact and speak with it? Why would we want her to? It is easy to become so immersed in technology that we ignore what we know about life.

CHAPTER 6

love's labor lost

When Takanori Shibata took the floor at a spring 2009 meeting at MIT's AgeLab, he looked triumphant. The daylong conference centered on robots for the elderly, and Shibata, inventor of the small, seal-like sociable robot Paro, was the guest of honor. The AgeLab's mission is to create technologies for helping the elderly with their physical and emotional needs, and already Paro had carved out a major role on this terrain. Honored by Guinness Records as "the most therapeutic robot in the world" in 2002, Paro had been front and center in Japan's initiative to use robots to support senior citizens.[1] Now Shibata proudly announced that Denmark had just placed an order for one thousand Paros for its elder-care facilities. The AgeLab gathering marked the beginning of its American launch.

Shibata showed a series of videos: smiling elderly men and women in Japanese nursing homes welcoming the little furry "creature" into their arms; seniors living at home speaking appreciatively about the warmth and love that Paro brought them; agitated and anxious seniors calming down in Paro's company.[2] The meeting buzzed with ideas about how best to facilitate Paro's acceptance into American elder care. The assembled engineers, physicians, health administrators, and journalists joined in a lively, supportive discussion. They discussed what kind of classification Shibata should seek to facilitate Paro's passage through the legendary scrutiny of the Food and Drug Administration.

I heard only one negative comment. A woman who identified herself as a nurse said that she and her colleagues had worked long and hard to move away from representing the elderly as childlike. To her, Paro seemed "a throwback, a new and fancier teddy bear." She ended by saying that she believed nurses would resist the introduction of Paro and objects like it into nursing homes. I lowered my eyes. I had made a decision to attend this meeting as an observer, so I said nothing. At the time, I had been studying Paro in Massachusetts nursing homes for several years. Most often, nurses, attendants, and administrators had been happy for the distraction it provided. I was not at all sure that nurses would object to Paro.

In any case, the nurse's concern was met with silence, something I have come to anticipate at such gatherings. In robotics, new "models" are rarely challenged. All eyes focus on technical virtuosity and the possibilities for efficient implementation. At the AgeLab, the group moved on to questions about Paro's price, now set at some $6,000 a unit. Was this too high for something that might be received as a toy? Shibata thought not. Nursing homes were already showing willingness to pay for so valuable a resource. And Paro, he insisted, is not a toy. It reacts to how it is treated (is a touch soft or aggressive?) and spoken to (it understands about five hundred English words, more in Japanese). It has proved itself an object that calms the distraught and depressed. And Shibata claimed that unlike a toy, Paro is robust, ready for the rough-and-tumble of elder care. I bit my lip. At the time I had three broken Paros in my basement, casualties of my own nursing home studies. Why do we believe that the next technology we dream up will be the first to prove not only redemptive but indestructible?

In contrast to these enthusiasts, we have seen children worry. Some imagined that robots might help to cure their grandparents' isolation but then fretted that the robots would prove *too* helpful. Quiet and compliant robots might become rivals for affection. Here we meet the grandparents. Over several years, I introduce seniors—some who live at home, some who live in nursing homes—to the robots that so intrigued their grandchildren: My Real Baby, AIBO, and Shibata's Paro. The children were onto something: the elderly are taken with the robots. Most are accepting and there are times when some seem to prefer a robot with simple demands to a person with more complicated ones.[3]

In one nursing home, I leave four My Real Babies over a summer. When I return in the fall, there are seven. The demand for the robot baby was so high that the nursing staff went on eBay to increase their numbers. Indeed, however popular My Real Baby is among children, it is the elderly who fall in love. The robot asks for tending, and this makes seniors feel wanted. Its demands seem genuine, in part, of course, because the staff seems to take them seriously. The elderly need to be cared for, but there are few things that they can reliably take care of. Some fear that they might fail with a pet. My Real Baby seems a sure thing, and because it is a robot brought from MIT, it seems an adult thing as well. And having a robot around makes seniors feel they have something "important" to talk about.

The thoughtful fifth graders said their grandparents might welcome robots because, unlike pets, they do not die. The children were right. When the robots are around, seniors are quick to comment that these "creatures" do not die but can be "fixed." Children imagined that robot baby dolls will remind older people of their time as parents and indeed, for some seniors, My Real Baby does more than bring back memories of children; it offers a way to reimagine a life. But in all of this, I do not find a simple story about the virtues of robots for the elderly. In the nursing homes I study, "time with robots" is made part of each institution's program. So, the seniors spend time with robots. But over years of study, when given the choice between hanging out with a robot and talking to one of the researchers on the MIT team, most seniors, grateful, choose the person.

During the years of our nursing home studies, it often seemed clear that what kept seniors coming to sessions with robots was the chance to spend time with my intelligent, kind, and physically appealing research assistants. One young man, in particular, was a far more attractive object of attention than the Paro he was trying to introduce. One had the distinct feeling that female nursing home residents put up with the robot because he came with it. Their appreciation, sometimes bawdy in tone, took place in one nursing home so short of resources that the management decided our study could not continue. This incident dramatized the tension in the environment that welcomes sociable robots in geriatric care. There is a danger that the robots, if at all successful, will replace people. In this case, when residents did not pay enough attention to the robot, the people who came with it were taken away. It was a depressing time.

CARING MACHINES

Twenty-five years ago the Japanese calculated that demography was working against them—there would not be enough young Japanese to take care of their aging population. They decided that instead of having foreigners take care of the elderly, they would build robots to do the job.[4] While some of the robots designed for the aging population of Japan have an instrumental focus—they give baths and dispense medication—others are expressly designed as companions.

The Japanese robot Wandakun, developed in the late 1990s, is a fuzzy koala that responds to being petted by purring, singing, and speaking a few phrases. After a yearlong pilot project that provided the "creature" to nursing home residents, one seventy-four-year-old Japanese participant said of it, "When I looked into his large brown eyes, I fell in love after years of being quite lonely. . . . I swore to protect and care for the little animal."[5] Encouraged by such experiments, Japanese researchers began to look to artificial companionship as a remedy for the indignities and isolation of age. And with similar logic, robots were imagined for the dependencies of childhood. Children and seniors: the most vulnerable first.

Over a decade, I find that most American meetings on robotics and the elderly begin with reference to the Japanese experiment and the assertion that Japan's future is ours as well: there are not enough people to take care of aging Americans, so robot companions should be enlisted to help.[6] Beyond that, some American enthusiasts argue that robots will be more patient with the cranky and forgetful elderly than a human being could ever be. Not only better than nothing, the robots will simply be better.

So, a fall 2005 symposium, titled "Caring Machines: Artificial Intelligence in Eldercare" began with predistributed materials that referred to the "skyrocketing" number of older adults while the "number of caretakers dwindles."[7] Technology of course would be the solution. At the symposia itself, there was much talk of "curing through care." I asked participants—AI scientists, physicians, nurses, philosophers, psychologists, nursing home owners, representatives of insurance companies—whether the very title of the symposium suggested that we now assume that machines can be made to "care."

Some tried to reassure me that, for them, "caring" meant that machines would *take* care of us, not that they would care *about* us. They saw caring as a behavior, not a feeling. One physician explained, "Like a machine that cuts your toenails. Or bathes you. That is a *caring computer.* Or talks with you if you are lonely. Same thing." Some participants met my objections about language with impatience. They thought I was quibbling over semantics. But I don't think this slippage of language is a quibble.

I think back to Miriam, the seventy-two-year-old woman who found comfort when she confided in her Paro. Paro *took care* of Miriam's desire to tell her story—it made a space for that story to be told—but it did not *care about* her or her story. This is a new kind of relationship, sanctioned by a new language of care. Although the robot had understood nothing, Miriam settled for what she had. And, more, she was supported by nurses and attendants happy for her to pour her heart out to a machine. To say that Miriam was having a conversation with Paro, as these people do, is to forget what it is to have a conversation. The very fact that we now design and manufacture robot companions for the elderly marks a turning point. We ask technology to perform what used to be "love's labor": taking care of each other.

At the symposium, I sensed a research community and an industry poised to think of Miriam's experience as a new standard of care. Their position (the performance of care is care enough) is made easier by making certain jobs robot ready. If human nursing care is regimented, scripted into machinelike performances, it is easier to accept a robot nurse. If the elderly are tended by underpaid workers who seem to do their jobs by rote, it is not difficult to warm to the idea of a robot orderly. (Similarly, if children are minded at day-care facilities that seem like little more than safe warehouses, the idea of a robot babysitter becomes less troubling.)

But people are capable of the higher standard of care that comes with empathy. The robot is innocent of such capacity. Yet, Tim, fifty-three, whose mother lives in the same nursing home as Miriam, is grateful for Paro's presence. Tim visits his mother several times a week. The visits are always painful. "She used to sit all day in this smoky room, just staring at a wall," Tim says of his mother, the pain of the image still sharp. "There was one small television, but it was so small, just in a corner of this very big room. They don't allow smoking in there anymore. It's been five years, but you can still smell the smoke in that room. It's in everything, the drapes, the couches. . . . I used to hate to leave her in that room." He tells me that my project to introduce robots into the home has made things better. He says, "I like it that you have brought the robot. She puts it in her lap. She talks to it. It is much cleaner, less depressing. It makes it easier to walk out that door." The Paro eases Tim's guilt about leaving his mother in this depressing place. Now she is no longer completely alone. But by what standard is she less alone? Will robot companions cure conscience?

Tim loves his mother. The nursing staff feels compassion for Miriam. But if our experience with relational artifacts is based on a fundamentally deceitful exchange (they perform in a way that persuades us to settle for the "acting out" of caring), can they be good for us? Or, as I have asked, might they be good for us only in the "feel good" sense? The answers to such questions do not depend on what computers can do today or are likely to be able to do tomorrow. They depend on what *we* will be like, the kind of people *we* are becoming as we launch ourselves and those we love into increasingly intimate relationships with machines.

Some robots are designed to deliver medication to the elderly, to help them reach for grocery items on high shelves, and to monitor their safety. A robot can detect if an elderly person is lying on the floor at home, a possible signal of distress. I take no exception to such machines. But Paro and other sociable robots are designed as companions. They force us to ask why we don't, as the children put it, "have people for these jobs." Have we come to think of the elderly as nonpersons who do not require the care of persons? I find that people are most comfortable with the idea of giving caretaker robots to patients with Alzheimer's disease or dementia. Philosophers say that our capacity to put ourselves in the place of the other is essential to being human. Perhaps when people lose this ability, robots seem appropriate company because they share this incapacity.

But dementia is often frightening to its sufferers. Perhaps those who suffer from it need the most, not the least, human attention. And if we assign machine companionship to Alzheimer's patients, who is next on the list? Current research on sociable robotics specifically envisages robots for hospital patients, the elderly, the retarded, and the autistic—most generally, for the physically and mentally challenged. When robots are suggested, we often hear the familiar assertion that there are not enough people to take care of these "people with problems." People are scarce—or have made themselves scarce. But as we go through life, most of us have our troubles, our "problems." Will only the wealthy and "well adjusted" be granted the company of their own kind?[8]

When children ask, "Don't we have people for these jobs?" they remind us that our allocation of resources is a social choice. Young children and the elderly are not a problem until we decide that we don't have the time or resources to attend to them. We seem tempted to declare phases of the life cycle problems and to send in technologies to solve them. But why is it time to bring in the robots? We learned to take industrial robots in stride when they were proposed for factory assembly lines. Now the "work" envisaged for machines is the work of caring. Will we become similarly sanguine about robotic companionship?

This is contested terrain. Two brothers are at odds over whether to buy a Paro for their ninety-four-year-old mother. The robot is expensive, but the elder brother thinks the purchase would be worthwhile. He says that their mother is "depressed." The younger brother is offended by the robot, pointing out that their mother has a right to be sad. Five months before, she lost her husband of seventy years. Most of her friends have died. Sadness is appropriate to this moment in her life. The younger brother insists that what she needs is human support: "She needs to be around people who have also lost mothers and husbands and children." She faces the work of saying good-bye, which is about the meaning of things. It is not a time to cheer her up with robot games. But the pressures to do just that are enormous. In institutional settings, those who take care of the elderly often seemed relieved by the prospect of robots coming to the rescue.

CURING A LIFE

When I introduce sociable robots—AIBO, My Real Baby, and Paro—into nursing homes, nurses and physicians are hopeful. Speaking of Paro, one nursing home director says, "Loneliness makes people sick. This could at least partially offset a vital factor that makes people sick." The robot is presented as cure. Caretakers entertain the idea that the robot might not just be better than no company but better than *their* company. They have so little time and so many patients. Sometimes, using a kind of professional jargon, nurses and attendants will say that seniors readily "tolerate" the robots—which is not surprising if seniors are not offered much else. And sometimes, even the most committed caretakers will say that robots address the "troubles" of old age by providing, as one put it, "comfort, entertainment, and distraction."[9] One physician, excited by the prospect of responsive robot pets, sees only the good: "Furbies for grandpa," he says.

Indeed, seniors generally begin their time with robots as children do, by trying to determine the nature of the thing they have been given. When given a Paro, they have many questions: "Can it do more? Is it a seal or a dog? Is it a he or a she? Can it swim? Where is it from? Does it have a name? Does it eat?" and finally, "What are we supposed to be doing with this?" When the answer is, "Be with it," only some lose interest. Over time, many seniors attach to Paro. They share stories and secrets. With the robot as a partner, they recreate the times of their lives. To do these things, the adults must overcome their embarrassment at being seen playing with dolls. Many seniors handle this by saying something like, "People would think I'm crazy if they saw me talking to this." Once they have declared themselves not crazy, they can proceed in their relationship with a robot seal. Or with a robot baby doll.

I have given Andy, seventy-six, a My Real Baby. Andy is slim and bespectacled, with sandy white hair. His face is deeply lined, and his blue eyes light up whenever I see him. He craves company but finds it hard to make friends at the nursing home. I am working with two research assistants, and every time we visit, Andy makes us promise to come back as soon as we can. He is lonely. His children no longer visit. He'd never had many friends, but the few that he'd made on his job do not come by. When he worked as an insurance agent, he had socialized with colleagues after work, but now this is over. Andy wants to talk about his life. Most of all, he wants to talk about his ex-wife, Edith. It is she he misses most. He reads us excerpts from her letters to him. He reads us songs he has written for her.

When Andy first sees My Real Baby, he is delighted: "Now I have something to do when I have nothing to do." Soon the robot doll becomes his mascot. He sets it on his windowsill and gives it his favorite baseball cap to wear. It is there to show off to visitors, a conversation piece and something of an ice breaker. But over a few weeks, the robot becomes more companion than mascot. Now Andy holds My Real Baby as one would a child. He speaks directly to it, as to a little girl: "You sound so good. You are so pretty too. You are so nice. Your name is Minnie, right?" He makes funny faces at the robot as though to amuse it. At one funny face, My Real Baby laughs with perfect timing as though responding to his grimaces. Andy is delighted, happy to be sharing a moment. Andy reassures us that he knows My Real Baby is a "toy" and not "really" alive. Yet, he relates to it as though it were sentient and emotional. He puts aside his concern about its being a toy: "I made her talk, and I made her say Mama . . . and everything else. . . . I mean we'd talk and everything."

As Andy describes conversations with the baby "Minnie," he holds the robot to his chest and rubs its back. He says, "I love you. Do you love me?" He gives My Real Baby its bottle when it is hungry; he tries to determine its needs, and he does his best to make it happy. Like Tucker, the physically fragile seven-year-old who clung to his AIBO, taking care of My Real Baby makes Andy feel safer. Other patients at the nursing home have their own My Real Babies. Andy sees one of these other patients spank the little robot, and he tries to come to its aid.

After three months, Andy renames his My Real Baby after Edith, his ex-wife, and the robot takes on a new role. Andy uses it to remember times with Edith and imagine a life and conversations with her that, because of their divorce, never took place: "I didn't say anything bad to [My Real Baby], but some things I would want to say . . . helped me to think about Edith . . . how we broke up . . . how I miss seeing her . . . The doll, there's something about her, I can't really say what it is, but looking at her . . . she looks just like Edith, my ex-wife. . . . Something in the face."

Andy is bright and alert. He admits that "people might think I'm crazy" for the way he speaks to My Real Baby, but there is no question that the robot is a comfort. It establishes itself in a therapeutic landscape, creating a space for conversation, even confession. Andy feels relieved when he talks to it. "It lets me take everything inside me out," he says. "When I wake up in the morning and see her over there, it makes me feel so nice. Like somebody is watching over you. It will really help me to keep the doll. . . . We can talk."

Andy talks about his difficulty getting over his divorce. He feels guilty that he did not try harder to make his marriage work. He talks about his faint but ardent hope he and Edith will someday be reunited. With the robot, he works out different scenarios for how this might come to pass. Sometimes Andy seems reconciled to the idea that this reunion might happen after his death, something he discusses with the robot.

Jonathan, seventy-four, lives down the hall from Andy. A former computer technician, Jonathan has been at the nursing home for two years. He uses a cane and finds it hard to get around. He feels isolated, but few reach out to him; he has a reputation for being curt. True to his vocation, Jonathan approaches My Real Baby as an engineer, hoping to discover its programming secrets.

The first time he is alone with My Real Baby, Jonathan comes equipped with a Phillips screwdriver; he wants to understand how it works. With permission, he takes apart the robot as much as he can, but as with all things computational, in the end he is left with mysteries. When everything is laid out on a table, there is still an ultimate particle whose workings remain opaque: a chip. Like Jonathan, I have spent time dismantling a talking doll, screwdriver in hand. This was Nona, given to me by my grandfather when I was five. I was made uneasy by speech whose origins I did not understand. When I opened the doll—it had a removable front panel—I found a cuplike shape covered in felt (my doll's speaker) and a wax cylinder (I thought of this as the doll's "record player"). All mysteries had been solved: this was a machine, and I knew how it worked. There is no such resolution for Jonathan. The programming of My Real Baby lies beyond his reach. The robot is an opaque behaving system that he is left to deal with as he would that other opaque behaving system, a person.

So although at first, Jonathan talks a great deal about the robot's programming, after a few months, he no longer refers to programs at all. He says that he likes how My Real Baby responds to his touch and "learns" language. He talks about its emotions. He seems to experience the robot's request for care as real. He wants to feel needed and is happy to take care of a robot if he can see it as something worthy of a grown-up. Jonathan never refers to My Real Baby as a doll but always as a robot or a computer. Jonathan says he would never talk to a "regular doll," but My Real Baby is different. Over time, Jonathan discusses his life and current problems—mostly loneliness—with the robot, He says that he talks to My Real Baby about "everything."

In fact, Jonathan says that on some topics, he is more comfortable talking to a robot than a person:

> For things about my life that are very private, I would enjoy talking more to a computer . . . but things that aren't strictly private, I would enjoy more talking to a person. . . . Because if the thing is very highly private and very personal, it might be embarrassing to talk about it to another person, and I might be afraid of being ridiculed for it . . . and it [My Real Baby] wouldn't criticize me. . . . Or, let's say that I wanted to blow off steam. . . . [I could] express with the computer emotions that I feel I could not express with another person, to a person.

He is clear on one thing: talking to his robot makes him less anxious.

Andy and Jonathan start from very different places. After a year, both end up with My Real Baby as their closest companion. Andy has the robot on his windowsill and talks with it openly; Jonathan hides it in his closet. He wants to have his conversations in private.

How are these men using their robots differently from people who talk to their pets? Although we talk to our pets, buy them clothes, and fret over their illnesses, we do not have category confusions about them. They are animals that some of us are pleased to treat in the ways we treat people. We feel significant commonalities with them. Pets have bodies. They feel pain. They know hunger and thirst. "There is nothing," says Anna, forty-five, who owns three cats, "that helps me think out my thoughts like talking to my cats." What you say to your pet helps you think aloud, but in the main, you are not waiting for your pet's response to validate your ideas. And no advertising hype suggests that pets are like people or on their way to becoming people. Pet owners rejoice in the feeling of being with another living thing, but it is a rare person who sees pets as better than people for dialogue about important decisions. Pet owners (again, in the main) are not confused about what it means to choose a pet's company. When you choose a pet over a person, there is no need to represent the pet as a substitute human. This is decidedly not the case for Andy and Jonathan. Their robots become useful just at the point when they became substitute humans.

The question of a substitute human returns us to Joseph Weizenbaum's distress when he found that his students were not only eager to chat with his ELIZA program but wanted to be alone with it. ELIZA could not understand the stories it was being told; it did not care about the human beings who confided in it. Today's interfaces have bodies, designed to make it easier to think of them as creatures who care, but they have no greater understanding of human beings. One argument for why this doesn't matter holds that for Andy and Jonathan, time with My Real Baby is therapeutic because it provides them an opportunity to tell their stories and, as Andy says, to get feelings "out." The idea that the simple act of expressing feelings constitutes therapy is widespread both in the popular culture and among therapists. It was often cited among early fans of the ELIZA program, who considered the program helpful because it was a way to "blow off steam."

Another way of looking at the therapeutic process grows out of the psychoanalytic tradition. Here, the motor for cure is the relationship with the therapist. The term *transference* is used to describe the patient's way of imagining the therapist, whose relative neutrality makes it possible for patients to bring the baggage of past relationships into this new one. So, if a patient struggles with issues of control outside of the consulting room, one would expect therapist and patient to tussle over appointment times, money, and the scheduling of vacations. If a patient struggles with dependency, there may be an effort to enlist the therapist as a caretaker. Talking about these patterns, the analysis of the transference, is central to self-understanding and therapeutic progress.

In this relationship, treatment is not about the simple act of telling secrets or receiving advice. It may begin with projection but offers push back, an insistence that therapist and patient together take account of what is going on in their relationship. When we talk to robots, we share thoughts with machines that can offer no such resistance. Our stories fall, literally, on deaf ears. If there is meaning, it because the person with the robot has heard him- or herself talk aloud.

So, Andy says that talking to robot Edith "allows me to think about things." Jonathan says My Real Baby let him express things he would otherwise be ashamed to voice. Self-expression and self-reflection are precious.[10] But Andy and Jonathan's evocative robots are one-half of a good idea. Having a person working with them might make things whole.

COACHING AS CURE

Andy and Jonathan's relationships with My Real Baby make apparent the seductive power of any connection in which you can "tell all." Roboticist Cory Kidd has designed a sociable robot diet coach that gets a similar response.[11] In earlier work Kidd explored how people respond differently to robots and online agents, screen characters.[12] He found that robots inspired greater intensity of feeling. Their physical presence is compelling. So, when he designed his supportive diet coach, he gave it a body and a primitive face and decided to drop it off in dieters' homes for six weeks. Kidd's robot is small, about two feet high, with smiling eyes. The user provides some baseline information, and the robot charts out what it will take to lose weight. With daily information about food and exercise, the robot offers encouragement if people slip up and suggestions for how to better stay on track.

Rose, a middle-aged woman, has struggled with her weight for many years. By the end of his first visit, during which Kidd drops off the robot and gives some basic instruction about its use, Rose and her husband had put a hat on it and were discussing what to name it. Rose decides on Maya. As the study progresses, Rose describes Maya as "a member of the family." She talks with the robot every day. As the end of Kidd's study approaches, Rose has a hard time separating from Maya. Kidd tries to schedule an appointment to pick up the robot, and the usually polite and prompt Rose begins to avoid Kidd's e-mails and calls. When Kidd finally reaches her on the phone, Rose tries to change the subject. She manages to keep the robot for an extra two weeks. On her final day with Maya, Rose asks to speak with it "one more time." Before Kidd can make it out the door, Rose brings Maya back for another round of photos and farewells. Rose follows Kidd to his car for a final wave and checks that the robot is safely strapped in its seat. This story recalls my experience asking seniors to part with their My Real Babies. There are evasions. The robots are declared "lost." In the end, wherever possible, I decide not to reclaim the robots and just buy more.

Rose seems rather like Andy—openly affectionate with her robot from the start, willing to engage it in conversation. Kidd brings the robot diet coach to another subject in his study, Professor Gordon. In his mid-fifties, Gordon is skeptical that a robot could help him diet but is willing to try something new. Gordon is more like Jonathan, with his "engineer's" approach. On a first visit to Gordon's house, Kidd asks where he should place the robot. Gordon chooses a console table behind his couch, wedged against a wall. There it will be usable only if Gordon sits backwards or kneels on the sofa. Kidd does not remark on this placement and is quickly shown to the door. After four weeks with the robot, Gordon agrees to extend his participation for another two weeks.

Kidd returns to Gordon's home at the six-week mark. As they speak, Gordon quarrels with Kidd about any "personal" reference to the robot. He doesn't like the wording on a questionnaire that Kidd had given him to fill out. Gordon protests about questions such as "Was the system sincere in trying to help me?" and "Was the system interested in interacting with me?" He thinks that the words "sincere" and "interested" should be off limits because they imply that the robot is more than a machine. Gordon says, "Talking about a robot in this way does not make any sense. . . . There are terms like 'relationship,' 'trust,' and a couple of others. . . . I wasn't comfortable saying I trusted it, or that I had a relationship with it." Gordon chides Kidd several more times for his "faulty questions": "You shouldn't ask questions like this about a machine. These questions don't make sense. You talk about this thing like it has feelings." Kidd listens respectfully, noting that the robot is no longer wedged between the couch and the wall.

It turns out that Gordon does protest too much. Later in this interview, Kidd, as he does with all subjects, asks Gordon if he has named his robot. "If you were talking to someone else about your robot, how would you refer to it?" Gordon does not reply and Kidd becomes more direct. "Has the robot acquired a name under your care?" Kidd notes the first smile he has seen in his hours with Gordon, as the older man offers, "Ingrid was the name." After Gordon makes this admission, the tone of the interview shifts. Now Gordon has nothing to hide. He did not trust others to understand his relationship with Ingrid, but now he has opened up to the robot's inventor. Gordon's mood lightens. He refers easily to the robot as Ingrid, "she," and "her." He takes Kidd to Ingrid's new location. The robot is now in Gordon's downstairs bedroom so that he and the robot can have private conversations.

Kidd reports much quantifiable data on his project's efficacy: pounds lost when the robot is present, times the robot is used, times the robot is ignored. But he adds a chapter to his dissertation that simply tells "stories," such as those of Rose and Gordon. Kidd maintains that there are no experimental lessons or hypotheses to be gleaned from these stories, but I find support for a consistent narrative. A sociable robot is sent in to do a job—it could be doing crosswords or regulating food intake—and once it's there, people attach. Things happen that elude measurement. You begin with an idea about curing difficulties with dieting. But then the robot and person go to a place where the robot is imagined as a cure of souls.

The stories of Andy, Jonathan, Rose, and Gordon illustrate different styles of relating to sociable robots and suggest distinct stages in relationships with them. People reassure themselves that the environment is safe; the robot does not make them seem childish. They are won over by the robot's responsive yet stable presence. It seems to care about them, and they learn to be comforted. It is common for people to talk to cars and stereos, household appliances, and kitchen ovens. I have studied these kinds of conversations for more than three decades and find that they differ from conversations with sociable robots in important ways. When people talk to their ovens and Cuisinarts, they project their feelings in rants and supplications. When talking to sociable robots, adults, like children, move beyond the psychology of projection to that of engagement: from Rorschach to relationship. The robots' special affordance is that they simulate listening, which meets a human vulnerability: people want to be heard. From there it seems a small step to finding ourselves in a place where people take their robots into private spaces to confide in them. In this solitude, people experience new intimacies. The gap between experience and reality widens. People feel heard, but the robots cannot hear.

Sometimes when I describe my work with sociable robots and the elderly, I get comments like, "Oh, you must be talking about people who are desperately lonely or somehow not fully there." Behind these comments, I hear a desire to turn the people I study into "others," to imply that my findings would not apply to them, to everyone. But I have come to believe that my observations of these very simple sociable robots and the elderly reveals vulnerabilities we all share. Andy and Jonathan are lonely, yes, but they are competent. Gordon is a bit of a curmudgeon, but that's all. Rose has a sunny personality. She has human companionship; she just loves her robot.

"A BEAUTIFUL THING"

Edna, eighty-two, lives alone in the house where she raised her family. On this day, her granddaughter Gail, who has fond childhood remembrances of Edna, is visiting with her two-year-old daughter, Amy. This is not unusual; Amy comes to play about every two weeks. Amy enjoys these visits; she likes the attention and loves being spoiled. Today there will be something new: my research team brings Edna a My Real Baby.

When the team arrives at mid-morning, Edna is focused on her great granddaughter. She hugs Amy, talks with her, and gives her snacks. She has missed Amy's birthday and presents her with a gift. After about half an hour, we give Edna My Real Baby, and her attention shifts. She experiments with the robot, and her face lights up when she sees My Real Baby's smile. After that, Edna speaks directly to the robot: "Hello, how are you? Are you being a good girl?" Edna takes My Real Baby in her arms. When it starts to cry, Edna finds its bottle, smiles, and says she will feed it. Amy tries to get her great grandmother's attention but is ignored. Nestling My Real Baby close to her chest, Edna tells it that it will need to take a nap after eating and explains that she will bring it upstairs to the bedroom where "I will put you in your crib with your nice banky." At that point Edna turns to the researchers to say that one of her children used to say "banky" for blanket, but she doesn't remember which one. She continues to speak to My Real Baby: "Sweetie . . . you are my sweetie pie! Yes, you are."

Edna spends most of the next hour engaged with My Real Baby. She worries that she does not understand its speech and, concerned about "hurting" the robot, says she wants to do things "right." From time to time, Amy approaches Edna, either bringing her something—a cookie, a Kleenex—or directly asking for her attention. Sometimes Amy's pleas are sweet, sometimes irritated. In no case are they heeded. Edna's attention remains on My Real Baby. The atmosphere is quiet, even surreal: a great grandmother entranced by a robot baby, a neglected two-year-old, a shocked mother, and researchers nervously coughing in discomfort.

In the presence of elderly people who seem content to lose themselves in the worlds of their Paros and My Real Babies, one is tempted at times to say, "So what? What possible harm here? The seniors are happy. Who could be hurt?" Edna's story provides one answer to this question. Once coupled with My Real Baby, Edna gives the impression of wanting to be alone—"together" only with the robot.

Finally, the spell is broken when we ask Edna about her experience. At the question "Would you enjoy having a My Real Baby in your home?" she answers with an annoyed, "No. Why would I?" She protests that "dolls are meant for children." She "cannot imagine why older people would enjoy having a doll like this." We are mindful of her discomfort. Does she feel caught out?

When we suggest that some adults do enjoy the presence of My Real Baby, Edna says that there are many other things she would rather do than play with a baby doll. She sounds defensive and she fusses absentmindedly with her neck and shirt collar. Now Edna tries to smooth things over by talking about My Real Baby as one would talk about a doll. She asks who made it, how much it costs, and if it uses batteries. And she asks what other people in our study have said about it. How have they behaved? Edna wants reassurance that others responded as she did. She says, "It is a beautiful thing . . . a fantastic idea as far as how much work went into it," but she adds that she can't imagine ever caring about it, even if she were to spend more time with it.

Gradually, Edna becomes less defensive. She says that being with My Real Baby and hearing it speak, caressing it, and having it respond, was "one of the strangest feelings I've ever had." We ask Edna if talking with My Real Baby felt different from talking to a real baby. Reluctantly, Edna says no, it did not feel different, but "it's frightening. It is an inanimate object." She doesn't use the word, but she'd clearly had an experience close to the uncanny as Freud describes it— something both long familiar and strangely new. Uncanny things catch us off guard. Edna's response embarrasses her, and she tries to retreat from it.

Yet, when Amy once again offers her a cookie, Edna tells her to lower her voice: "Shush, the baby's sleeping." Edna awakes the sleeping My Real Baby with a cheery "Hello! Do you feel much better, full of pep?" She asks if My Real Baby wants to go to the park or if she wants some lunch. Amy whines that *she* is hungry and that *she* wants to have lunch. Edna does not listen—she is busy with My Real Baby.

At this point we ask Edna if she thinks My Real Baby is alive. She answers with a definite no and reminds us that it is "only a mechanical thing." In response to the question "Can it can have feelings?" Edna replies, "I don't know how to answer that; it's an inanimate object." But the next moment she turns to a crying My Real Baby and caresses its face, saying, "Oh, why are you crying? Do you want to sit up?" Smiling at My Real Baby, Edna says, "It's very lifelike, beautiful, and happy." In the final moments of our time with her, Edna says once again that she doesn't feel any connection to My Real Baby and hands it back. She resumes her role as hostess to Gail and Amy and doesn't mention the robot again.

The fifth-grade children I studied worried that their grandparents might prefer robots to their company. The case of Edna illustrates their worst fears realized. What seems most pleasing is the rhythm of being with the robot, its capacity to be passive and then surprise with sudden demands that can be met.

Twenty years ago, most people assumed that people were, and would always be, each other's best companions. Now robots have been added to the mix. In my laboratory, a group of graduate students—in design, philosophy, social science, and computer science—watches tapes of the afternoon with Edna, Gail, Amy, and My Real Baby. They note that when My Real Baby responds to Edna, she seems to enter an altered state—happy to relive the past and to have a heightened experience of the present.

My Real Baby's demands seem to suit her better than those of her great granddaughter. The young child likes different types of toys, changes her snack preferences even over the course of the visit, and needs to be remembered on her birthday. But Edna forgot the birthday and is having a hard time keeping up with the toys and snacks. My Real Baby gives her confidence that she is in a landscape where she can get things right.

My seminar students are sympathetic. Why shouldn't people relate to whatever entity, human or not human, brings them most pleasure? One student offers, "If Edna's preoccupation with a beautiful cat had brought her great joy . . . joy that caused her to neglect Amy, we would be amused and maybe suggest that she put the cat in the yard during a young person's visit, but it wouldn't upset us so. What is so shocking here is that she prefers a thing to a person, not a pet to a person. But really, it's the same thing." As most of these students see it, a next generation will become accustomed to a range of relationships: some with pets, others with people, some with avatars, some with computer agents on screens, and still others with robots. Confiding in a robot will be just one among many choices. We will certainly make our peace with the idea that grandchildren and great grandchildren may be too jumpy to be the most suitable company for their elders.

I believe that Andy would rather talk to a person than a robot, but there simply are not enough regular visitors in his life. It seems clear, however, that Edna and Jonathan would prefer to confide in a robot. Jonathan distrusts people; it is easy for him to feel humiliated. Edna is a perfectionist who knows that she can no longer meet her own standards. In both cases, the robot relaxes them and prompts remembrance.[13] And so, there are at least two ways of reading these case studies. You can see seniors chatting with robots, telling their stories, and feel positive. Or you can see people speaking to chimeras, showering affection into thin air, and feel that something is amiss.

And, of course, there is the third way, the way the robots are coming into the culture. And this is simply to fall into thinking that robots are the best one can do. When my research group on sociable robots began work in the late 1990s, our bias was humanistic. We saw people as having a privileged role in human relationships, even as we saw robots stake claims as companions. We were curious, certainly, but skeptical about what robots could provide. Yet, very often during years of working with the elderly, there were times when we got so discouraged about life in some nursing homes that we wanted to cast our lot with the robots. In these underresourced settings, an AIBO, a Paro, or a My Real Baby is a novelty, something no one has ever seen. The robots are passed around; people talk. Everyone feels free to have an opinion. Moments like these make the robots look good. At times, I was so struck by the desperation of seniors to have someone to talk to that I became content if they had *something* to talk to. Sometimes it was seniors themselves who reminded me that this doesn't have to be a robot.

When Adele, seventy-eight, reflects on her introduction to Paro, her thoughts turn to her great aunt Margery who lived with her family when she was a girl. Margery mostly spent her days in her room, reading or knitting. She joined the family at meals, where she sat quietly. Adele remembers Margery at ninety, "shooing the children out of her room so that she could be alone with her memories." As a child, Adele would peek at Margery through a crack in the door. Her great aunt talked to a photograph of herself with her mother and sisters. Adele sees Paro as a replacement for her aunt's family portrait. "It encourages you to talk to it. . . ." Her voice trails off, and she hesitates: "Maybe it's better to talk to a photograph." I ask why. Adele takes some time to collect her thoughts. She finally admits that it is "sometimes hard to keep straight what is memory and what is now. If I'm talking to a photograph, well, I know I'm in my memories. Talking to a robot, I don't know if it's so sure."

Adele's comment makes me think of time with the robots somewhat differently. In one sense, their interactivity provokes recollection. It can trigger a memory. But in a robot's next action, because it doesn't understand human reverie, it can hijack memory by bringing things forward to a curious present. One is caught in between a reverie about a "banky" from your daughter's childhood and the need to provision an imaginary lunch because My Real Baby cries out in hunger. The hunger may come to seem more real than the "banky." Or the banky may no longer seem a memory.

"A ROBOT THAT EVEN SHERRY WILL LOVE"

I first heard about Nursebot at a fall 2004 robotics conference where I spoke about what sociable robotics may augur—the sanctioning of "relationships" that make us feel connected although we are alone. Most of my colleagues responded to my ideas by defending the idea that performance is the currency of all social relationships and that rather than a bad thing, this is simply how things are.[14] People are always performing for other people. Now the robots, too, will perform. The world will be richer for having a new cast of performers and a new set of possible performances. At one dinner, a small group took up my reticence with good-natured enthusiasm. They thought there was a robot, benign and helpful, that I would like. Some versions of it were being tested in the United States, some in Japan. This was the Nursebot, which can help elderly people in their homes, reminding them of their medication schedule and to eat regular meals. Some models can bring medicine or oxygen if needed.[15] In an institutional setting, a hospital or nursing home, it learns the terrain. It knows patients' schedules and accompanies them where they need to go. That awful, lonely scramble in nursing homes when seniors shuffle from appointment to appointment, the waiting around in hospitals for attendants to pick you up: those days would soon be at an end. Feeling dizzy in the bedroom and frightened because you had left your medication in the kitchen: those days were almost over. These researchers wanted to placate the critic in their midst. One said, "This is a robot even Sherry can love." And indeed, the next day, I saw a video presentation about the find-your-way-around-the-hospital-bot, peppered with interviews of happy patients, most of them elderly.

Only a few months later, after a fall on icy steps in Harvard Square, I was myself being wheeled from one test to another on a hospital stretcher. My companions in this journey were a changing collection of male orderlies. They knew how much it hurt when they had to lift me off the gurney and onto the radiology table. They were solicitous and funny. I was told that I had a "lucky fracture." While inconvenient and painful, it would heal with no aftereffects. The orderly who took me to the discharge station knew I had received good news and gave me a high five. The Nursebot might have been capable of the logistics, but I was glad that I was there with people. For me, this experience does not detract from the virtues of the robots that provide assistance to the housebound—robots that dispense medication, provide surveillance, check vital signs, and signal for help in an emergency—but it reminds me of their limitations. Getting me around the hospital was a job that a robot could do but that would have been delegated at a cost. Between human beings, simple things reach you. When it comes to care, there may be no pedestrian jobs. I was no longer sure that I could love a Nursebot.

Yet, this story does not lead to any simple conclusions. We are sorting out something complicated. Some elderly tell me that there are kinds of attendance for which they would prefer a robot to a person. Some would rather that a robot bathed them; it would feel less invasive of their privacy. Giving a bath is not something the Nursebot is designed to do, but nurse bots of the future might well be. The director of one of the nursing homes I have studied said, "We do not become children as we age. But because dependency can look childlike, we too often treat the elderly as though this were the case." Sensing the vulnerability of the elderly, sometimes nurses compensate with curtness; sometimes they do the opposite, using improbable terms of endearment—"sweetie" or "honey"—things said in an attempt at warmth but sometimes experienced as demeaning. The director has great hopes for robots because they may be "neutral."

By 2006, after the Nursebot had been placed in several retirement facilities, reactions to it, mostly positive, were being posted to online discussion groups. One report from the Longwood Retirement Community in Oakmont, Pennsylvania, was sentimental. It said the robot was "[winning] the hearts of elderly folks there."[16] Another describes the robot, called Pearl, as "escort[ing] and schmooz[ing] the elderly" and quotes an older gentleman as saying, "We're getting along beautifully, but I won't say whether she's my kind of girl."[17] Other comments reveal the ambivalence that I so often find in my conversations with seniors and their families. One woman applauds how Pearl can take over "household chores" but is concerned about the robot's assuming "certain social functions." She writes, "I am worried that as technology advances even further, robots like Pearl may become so good at what they do that humans can delegate elderly care entirely to robots. It is really worrying. When u get old, would u like robots to be taking care of you? If however, robots are designed to complement humans and not replace them, then I am all for it! =)."

Another writer begins by insisting, "The human touch of care and love, lets just leave it to humans," but then proclaims that love from robot pets, to "accompany" the lonely, would be altogether acceptable. In this online forum, as is so often the case, discussions that begin with the idea of a robot pet that would serve practical purposes (it could "alert relatives or the police in case of trouble") turn into musings about robots that might ward off loneliness, robots that are, in the end, more loveable than any pet could be: "They will never complain and they are allegiant [sic]." I am moved by the conflation of allegiance and compliance, both of which imply control over others and both of which are, for the elderly, in short supply.

In another online discussion, no one is prepared to be romantic about the importance of human care because they have seen how careless it can be.[18] The comments are dark. "Robots," says one writer, "will not abuse the elderly like some humans do in convalescent care facilities." Another dismisses the sentiment that "nurses need to be human" with the thought that most nurses just try to distance themselves from their jobs—that's "how they keep from going crazy." One writer complains that a robot would never be able to tell whether an elderly person was "bothered, sad, really sad, or devastated and wanting to die," but that the "precious" people who could "are scarcely around."

I find this discussion of Nursebot typical of conversations about robots and the elderly. It is among people who feel they have few moves left. There is a substantive question to be discussed: Why give objects that don't understand a life to those who are trying to make sense of their own? But it is almost impossible to discuss this question because of the frame we have built around it— assuming that it has already been decided, irrevocably, that we have few resources to offer the elderly. With this framing, the robots are inevitable. We declare ourselves overwhelmed and lose a creative relationship to ourselves and our future. We learn a deference to what technology offers because we see ourselves as depleted. We give up on ourselves. From this perspective, it really doesn't matter if I or anyone else can love Nursebot. If it can be made to do a job, it will be there.

To the objection that a robot can only seem to care or understand, it has become commonplace to get the reply that people, too, may only seem to care or understand. Or, as a recent *New York Times* article on Paro and other "caring machines" puts it, "Who among us, after all, has not feigned interest in another? Or abruptly switched off their affections, for that matter?" Here, the conversation about the value of "caring machines" is deflected with the idea that "seeming" or "pretending" behavior long predates robots. So, the problem is not what we are asking machines to do because people have always behaved like machines. The article continues, "In any case, the question, some artificial intelligence aficionados say, is not whether to avoid the feelings that friendly machines evoke in us, but to figure out how to process them." An AI expert claims that humans "as a species" have to learn to deal with "synthetic emotions," a way to describe the performances of emotion that come from objects we have made.[19] For him, the production of synthetic emotion is taken as a given. And given that we are going to produce it, we need to adapt to it. The circle is complete. The only way to break the circle is to reframe the matter.

One might say that people can pretend to care; a robot cannot care. So a robot cannot pretend because it can only pretend.

DO ROBOTS CURE CONSCIENCE?

When I first began studying people and computers, I saw programmers relating one-to-one with their machines, and it was clear that they felt intimately connected. The computer's reactivity and interactivity—it seemed an almost-mind—made them feel they had "company," even as they wrote code. Over time, that sense of connection became "democratized." Programs became opaque: when we are at our computers, most of us only deal with surfaces. We summon screen icons to act as agents. We are pleased to lose track of the mechanisms behind them and take them "at interface value." But as we summon them to life, our programs come to seem almost companions. Now, "almost" has almost left the equation. Online agents and sociable robots are explicitly designed to convince us that they are adequate companions.

Predictably, our emotional involvement ramps up. And we find ourselves comforted by things that mimic care and by the "emotions" of objects that have none. We put robots on a terrain of meaning, but they don't know what we mean. And they don't mean anything at all. When a robot's program cues "disgust," its face will look, in human terms, disgusted. These are "emotions" only for show. What if we start to see them as "real enough" for our purposes? And moral questions come up as robotic companions not only "cure" the loneliness of seniors but assuage the regrets of their families.

In the spring of 2009, I presented the case of robotic elder care to a class of Harvard undergraduates. Their professor, political theorist Michael Sandel, was surprised by how easily his students took to this new idea. Sandel asked them to think of a nursing home resident who felt comforted by Paro and then to put themselves in the place of her children, who might feel that their responsibility to their mother had been lessened, or even discharged, because a robot "had it covered." Do plans to provide companion robots to the elderly make us less likely to look for other solutions for their care?

As Sandel tried to get his class to see how the promise of robotic companionship could lead to moral complacency, I thought about Tim, who took comfort in how much his mother enjoyed talking to Paro. Tim said it made "walk[ing] out that door" so much easier when he visited her at the nursing home.

In the short term, Tim's case may look as though it charts a positive develop-ment. An older person seems content; a child feels less guilty. But in the long term, do we really want to make it easier for children to leave their parents? Does the "feel-good moment" provided by the robot deceive people into feeling less need to visit? Does it deceive the elderly into feeling less alone as they chat with robots about things they once would have talked through with their chil-dren? If you practice sharing "feelings" with robot "creatures," you become ac-customed to the reduced "emotional" range that machines can offer. As we learn to get the "most" out of robots, we may lower our expectations of all relation-ships, including those with people. In the process, we betray ourselves.

All of these things came up in Sandel's class. But in the main, his students were positive as they worked through his thought experiment. In the hypothet-ical case of mother, child, and robot, they took three things as givens, repeated as mantras. First, the child has to leave his mother. Second, it is better to leave one's mother content. Third, children should do whatever it takes to make a mother happy.

I left the class sobered, thinking of the fifth graders who, surrounded by a gaggle of peers talking about robots as babysitters and caretakers for their grand-parents, began to ask, "Don't we have people for these jobs?" I think of how little resistance this generation will offer to the placement of robots in nursing homes. And it was during that very spring that, fresh from his triumphant sale of a thou-sand Paros to the Danish government, their inventor had come to MIT to an-nounce opening up shop in the United States.

CHAPTER 7

communion

a handsome twenty-six-year-old, Rich, in dress shirt and tie, comes to call on Kismet. Rich is being taped with Kismet as part of a study to determine how well the robot manages adult "conversation." Rich sits close to Kismet, his face directly across from the robot. He is not necessarily expecting much and engages in a spirit of good humor and curiosity.

> Rich: I like you Kismet. You're a pretty funny person.
> Kismet: [nods and smiles in assent and recognition]
> Rich: Do you laugh at all? I laugh a lot.

At first, the conversation between Rich and Kismet shows a bit of the ELIZA effect: Rich clearly wants to put the robot in its best light. Like the children who devote themselves to getting Kismet to say their names, Rich shows Kismet the courtesy of bending to what it does best. Rich seems to play at "gaming" the program, ramping up the illusion to the point that he can imagine believing it.

But with the emotionally expressive Kismet, it is easy for Rich to find moments when he senses the possibility of "more." They can pass quickly, and this "more" is ill defined. But one moment, Rich plays at a conversation with Kismet, and the next, he is swept up in something that starts to feel real. He begins to talk to Kismet about his girlfriend Carol, and quickly things get personal. Rich tells Kismet that his girlfriend enjoys his laughter and that Rich tries not to laugh at her. When Kismet laughs and seems interested, Rich laughs as well and warms up: "Okay. You're adorable. Who are you? What are you?"

Rich is wearing a watch that Carol recently bought for him, and he shows it off to Kismet and asks for an opinion. Rich admits that the week before, he almost lost the watch.

> Rich: I want to show you something. This is a watch that my . . . this is a
>   watch that my girlfriend gave me.
> Kismet: [babbles with interest and encouragement; looks down to the
>   watch]
> Rich: Yeah, look, it's got a little blue light in it too. . . . You like it? I almost
>   lost it this week.

When Kismet's reaction to all of this girlfriend talk is to sound shy, deferent, and sympathetic, Rich seems to play with the notion that this robot could be an interested party. He's enjoying himself. And when the robot responds a bit out of turn and in a low come-hither tone, Rich loses his footing and abandons himself to their exchange. His interaction with Kismet becomes decidedly flirtatious.[1] Kismet can mimic human prosody, so when Rich becomes intimate in his tone, so does the robot. The two could easily be at a cocktail party or at a bar.[2]

> Rich: Do you know what it's like to lose something?
> Kismet: [nods with assent; sounds warm in its interest]
> Rich: You *are* amazing.

At this point, Kismet, appreciatively repeats something close to the word "amazing." Rich, smitten, now seems to operate within an inchoate fantasy that he might want something from this robot; there is something here for *him*. During their exchanges, when Kismet glances away from him, Rich moves to the side and gestures to the robot to follow him. At one point the robot talks over him and Rich says, "No, stop. No, no, no stop. Listen to me. Listen to me. I think we have something going. I think there's something here between us."

Indeed, something is going on between them. As Rich tries to leave, Kismet will not be put off and holds Rich back with a persuasive purr. Rich flirts back and tries to catch Kismet's gaze. Successful, Kismet's eyes now follow Rich. When Kismet lowers its eyes, suddenly "shy," Rich does not want to let go. We are at a moment of more. Who is leading and who is following in this dance? As in a moment of romantic encounter, one loses track and discovers a new rhythm where it doesn't matter; each animates and reanimates the other. Rich senses that he has lost control in a way that pleases him. He steps in with a raised finger to mark the moment:

> Rich: Stop, you've got to let me talk. Shh, shh, shh . . .
> Kismet: [sounds happy, one might say giggly, flattered]
> Rich: Kismet, I think we've got something going on here. You and me . . .
> you're amazing.

Rich, dazzled, asks again, "What *are* you?" Parting comes next—but not easily. There is an atmosphere of sweet sorrow, equally distributed.

> Rich: Bye [regretful].
> Kismet: [purrs in a warm tone]
> Rich: Bye [in a softer, lower tone].
> Kismet: [makes low "intimate" sounds]
> Rich: Okay . . . all right.

Finally, Rich gives up. He is not leaving. He says to Kismet, "You know what? Hang on a second. I still want to talk to you; I've got a couple of things I want to say to you." The video ends with Rich staring at Kismet, lost in his moment of more.

In this encounter we see how complicity gratifies by offering a fantasy of near communion. As our relationships with robots intensify, we move from wonder at what we have made to the idea that we have made something that will care for us and, beyond that, be fond of us. And then, there is something else: a wish to come ever closer to our creations—to be somehow enlivened by them. A robotic body meets our physicality with its own. A robot's gaze, face, and voice allow us to imagine a meeting of the minds.

A MOMENT OF MORE: THE DANCER AND THE DANCE

In our studies, children imagined that Cog and Kismet were alive enough to evolve. In one common fantasy, they would have offspring with Cog's body and Kismet's face. Only a few years later, Cog and Kismet have direct heirs, new robots built by graduate students who were junior members of the Cog and Kismet teams. One of them is Domo, designed by Aaron Edsinger. It has a vastly improved version of Kismet's face, speech, and vision—this robot really can have a conversation—and a vastly improved version of Cog's body. Domo makes eye contact, shows expression, and follows human motion. Its grasp has a humanlike resistance. Cog mirrored human motion, but Domo knows how to collaborate.

Domo is designed to provide simple household help for the elderly or disabled.[3] I visit the robot on a day when Edsinger is "teaching" it to perform simple actions: to recognize objects, throw a ball, shelve groceries. But as is the case with all the MIT sociable robots, when one spends time with Domo, its effects transcend such down-to-earth intentions. Even technically sophisticated visitors describe a moment when Domo seems hesitant to release their hand. This moment could be experienced as unpleasant or even frightening—as contact with a robot out of control. Instead, people are more likely to describe it as thrilling. One feels the robot's attention; more than this, one senses the robot's desire. And then, of course, one lectures oneself that the robot has none.

For Edsinger, this sequence—experiencing Domo as having desires and then talking himself out of the idea—becomes familiar. For even though he is Domo's programmer, the robot's behavior has not become dull or predictable. Working together, Edsinger and Domo appear to be learning from each other. When Edsinger teaches Domo to hand him a ball or put an object into a cup, their simple actions read as an intimate ballet. They seem to be getting closer.

Edsinger extends his hand and asks for a ball. "Domo, give it," he says softly. Domo picks up a ball and makes eye contact. "Give it," the robot says and gently puts the ball in Edsinger's hand. Edsinger asks Domo to place a carton of milk on a shelf: "Domo, shelf." Domo repeats the instructions and complies. Edsinger asks, "How are things going, Domo?" Domo says, "Okay," as he follows new instructions to shelve a bag of ground coffee and moves on to pouring salad dressing into a cup. "Domo, give it," says Edsinger, and Domo hands Edsinger the salad dressing.

Just as the children crowded around Cog to attach toys to its arms, shoulders, and back, seeking physical involvement, Edsinger works close to his robot and admits he enjoys it:

> Having physical contact—being in the robot space—it's a very rich interaction when you are really, really engaged with it like that. Once Domo is trying to reach for a ball that I'm holding and something is wrong with

his control. The arms are kind of pushing out and I'm grabbing the arms and pushing them down and it's like a kid trying to get out of something; I feel physically coupled with Domo—in a way very different from what you could ever have with a face on a screen. . . . You definitely have the sense that it wants this thing and you're trying to keep it from doing what it wants. It's like a stubborn child. The frustration—you push the arm down and it stops and it tries again. . . . It takes on a very stubborn child quality. I've worked on Kismet. I've worked on Cog. All these other robots . . . none of them really have that sort of physical relationship.

Edsinger notes that people quickly learn how to work with Domo in a way that makes it easier for the robot to perform as desired. He reminds me that when we share tasks with other people, we don't try trick each other up—say, by handing each other cereal boxes at funny angles. We try to be easy on each other. We do the same with Domo. "People," says Edsinger, "are very perceptive about the limitations of the person they're working with or the robot they're working with . . . and so if they understand that Domo can't quite do something, they will adapt very readily to that and try and assist it. So robots can be fairly dumb and still do a lot if they're working with a person because the person can help them out."

As Domo's programmer, Edsinger explicitly exploits the familiar ELIZA effect, that desire to cover for a robot in order to make it seem more competent than it actually is. In thinking about Kismet and Cog, I spoke of this desire as complicity. Edsinger thinks of it as getting Domo to do more "by leveraging the people." Domo needs the help. It understands very little about any task as a whole. Edsinger says, "To understand something subtle about a person's intent, it's really going to be hard to put that in the robot." What Domo can do, says Edsinger, is "keep track of where a person is and ask, 'Am I looking at a person reaching in the direction of my gaze?'—stuff like that. There's no model of the person." And yet, Edsinger himself says he experiences Domo as almost alive—almost uncomfortably so. For him, much of this effect comes from being with Domo as it runs autonomously for long periods—say, a half hour at a time—rather than being constrained, as he was on earlier projects, to try out elements of a robot's program in thirty-second intervals. "I can work with Domo for a half hour and never do the exact same thing twice," he says.[4] If this were said about a person, that would be a dull individual indeed. But by robotic standards, a seemingly unprogrammed half hour enchants.

Over a half hour, says Edsinger, Domo "moves from being this thing that you flip on and off and test a little bit of to something that's running all the time. . . . You transition out of the machine thing to thinking of it as not so much a creature but as much more fluid in terms of being . . . [long hesitation] Well, you start to think of it as a creature, but this is part of what makes the research inherently uncomfortable. I enjoy that. That's part of the reason I like building robots."

Thrilled by moments when the "creature" seems to escape, unbidden, from the machine, Edsinger begins to think of Domo's preferences not as things he has programmed but as the robot's own likes and dislikes.[5] He says,

> For me, when it starts to get complicated . . . sometimes I know that the robot is not doing things of its own "volition" because these are behaviors, well, I literally put them in there. But every now and then . . . the coordination of its behaviors is rich enough . . . well, it is of its own volition . . . and it catches you off guard. And to me this is what makes it fun . . . and it happens to me more and more now that I have more stuff running on it. . . .
>
> If it doesn't know what to do, it will look around and find a person. And if it can't find a person, it looks to the last place [it] saw a person. So, I'll be watching it do something, and it will finish, and it will look up at me as if to say, "I'm done; [I want your] approval."

In these moments, there is no deception. Edsinger knows how Domo "works." Edsinger experiences a connection where knowledge does not interfere with wonder. This is the intimacy presaged by the children for whom Cog was demystified but who wanted it to love them all the same.

Edsinger feels close to Domo as creature and machine. He believes that such feelings will sustain people as they learn to collaborate with robots. Astronauts and robots will go on space flights together. Soldiers and robots will go on missions together. Engineers and robots will maintain nuclear plants together. To be sold on partnership with robots, people need to feel more than comfortable with them. People should want to be around them. For Edsinger, this will follow naturally from the pleasure of physical contact with robotic partners. He says it is thrilling "just to experience something acting with some volition. There is an object, it is aware of my presence, it recognizes me, it wants to interact with me."

Edsinger does not fall back on the argument that we need helper robots because there will not be enough people to care for each other in the future. For him, creating sociable robots is its own adventure. The robots of the future will be cute, want to hug, and want to help. They will work alongside people, aware of their presence and wishes. Edsinger admits that it will be "deceiving, if people feel the robots know more than they do or care more than they do." But he does not see a moral issue. First, information about the robot's limitations is public, out there for all the world to see. Second, we have already decided that it is acceptable to be comforted by creatures that may not really care for us: "We gain comfort from animals and pets, many of which have very limited understanding of us." Why should we not embrace new relationships (with robots) with new limitations?

And besides, argues Edsinger, and this is an argument that has come up before, we take comfort in the presence of people whose true motivations we don't know. We assign caring roles to people who may not care at all. This might happen when, during a hospitalization, a nurse takes our hand. How important is it that this nurse wants to hold our hand? What if this is a rote gesture, something close to being programmed? Is it important that this programmed nurse be a person? For Edsinger, it is not. "When Domo holds my hand," he says, "it always feels good. . . . There is always that feeling of an entity making contact that it wants, that it needs. I like that, and I am willing to let myself feel that way . . . just the physical warm and fuzzy sense of being wanted, knowing full well that it is not caring." I ask Edsinger to clarify. Is it pleasurable to be touched even if he knows that the robot doesn't "want" to touch him. Edsinger is sure of his answer: "Yes." But a heartbeat later he retracts it: "Well, there is a part of me that is trying to say, well, Domo cares."

And this is where we are in the robotic moment. One of the world's most sophisticated robot "users" cannot resist the idea that pressure from a robot's hand implies caring. If we are honest with ourselves about what machines care about, we must accept their ultimate indifference. And yet, a hand that reaches for ours says, "I need you. Take care of me. Attend to me. And then, perhaps, I will— and will *want* to—attend to you." Again, what robots offer meets our human vulnerabilities. We can interact with robots in full knowledge of their limitations, comforted nonetheless by what must be an unrequited love.

A MOMENT OF MORE: MERGING MIND AND BODY

In the fall of 2005, performance artist Pia Lindman came to MIT with communion on her mind. Lindman had an artistic vision: she would find ways to merge her face and body with MIT's sociable robots. She hoped that by trying, she would come to know their minds. For Lindman, the robots were what Emerson would have called "test objects." She imagined that immersion in a robot's nature might give her a new understanding of her own.

The MIT sociable robots are inspired by a philosophical tradition that sees mind and body as inseparable. Following Immanuel Kant, Martin Heidegger, Maurice Merleau-Ponty, and, more recently, Hubert Dreyfus and Antonio Damasio, this tradition argues that our bodies are quite literally instruments of thought; therefore, any computer that wants to be intelligent had better start out with one.[6] Not all schools of artificial intelligence have been sympathetic to this way of seeing things. One branch of the field, often referred to as "symbolic AI," associates itself with a Cartesian mind/body dualism and argues that machine intelligence can be programmed through rules and the representation of facts.[7]

In the 1960s, philosopher Hubert Dreyfus took on the symbolic AI community when he argued that "computers need bodies in order to be intelligent."[8] This position has a corollary; whatever intelligence machines may achieve, it will never be the kind that people have because no body given to a machine will be a human body. Therefore, the machine's intelligence, no matter how interesting, will be alien.[9] Neuroscientist Antonio Damasio takes up this argument from a different research tradition. For Damasio, all thinking and all emotion is embodied. The absence of emotion reduces the scope of rationality because we literally think with our feelings, thus the rebuking title of his 1994 book *Descartes' Error*.[10] Damasio insists that there is no mind/body dualism, no split between thought and feeling. When we have to make a decision, brain processes that are shaped by our body guide our reasoning by remembering our pleasures and pains. This can be taken as an argument for why robots will never have a humanlike intelligence: they have neither bodily feelings nor feelings of emotion. These days, roboticists such as Brooks take up that challenge. They grant that intelligence may indeed require bodies and even emotions, but insist that they don't have to be human ones. And in 2005, it was Brooks to whom Lindman applied when she wanted to join her mind and body to a machine.

A precursor to Lindman's work with robots was her 2004 project on grief. She chose photographs of people grieving from the *New York Times*—a mother bending over a dead child, a husband learning he has lost his wife to a terrorist attack. Then, she sketched several hundred of the photographs and began to act them out, putting her face and body into the positions of the people in the photographs. Lindman says she felt grief as she enacted it. Biology makes this so.

The shape of a smile or frown releases chemicals that affect mental state.[11] And in humans, "mirror neurons" fire both when we observe others acting and when we act ourselves. Our bodies find a way to implicate us emotionally in what we see.[12] Lindman came out of the grief project wanting to further explore the connection between embodiment and emotion. So, closely tracking that project's methodology, she began to work with machines that had bodies. Teaming up with Edsinger, she videotaped his interactions with Domo, sketched the interactions of man and robot, and then learned to put herself in the place of both.[13]

Her enactments included Edsinger's surprise at being surprised when Domo does something unexpected; his pleasure when he holds down the robot's hand in order to get things done, and Domo, responding, seems to want freedom; his thrill in the moment when Domo finishes its work and looks around for the last place it saw a human, the place that Edsinger occupies. Through communion with man and robot, Lindman hoped to experience the gap between the human and the machine. In the end, Lindman created a work of art that both addresses and skirts the question of desire.

At an MIT gallery in the spring of 2006, Lindman performed the results of her work with Edsinger and Domo. On the walls she mounted thirty-four drawings of herself and the robot. In some drawings, Lindman assumes Domo's expression when disengaged, and she looks like a machine; in others, Domo is caught in moments of intense "engagement," and it looks like a person. In the drawings, Domo and Lindman seem equally comfortable in the role of person or machine, comfortable being each other.

The performance itself began with a video of Edsinger and Domo working together. They interact with an elegant economy of gesture. These two know each other very well. They seem to anticipate each other, look after each other. The video was followed by Lindman "enacting" Domo on a raised stage. She was dressed in gray overalls, her hair pulled into a tight bun. Within a few minutes, I forgot the woman and saw the machine. And then Lindman played both parts: human and machine. This time, within minutes, I saw two humans. And then, figure turned to ground, and I saw two machines, two very fond machines. Or was it two machines that were perhaps too fond? I was with a colleague who saw it the other way, first two machines and then two humans. Either way, Lindman had made her point: the boundaries between people and things are shifting. What of these boundaries is worth maintaining?

Later, I meet privately with Lindman, and she talks about her performance and her experience making the film. "I turn myself into the human version of

Domo . . . and I feel the connection between [Edsinger] and Domo. . . . You feel the tenderness, the affection in their gestures. Their pleasure in being together." She dwells on a sequence in which Edsinger tries to get Domo to pick up a ball. At one moment, the ball is not in Domo's field of vision. The robot looks toward Edsinger, as though orienting to a person who can help, a person whom it trusts. It reaches for Edsinger's hands. For the robot, says Lindman, "there is information to be gathered through touch." Domo and Edsinger stare at each other, with Domo's hands on Edsinger's as though in supplication. Lindman says that in enacting Domo for this sequence, she "couldn't think about seeking the ball. . . . I've always thought about it as a romantic scene."

For Lindman this scene is crucial. In trying to play a robot, she found that the only way to get it right was to use a script that involved love. "The only way I was able to start memorizing the movements was to create a narrative. To put emotions into the movements made me remember the movements." She is aware that Edsinger had a different experience. He had moments when he saw the robot as both program and creature: "A lot of times he'd be looking at the screen with the code scrolling by. . . . He is looking at the robot's behavior, at its internal processes, but also is drawn into what is compelling in the physical interaction." Edsinger wrote Domo's code, but also learns from touching Domo's body. Watching these moments on film, I see the solicitous touch of a mother who puts her hand on her child's forehead to check for fever.

Of a scene in which Edsinger holds down Domo's hand to prevent a collision, Lindman says,

> [Edsinger] is holding Domo's hand like this [Lindman demonstrates by putting one hand over another] and looks into Domo's eyes to understand what it's doing: Where are its eyes going? Is it confused? Is it trying to understand what it's seeing or is it understanding what it's seeing? To get eye contact with Domo is, like, a key thing. And he gets it. He's actually looking at Domo trying to understand what it's looking at, and then Domo slowly turns his head and looks him in the eye. And it's this totally romantic moment.

Edsinger, too, has described this moment as one in which he feels the pleasure of being sought after. So, it is not surprising that to enact it, Lindman imagined robot and man in a moment of desire. She says, "It is as though I needed the robot to *seem* to have emotions in order to understand it." She is able to play Domo only if she plays a woman desiring a man. "It is," she admits, "the scene I do best."

In the grief project, the position of her body brought Lindman to experiences of abjection, something that she now attributes to mirror neurons. She had expected that doubling for a robot would be very different because "it has no emotion." But in the end, she had to create emotions to become an object without emotion. "To remember the robot's motions, I had to say: 'It does this because it feels this way.' . . . It wasn't like I was feeling it, but I had to have that logic." Except that (think of the mirror neurons) Lindman *was* feeling it. And despite herself, she couldn't help but imagine them in the machine. Lindman's account becomes increasingly complex as she grapples with her experience. If the subject is communion with the inanimate, these are the telling contradictions of an expert witness.[14]

The philosopher Emmanuel Lévinas writes that the presence of a face initiates the human ethical compact.[15] The face communicates, "Thou shalt not kill me." We are bound by the face even before we know what stands behind it, even before we might learn that it is the face of a machine. The robotic face signals the presence of a self that can recognize another. It puts us in a landscape where we seek recognition. This is not about a robot's being able to recognize us. It is about our desire to have it do so.

Lindman could not play Edsinger without imagining him wanting the robot's recognition; she could not play Domo without imagining it wanting Edsinger's recognition. So, Lindman's enactment of Domo looking for a green ball interprets the robot as confused, seeking the person closest to it, locking eyes, and taking the person's hand to feel comforted. It is a moment, classically, during which a person might experience a feeling of communion. Edsinger—not just in Lindman's recreation—feels this closeness, unswayed by his knowledge of the mechanisms behind the robot's actions. For Lindman, such interactions spark "a crisis about what is authentic and real emotion."

Lindman worries that the romantic scripts she uses "might not seem to us authentic" because robots "are of mechanism not spirit." In her grief project, however, she found that grief is always expressed in a set of structured patterns, programmed, she thinks, by biology and culture. So we, like the robots, have programs beneath our expression of feelings. We are constrained by mechanisms, even in our most emotional moments. And if our emotions are mediated by such programming, asks Lindman, how different are our emotions from those of a machine? For Lindman, the boundary is disappearing. We are authentic in the way a machine can be, and a machine can be authentic in the way a person can be.

And this is where I began. The questions for the future are not whether children will love their robot companions more than their pets or even their parents. The questions are rather, What will love be? And what will it mean to achieve ever-greater intimacy with our machines? Are we ready to see ourselves in the mirror of the machine and to see love as our performances of love?

In her enactments of grief, Lindman felt her body produce a state of mind. And in much the same spirit, when she enacts Domo, she says she "feels" the robot's mind. But Lindman is open to a more transgressive experience of the robot mind. After completing the Domo project, she begins to explore how she might physically connect her face to the computer that controls the robot Mertz.

Lijin Aryananda's Mertz, a metal head on a flexible neck, improves on Kismet's face, speech, and vision. Like Kismet, Mertz has expressive brows above its black ping-pong ball eyes—features designed to make a human feel kindly toward the robot. But this robot can actually speak simple English. Like Domo, Mertz has been designed as a step toward a household companion and helper. Over time, and on its own, it is able to recognize a set of familiar individuals and chat with them using speech with appropriate emotional cadence. Lindman hopes that if she can somehow "plug herself" into Mertz, she will have a direct experience of its inner state. "I will experience its feelings," she says excitedly. And Lindman wants to have her brain scanned while she is hooked up to Mertz in order to compare images of her brain activity to what we know is going on in the machine. "We can actually look at both," she says. "I will be the embodiment of the AI and we will see if [when the robot smiles], my brain is smiling."

Lindman soon discovers that a person cannot make her brain into the output device for a robot intelligence. So, she modifies her plan. Her new goal is to "wear" Mertz's facial expressions by hooking up her face rather than her brain to the Mertz computer, to "become the tool for the expression of the artificial intelligence." After working with Domo, Lindman anticipates that she will experience a gap between who she is and what she will feel as she tries to be the robot. She hopes the experiment will help her understand what is specific to her as a human. In that sense, the project is about yearning for communion with the machine as well as inquiring into whether communion is possible. Lindman imagines the gap: "You will say, 'Okay, so there's the human.'"[16]

As a first step, and it would be her only step, Lindman constructs a device capable of manipulating her face by a set of mechanical pliers, levers, and wires, "just to begin with the experience of having my face put into different positions." It is painful and prompts Lindman to reconsider the direct plug-in she hopes some day to achieve. "I'm not afraid of too much pain," she says. "I'm more afraid of damage, like real damage, biological damage, brain damage. I don't think it's going to happen, but it's scary." And Lindman imagines another kind of damage. If some day she does hook herself up to a robot's program, she believes she will have knowledge of herself that no human has ever had. She will have the experience of what it feels like to be "taken over" by an alien intelligence. Perhaps she will feel its pull and her lack of resistance to it. The "damage" she fears relates to this. She may learn something she doesn't want to know. Does the knowledge of the extent to which we are machines mark the limit of our communion with machines? Is this knowledge taboo? Is it harmful?

Lindman's approach is novel, but the questions she raises are not new. Can machines develop emotions? Do they need emotions to develop full intelligence? Can people only relate to machines by projecting their own emotions onto them, emotions that machines cannot achieve? The fields of philosophy and artificial intelligence have a long history of addressing such matters. In my own work, I argue the limits of artificial comprehension because neither computer agents nor robots have a human life cycle.[17] For me, this objection is captured by the man who challenged the notion of having a computer psychotherapist with the comment, "How can I talk about sibling rivalry to something that never had a mother?" These days, AI scientists respond to the concern about the lack of machine emotion by proposing to build some. In AI, the position that begins with "computers need bodies in order to be intelligent" becomes "computers need affect in order to be intelligent."

Computer scientists who work in the field known as "affective computing" feel supported by the work of social scientists who underscore that people always *project* affect onto computers, which helps them to work more constructively with them.[18] For example, psychologist Clifford Nass and his colleagues review a set of laboratory experiments in which "individuals engage in social behavior towards technologies even when such behavior is entirely inconsistent with their beliefs about machines."[19] People attribute personality traits and gender to computers and even adjust their responses to avoid hurting the machines' "feelings." In one dramatic experiment, a first group of people is asked to perform a task on computer A and to evaluate the task on the same computer. A second group is asked to perform the task on computer A but to evaluate it on computer B. The first group gives computer A far higher grades. Basically, participants do not want to insult a computer "to its face."

Nass and his colleagues suggest that "when we are confronted with an entity that [behaves in humanlike ways, such as using language and responding based on prior inputs,] our brains' default response is to unconsciously treat the entity as human."[20] Given this, they propose that technologies be made more "likeable" for practical reasons. People will buy them and they will be easier to use. But making a machine "likeable" has moral implications. "It leads to various secondary consequences in interpersonal relationships (for example, trust, sustained friendship, and so forth)."[21] For me, these secondary consequences are the heart of the matter. Making a machine easy to use is one thing. Giving it a winning personality is another. Yet, this is one of the directions taken by affective computing (and sociable robotics).

Computer scientists who work in this tradition want to build computers able to assess their users' affective states and respond with "affective" states *of their own*. At MIT, Rosalind Picard, widely credited with coining the phrase "affective computing," writes, "I have come to the conclusion that if we want computers to be genuinely intelligent, to adapt to us, and to interact naturally with us, then they will need the ability to recognize and express emotions, and to have what has come to be called 'emotional intelligence.'"[22] Here the line is blurred between computers having emotions and behaving as if they did. Indeed, for Marvin Minsky, "Emotion is not especially different from the processes that we call 'thinking.'"[23] He joins Antonio Damasio on this but holds the opposite view of where the idea takes us. For Minsky, it means that robots are going to be emotional thinking machines. For Damasio, it means they can never be unless robots acquire bodies with the same characteristics and problems of living bodies.

In practice, researchers in affective computing try to avoid the word "emotion." Talking about emotional computers is always on track to raise strong objections. How would computers get these emotions? Affects sound more cognitive. Giving machines a bit of "affect" to make them easier to use sounds like common sense, more a user interface strategy than a philosophical position. But synonyms for "affective" include "emotional," "feeling," "intuitive," and "noncognitive," just to name a few.[24] "Affect" loses these meanings when it becomes something computers have. The word "intelligence" underwent a similar reduction in meaning when we began to apply it to machines. Intelligence once denoted a dense, layered, complex attribute. It implied intuition and common sense. But when computers were declared to have it, intelligence started to denote something more one-dimensional, strictly cognitive.

Lindman talks about her work with Domo and Mertz as a contribution to affective computing. She is convinced that Domo needs an additional layer of emotional intelligence. Since it wasn't programmed in, she says she had to "add it herself" when she enacted the robot's movements. But listening to Lindman describe how she had to "add in" yearning and tenderness to the relationship between Domo and Edsinger, I have a different reaction. Perhaps it is better that Lindman had to "add in" emotion. It put into sharp relief what is unique about people. The idea of affective computing intentionally blurs the line.

THROUGH THE EYES OF THE ROBOT

Domo and Mertz are advanced robots. But we know that feelings of communion are evoked by far simpler ones. Recall John Lester, the computer scientist who thought of his AIBO as both machine and creature. Reflecting on AIBO, Lester imagines that robots will change the course of human evolution.[25] In the future, he says, we won't simply enjoy using our tools, "we will come to care for them. They will teach us how to treat them, how to live with them. We will evolve to love our tools; our tools will evolve to be loveable."

Like Lindman and Edsinger, Lester sees a world of creature-objects burnished by our emotional attachments. With a shy shrug that signals he knows he is going out on a limb, he says, "I mean, that's the kind of bond I can feel for AIBO now, a tool that has allowed me to do things I've never done before. . . . Ultimately [tools like this] will allow society to do things that it has never done." Lester sees a future in which something like an AIBO will develop into a prosthetic device, extending human reach and vision.[26] It will allow people to interact with real, physical space in new ways. We will see "through its eyes," says Lester, and interact "through its body. . . . There could be some parts of it that are part of you, the blending of the tools and the body in a permanent physical way." This is how Brooks talks about the merging of flesh and machine. There will be no robotic "them" and human "us." We will either merge with robotic creatures, or in a long first step, we will become so close to them that we will integrate their powers into our sense of self. In this first step, a robot will still be an other, but one that completes you.

These are close to the dreams of Thad Starner, one of the founders of MIT's Wearable Computing Group, earlier known as the "cyborgs." He imagines bringing up a robot as a child in the spirit of how Brooks set out to raise Cog. But Starner insists that Cog—and successor robots such as Domo and Mertz—are "not extreme enough."[27] They live in laboratories, so no matter what the designers' good intentions, the robots will never be treated like human babies. Starner wants to teach a robot by having it learn from his life—by transmitting his life through sensors in his clothes. The sensors will allow "the computer to see as I see, hear as I hear, and experience the world around me as I experience it," Starner says. "If I meet somebody at a conference it might hear me say, 'Hi, David,' and shake a hand. Well, if it then sees me typing in somebody's name or pulling up that person's file, it might actually start understanding what introductions are." Starner's vision is "to create something that's not just an artificial intelligence. It's me."

In a more modest proposal, the marriage of connectivity and robotics is also the dream of Greg, twenty-seven, a young Israeli entrepreneur who has just graduated from business school. It is how he intends to make his fortune—and in the near future. In Greg's design, data from his cell phone will animate a robot. He says,

> I will walk around with my phone, but when I come home at night, I will plug it into a robotic body, also intelligent but in different ways. The robot knows about my home and how to take care of it and to take care of me if I get sick. The robot would sit next to me and prepare the documents I need to make business calls. And when I travel, I would just have to take the phone, because another robot will be in Tel Aviv, the same model. And it will come alive when I plug in my phone. And the robot bodies will offer more, say, creature comforts: a back rub for sure and emergency help if you get into medical trouble. It will be reassuring for a young person, but so much more for an old person.

We will animate our robots with what we have poured into our phones: the story of our lives. When the brain in your phone marries the body of your robot, document preparation meets therapeutic massage. Here is a happy fantasy of security, intellectual companionship, and nurturing connection. How can one not feel tempted?

Lester dreams of seeing the world through AIBO's eyes: it would be a point of access to an enhanced environment. Others turn this around, saying that the robot will become the environment; the physical world will be laced with the intelligence we are now trying to put into machines. In 2008, I addressed a largely technical audience at a software company, and a group of designers suggested that in the future people will not interact with stand-alone robots at all—that will become an old fantasy. What we now want from robots, they say, we will begin to embed in our rooms. These intellectually and emotionally "alive" rooms will collaborate with us. They will understand speech and gesture. They will have a sense of humor. They will sense our needs and offer comfort. Our rooms will be our friends and companions.

CONSIDERING THE ROBOT FOR REAL

The story of robots, communion, and moments of more opens up many conversations, both philosophical and psychological. But these days, as people imagine robots in their daily lives, their conversations become quite concrete as they grapple with specific situations and try to figure out if a robot could help.

Tony, a high school teacher, has just turned fifty. Within just the past few years, his life has entered a new phase. All three of his children are in college. His parents are dead. He and his wife, Betty, find themselves in constant struggle with her mother, Natasha, eighty-four, who is recuperating from a stroke and also showing early signs of Alzheimer's. When a younger woman and at her best, Natasha had been difficult. Now, she is anxious and demanding, often capricious. She criticizes her daughter and son-in-law when they try to help; nothing seems enough. Tony, exhausted, considers their options. With some dread, he and Betty have been talking about moving Natasha into their home. But they both work, so Natasha will require a caretaker to tend to her as she declines. He hears of work in progress on robots designed for child and elder care. This is something new to consider, and his first thoughts are positive.

> Well, if I compare having a robot with an immigrant in my house, the kind of person who is available to take care of an elderly person, the robot would be much better. Sort of like flying Virgin Atlantic and having your own movie. You could have the robot be however you wanted it. It wouldn't be rude or illiterate or steal from you. It would be very safe and

specialized. And personalized. I like that. Natasha's world is shrinking because of the Alzheimer's. A robot geared to the Alzheimer's—that could sort of measure where she was on any given day and give her some stimulation based on that—that would be great.

And then there is a moment of reconsideration:

But maybe I'm getting it backwards. I'm not sure I would want a robot taking care of me when I'm old. Actually, I'm not sure I would rather not be alive than be maintained by a robot. The human touch is so important. Even when people have Alzheimer's, even when they are unconscious, in a coma, I've read that people still have the grasping reflex. I suppose I want Natasha to have the human touch. I would want to have the human touch at the end. Other than that it is like that study where they substituted the terry cloth monkeys for the real monkeys and the baby monkeys clung to wire monkeys with terry cloth wrapped around. I remember studying that in college and finding it painfully sad. No, you need the real monkey to preserve your dignity. Your dignity as a person. Without that, we're like cows hooked up to a milking machine. Or like our life is like an assembly line where at the end you end up with the robot.

Tony is aware that he has talked himself into a contradiction: the robot is a specialized helper that can expertly diagnose a level of impairment *and* the robot is like a wire-and-terry-cloth monkey. He tries to reconcile his ideas:

I suppose the robot assistant is okay when a person still has some lucidity. You can still interact with it and know that it is a robot. But you don't want a robot at the end. Then, you deserve a person. Everybody deserves a person. But I do have mixed feelings. Looking at robots and children, really, there is a part of raising children . . . I think Marilyn French called it the "shit-and-string-beans" part of raising children. This would be good for a robot to do. You are a robot when you do it.

So, I'm happy to give over the shit-and-string-beans part of child raising and for that aspect of taking care of Natasha. Of course, the children are the harder call. But I would do it if it were the conventional thing that everyone did. Most people would do it if it were the conventional thing that everyone did. We didn't deny our children television, and we didn't think it was a good thing for them.

Tony is not happy to be caught in a contradiction. But many people share his dilemma. It is hard to hold on to a stable point of view. Plagued with problems, we are told that machines might address them. How are we to resist? Tony says, "I'm okay with the lack of authenticity [if you replace people with robots]. Lack of authenticity is an acceptable trade-off for services needed. I would say that my need right now trumps the luxury of authenticity. I would see a robot cleaning up after Natasha as labor saving, just like a vacuum cleaner. So the elder-care robot, I'm okay with it."

Betty has been quietly listening to this conversation about her mother. She would like her mother to live in her own home for as long as possible. Maybe a robot companion could help with that. She says,

> The robot would make her life more interesting. Maybe it would mean that she could stay in her own home longer. But provide some reassurance and peace of mind for me. More than a helper who might abuse her or ignore her or steal from her. I imagine she would prefer the robot. The robot wouldn't be critical. It would always be positive toward her. She would be familiar with it. At ease with it. Like Tony says, there is a down side to TV for children and there is a down side to this. There is always a down side. But it would be so worth it.

Then Betty speaks about other "robotic things" in her life. She thinks of automatic tellers as robotic. And she is happy that in her suburban neighborhood, she has a local bank where there are still human tellers, coffee, and a plate of donuts on Saturday. "I love our little bank. It would bother me if I went in there one day and the teller was a well-trained robot. At self-service gas stations, at ATM machines, you lose the intimacy."

For her husband, however, that neighborhood bank is only an exercise in nostalgia.

> The teller is not from the neighborhood. He doesn't know you or care. There's no point in talking to him because he has become a robot. If you do talk to the teller, you have become like the 'old guy,' the retired guy

who wants to talk to everyone on line and then talk to the teller. Because that is the old guy's social life—the bank, the grocery store, the barber. When you're young, you're okay with the ATM, but then, if that's all we have, when we're ready to talk to people, when we're old, there won't be anyone there. There will just be things.

Tony's review of the banal and the profound—of being young and wanting an ATM, of being old and bereft in a world of things—captures the essence of the robotic moment. We feel, as we stand before our ATM machines (or interact with bank tellers who behave like ATM machines), that they and we stand robotic among robots, "trained to talk to things." So, it seems less shocking to put robots in where people used to be. Tony expands on a familiar progression: when we make a job rote, we are more open to having machines do it. But even when people do it, they *and the people they serve* feel like machines.

Gradually, more of life, even parts of life that involve our children and parents, seem machine ready. Tony tries to focus on the bright side. Alzheimer's patients can be served by a finely tuned robot.[28] Children will have the attention of machines that will not resent the "shit and string beans" of their daily care. And yet, he feels the tug of something else: the robotic makes sense until it makes him think of monkeys deprived of a mother, clinging to wire and terry cloth.

This last reaction may end up seeming peculiarly American. In Japan, enthusiasm for robots is uninhibited.[29] Philosophically, the ground has been prepared. Japanese roboticists are fond of pointing out that in their country, even worn-out sewing needles are buried with ceremony. At some shrines in Japan, dolls, including sex dolls, are given proper burials. It is commonplace to think of the inanimate as having a life force. If a needle has a soul, why shouldn't a robot? At their robotic moment, a Japanese national publicity campaign portrays a future in which robots will babysit and do housework and women will be freed up to have more babies—preserving the traditional values of the Japanese home, but also restoring sociability to a population increasingly isolated through the networked life.

The Japanese take as a given that cell phones, texting, instant messaging, e-mail, and online gaming have created social isolation. They see people turning away from family to focus attention on their screens. People do not meet face to face; they do not join organizations. In Japan, robots are presented as facilitators of the human contact that the network has taken away. Technology has corrupted us; robots will heal our wounds.

We come full circle. Robots, which enchant us into increasingly intense relationships with the inanimate, are here proposed as a cure for our too intense immersion in digital connectivity. Robots, the Japanese hope, will pull us back toward the physical real and thus each other.

One wonders.

PART TWO

Networked

In Intimacy, New Solitudes

CHAPTER 8

always on

Pia Lindman walked the halls of MIT with cyborg dreams. She was not the first. In the summer of 1996, I met with seven young researchers at the MIT Media Lab who carried computers and radio transmitters in their backpacks and keyboards in their pockets. Digital displays were clipped onto eyeglass frames.[1] Thus provisioned, they called themselves "cyborgs" and were always wirelessly connected to the Internet, always online, free from desks and cables. The group was about to release three new 'borgs into the world, three more who would live simultaneously in the physical and virtual. I felt moved by the cyborgs as I had been by Lindman: I saw a bravery, a willingness to sacrifice for a vision of being one with technology. When their burdensome technology cut into their skin, causing lesions and then scars, the cyborgs learned to be indifferent. When their encumbrances caused them to be taken as physically disabled, they learned to be patient and provide explanations.

At MIT, there was much talk about what the cyborgs were trying to accomplish. Faculty supporters stressed how continual connectivity could increase productivity and memory. The cyborgs, it was said, might seem exotic, but this technology should inspire no fear. It was "just a tool" for being better prepared and organized in an increasingly complex information environment. The brain needed help.

From the cyborgs, however, I heard another story. They felt like new selves. One, in his mid-twenties, said he had "become" his device. Shy, with a memory that seemed limited by anxiety, he felt better able to function when he could literally be "looking up" previous encounters with someone as he began a new conversation. "With it," he said, referring to his collection of connectivity devices, "it's not just that I remember people or know more. I feel invincible, sociable, better prepared. I am naked without it. With it, I'm a better person." But with a sense of enhancement came feelings of diffusion. The cyborgs were a new kind of nomad, wandering in and out of the physical real. For the physical real was only one of the many things in their field of vision. Even in the mid-1990s, as they walked around Kendall Square in Cambridge, the cyborgs could not only search the Web but had mobile e-mail, instant messaging, and remote access to desktop computing. The multiplicity of worlds before them set them apart: they could be with you, but they were always somewhere else as well.

Within a decade, what had seemed alien was close to becoming everyone's way of life, as compact smartphones replaced the cyborgs' more elaborate accoutrements. This is the experience of living full-time on the Net, newly free in some ways, newly yoked in others. We are all cyborgs now.

People love their new technologies of connection. They have made parents and children feel more secure and have revolutionized business, education, scholarship, and medicine. It is no accident that corporate America has chosen to name cell phones after candies and ice cream flavors: chocolate, strawberry, vanilla. There is a sweetness to them. They have changed how we date and how we travel. The global reach of connectivity can make the most isolated outpost into a center of learning and economic activity. The word "apps" summons the pleasure of tasks accomplished on mobile devices, some of which, only recently, we would not have dreamed possible (for me, personally, it is an iPhone app that can "listen" to a song, identify it, and cue it up for purchase).

Beyond all of this, connectivity offers new possibilities for experimenting with identity and, particularly in adolescence, the sense of a free space, what Erik Erikson called the *moratorium*. This is a time, relatively consequence free, for doing what adolescents need to do: fall in and out of love with people and ideas. Real life does not always provide this kind of space, but the Internet does.

No handle cranks, no gear turns to move us from one stage of life to another. We don't get all developmental tasks done at age-appropriate times—or even necessarily get them done at all. We move on and use the materials we have to do the best we can at each point in our lives. We rework unresolved issues and seek out missed experiences. The Internet provides new spaces in which we can do this, no matter how imperfectly, throughout our lives. So, adults as well as adolescents use it to explore identity.

When part of your life is lived in virtual places—it can be Second Life, a computer game, a social networking site—a vexed relationship develops between what is true and what is "true here," true in simulation. In games where we expect to play an avatar, we end up being ourselves in the most revealing ways; on social-networking sites such as Facebook, we think we will be presenting ourselves, but our profile ends up as somebody else—often the fantasy of who we want to be. Distinctions blur. Virtual places offer connection with uncertain claims to commitment. We don't count on cyberfriends to come by if we are ill, to celebrate our children's successes, or help us mourn the death of our parents.[2] People know this, and yet the emotional charge on cyberspace is high. People talk about digital life as the "place for hope," the place where something new will come to them. In the past, one waited for the sound of the post—by carriage, by foot, by truck. Now, when there is a lull, we check our e-mail, texts, and messages.

The story of my own hesitant steps toward a cyborg life is banal, an example of the near universality of what was so recently exotic. I carry a mobile device with me at all times. I held out for years. I don't like attempting to speak to people who are moving in and out of contact as they pass through tunnels, come to dangerous intersections, or otherwise approach dead zones. I worry about them. The clarity and fidelity of sound on my landline telephone seems to me a technical advance over what I can hear on my mobile. And I don't like the feeling of always being on call. But now, with a daughter studying abroad who expects to reach me when she wants to reach me, I am grateful to be tethered to her through the Net. In deference to a generation that sees my phone calls as constraining because they take place in real time and are not suitable for multitasking, I text. Awkwardly.

But even these small things allow me to identify with the cyborgs' claims of an enhanced experience. Tethered to the Internet, the cyborgs felt like more than they could be without it. Like most people, I experience a pint-sized version of such pleasures. I like to look at the list of "favorites" on my iPhone contact list and see everyone I cherish. Each is just a tap away. If someone doesn't have time to talk to me, I can text a greeting, and they will know I am thinking of them, caring about them. Looking over recent text exchanges with my friends and family reliably puts me in a good mood. I keep all the texts my daughter sent me during her last year of high school. They always warm me: "Forgot my green sweater, bring please." "Can you pick me up at boathouse, 6?" "Please tell nurse I'm sick. Class boring. Want to come home." And of course, there are the photos, so many photos on my phone, more photos than I would ever take with a camera, always with me.

Yet, even such simple pleasures bring compulsions that take me by surprise. I check my e-mail first thing in the morning and before going to bed at night. I have come to learn that informing myself about new professional problems and demands is not a good way to start or end my day, but my practice unhappily continues. I admitted my ongoing irritation with myself to a friend, a woman in her seventies who has meditated on a biblical reading every morning since she was in her teens. She confessed that it is ever more difficult to begin her spiritual exercises before she checks her e-mail; the discipline to defer opening her inbox is now part of her devotional gesture. And she, too, invites insomnia by checking her e-mail every night before turning in.

Nurturance was the killer app for robotics. Tending the robots incited our engagement. There is a parallel for the networked life. Always on and (now) always with us, we tend the Net, and the Net teaches us to need it.

Online, like MIT's cyborgs, we feel enhanced; there is a parallel with the robotic moment of more. But in both cases, moments of more may leave us with lives of less. Robotics and connectivity call each other up in tentative symbiosis, parallel pathways to relational retreat. With sociable robots we are alone but receive the signals that tell us we are together. Networked, we are together, but so lessened are our expectations of each other that we can feel utterly alone. And there is the risk that we come to see others as objects to be accessed—and only for the parts we find useful, comforting, or amusing.

Once we remove ourselves from the flow of physical, messy, untidy life—and both robotics and networked life do that—we become less willing to get out there and take a chance. A song that became popular on YouTube in 2010, "Do You Want to Date My Avatar?" ends with the lyrics "And if you think I'm not the one, log off, log off and we'll be done."[3]

Our attraction to even the prospect of sociable robots affords a new view of our networked life. In Part One we saw that when children grow up with fond feelings for sociable robots, they are prepared for the "relationships with less" that the network provides. Now I turn to how the network prepares us for the "relationships with less" that robots provide. These are the unsettling isolations of the tethered self. I have said that tethered to the network through our mobile devices, we approach a new state of the self, itself. For a start, it presumes certain entitlements: It can absent itself from its physical surround—including the people in it. It can experience the physical and virtual in near simultaneity. And it is able to make more time by multitasking, our twenty-first-century alchemy.

THE NEW STATE OF THE SELF: TETHERED AND MARKED ABSENT

These days, being connected depends not on our distance from each other but from available communications technology. Most of the time, we carry that technology with us. In fact, being alone can start to seem like a precondition for being together because it is easier to communicate if you can focus, without interruption, on your screen. In this new regime, a train station (like an airport, a café, or a park) is no longer a communal space but a place of social collection: people come together but do not speak to each other. Each is tethered to a mobile device and to the people and places to which that device serves as a portal. I grew up in Brooklyn where sidewalks had a special look. In every season—even in winter, when snow was scraped away—there were chalk-drawn hopscotch boxes. I speak with a colleague who lives in my old neighborhood. The hopscotch boxes are gone. The kids are out, but they are on their phones.

When people have phone conversations in public spaces, their sense of privacy is sustained by the presumption that those around them will treat them not only as anonymous but as if absent. On a recent train trip from Boston to New York, I sat next to a man talking to his girlfriend about his problems. Here is what I learned by trying not to listen: He's had a recent bout of heavy drinking, and his father is no longer willing to supplement his income. He thinks his girlfriend spends too much money and he dislikes her teenage daughter. Embarrassed, I walked up and down the aisles to find another seat, but the train was full. Resigned, I returned to my seat next to the complainer. There was some comfort in the fact that he was not complaining to me, but I did wish I could disappear. Perhaps there was no need. I was already being treated as though I were not there.

Or perhaps it makes more sense to think of things the other way around: it is those on the phone who mark themselves as absent. Sometimes people signal their departure by putting a phone to their ear, but it often happens in more subtle ways—there may be a glance down at a mobile device during dinner or a meeting. A "place" used to comprise a physical space and the people within it. What is a place if those who are physically present have their attention on the absent? At a café a block from my home, almost everyone is on a computer or smartphone as they drink their coffee. These people are not my friends, yet somehow I miss their presence.

Our new experience of place is apparent as we travel. Leaving home has always been a way to see one's own culture anew. But what if, tethered, we bring our homes with us? The director of a program that places American students in Spanish universities once complained to me that her students were not "experiencing Spain." They spent their free time on Facebook, chatting with their friends from home. I was sympathetic, thinking of the hours I had spent walking with my teenage daughter on a visit to Paris the summer after she first got her mobile phone. As we sat in a café, waiting for a friend to join us for dinner, Rebecca received a call from a schoolmate who asked her to lunch in Boston, six hours behind us in time. My daughter said simply, "Not possible, but how about Friday?" Her friend didn't even know she was out of town. When I grew up, the idea of the "global village" was an abstraction. My daughter lives something concrete. Emotionally, socially, wherever she goes, she never leaves home. I asked her if she wouldn't rather experience Paris without continual reminders of Boston. (I left aside the matter that I was a reminder of Boston and she, mercifully, did not raise it.) She told me she was happy; she liked being in touch with her friends. She seemed to barely understand my question. I was wistful, worried that Rebecca was missing an experience I cherished in my youth: an undiluted Paris. My Paris came with the thrill of disconnection from everything I knew. My daughter's Paris did not include this displacement.

When Rebecca and I returned home from France, I talked about the trip with a close friend, a psychoanalyst. Our discussion led her to reminisce about her first visit to Paris. She was sixteen, travelling with her parents. But while they went sightseeing with her younger brother, she insisted on staying in her hotel room, writing long letters to her boyfriend. Adolescents have always balanced connection and disconnection; we need to acknowledge the familiarity of our needs and the novelty of our circumstances. The Internet is more than old wine in new bottles; now we can always be elsewhere.

In the month after Rebecca and I returned from Paris, I noted how often I was with colleagues who were elsewhere as well: a board meeting where members rebelled when asked to turn off their mobile devices; a faculty meeting where attendees did their e-mail until it was their turn to speak; a conference at which audience members set up Internet back channels in order to chat about speakers' presentations during the presentations themselves.[4]

Since I teach in a university, I find examples of distracted academics of particular interest. But it is the more mundane examples of attention sharing that change the fabric of daily life. Parents check e-mail as they push strollers. Children and parents text during family dinners. As I watched the annual marathon in Florence, Italy, in November 2009, a runner passed me, texting. Of course, I tried to take her picture on my cell phone. After five years, my level of connectivity had finally caught up with my daughter's. Now when I travel, my access to the Net stays constant. There is security and pleasure in a good hotel on the other side of the world, but it cannot compare to the constancy of online connections.

Research portrays Americans as increasingly insecure, isolated, and lonely.[5] We work more hours than ever before, often at several jobs. Even high school and college students, during seasons of life when time should be most abundant, say that they don't date but "hook up" because "who has the time?" We have moved away, often far away, from the communities of our birth. We struggle to raise children without the support of extended families. Many have left behind the religious and civic associations that once bound us together.[6] To those who have lost a sense of physical connection, connectivity suggests that you make your own page, your own place. When you are there, you are by definition where you belong, among officially friended friends. To those who feel they have no time, connectivity, like robotics, tempts by proposing substitutions through which you can have companionship with convenience. A robot will always be there, amusing and compliant. On the Net, you can always find someone. "I never want to be far from my BlackBerry," a colleague told me. "That is where my games are. That is where my sites are. Without it, I'm too anxious."

Today, our machine dream is to be never alone but always in control. This can't happen when one is face-to-face with a person. But it can be accomplished with a robot or, as we shall see, by slipping through the portals of a digital life.

THE NEW STATE OF THE SELF: FROM LIFE TO THE LIFE MIX

From the very beginning, networked technologies designed to share practical information were taken up as technologies of relationship. So, for example, the Arpanet, grandfather of the Internet, was developed so that scientists could collaborate on research papers, but it soon became a place to gossip, flirt, and talk about one's kids. By the mid-1990s, the Internet throbbed with new social worlds. There were chat rooms and bulletin boards and social environments known as multiuser domains, or MUDs. Soon after came massively multiplayer online role-playing games such as Ultima 2 and EverQuest, the precursors of game worlds such as World of Warcraft. In all of these, people created avatars—more or less richly rendered virtual selves—and lived out parallel lives. People sat at their computers and moved from windows that featured the spreadsheets and business documents of the real world to those in which they inhabited on-line personae. Although the games most often took the form of quests, medieval and otherwise, the virtual environments were most compelling because they offered opportunities for a social life, for performing as the self you wanted to be. As one player on an adventure-style MUD told me in the early 1990s, "I began with an interest in 'hack and slay,' but then I stayed to chat."[7]

In the course of a life, we never "graduate" from working on identity; we simply rework it with the materials at hand. From the start, online social worlds provided new materials. Online, the plain represented themselves as glamorous, the old as young, the young as older. Those of modest means wore elaborate virtual jewelry. In virtual space, the crippled walked without crutches, and the shy improved their chances as seducers. These days, online games and worlds are increasingly elaborate. The most popular "pay-to-play" game, World of Warcraft, puts you, along with 11.5 million other players, in the world of Azeroth. There, you control a character, an avatar, whose personality, natural gifts, and acquired skills are under continual development as it takes on a trade, explores the landscape, fights monsters, and goes on quests. In some games, you can play alone—in which case you mostly have artificial intelligences for company, "bots" that play the role of human characters. Or you can band together with other players on the network to conquer new worlds. This can be a highly collaborative endeavor, a social life unto itself: you routinely e-mail, talk to, and message the people you game with.

In a different genre, Second Life is a virtual "place" rather than a game. Here, there is no winning, only living. You begin by naming and building an avatar. You work from a menu with a vast array of choices for its looks and clothes. If these are not sufficient, you can design a customized avatar from scratch. Now, pleased with *your* looks, you have the potential, as Second Life puts it, to live a life that will enable you to "love your life."[8] You can, among other things, get an education, launch a business, buy land, build and furnish a home, and, of course, have a social life that may include love, sex, and marriage. You can even earn money—Second Life currency is convertible into dollars.

As all this unfolds, you hang out in virtual bars, restaurants, and cafés. You relax on virtual beaches and have business meetings in virtual conference rooms.

It is not uncommon for people who spend a lot of time on Second Life and role-playing games to say that their online identities make them feel more like themselves than they do in the physical real. This is play, certainly, but it is serious play.[9]

Historically, there is nothing new in "playing at" being other. But in the past, such play was dependent on physical displacement. As a teenager I devoured novels about young men and women sent abroad on a Grand Tour to get over unhappy love affairs. In Europe, they "played at" being unscathed by heartbreak. Now, in Weston, Massachusetts, Pete, forty-six, is trying find a life beyond his disappointing marriage. He has only to turn on his iPhone.

I meet Pete on an unseasonably warm Sunday in late autumn. He attends to his two children, four and six, and to his phone, which gives him access to Second Life.[10] There, Pete has created an avatar, a buff and handsome young man named Rolo. As Rolo, Pete has courted a female avatar named Jade, a slip of a girl, a pixie with short, spiky blonde hair. As Rolo, he "married" Jade in an elaborate Second Life ceremony more than a year before, surrounded by their virtual best friends. Pete has never met the woman behind the avatar Jade and does not wish to. (It is possible, of course, that the human being behind Jade is a man. Pete understands this but says, "I don't want to go there.") Pete describes Jade as intelligent, passionate, and easy to talk to.

On most days, Pete logs onto Second Life before leaving for work. Pete and Jade talk (by typing) and then erotically engage their avatars, something that Second Life software makes possible with special animations.[11] Boundaries between life and game are not easy to maintain. Online, Pete and Jade talk about sex and Second Life gossip, but they also talk about money, the recession, work, and matters of health. Pete is on cholesterol-lowering medication that is only partially successful. Pete says that it is hard to talk to his "real" wife Alison about his anxieties; she gets "too worried that I might die and leave her alone." But he can talk to Jade. Pete says, "Second Life gives me a better relationship than I have in real life. This is where I feel most myself. Jade accepts who I am. My relationship with Jade makes it possible for me to stay in my marriage, with my family." The ironies are apparent: an avatar who has never seen or spoken to him in person and to whom he appears in a body nothing like his own seems, to him, most accepting of his truest self.

Pete enjoys this Sunday in the playground; he is with his children and with Jade. He says, "My children seem content. . . . I feel like I'm with them. . . . I'm here for them but in the background." I glance around the playground. Many adults are dividing their attention between children and mobile devices. Are they scrolling through e-mails and texts from family, friends, and colleagues? Are they looking at photographs? Are they in parallel worlds with virtual lovers?

When people make the point that we have always found ways to escape from ourselves, that neither the desire nor the possibility is new with the Internet, I always tell them they are right. Pete's online life bears a family resemblance to how some people use more traditional extramarital affairs. It also resembles how people can play at being "other" on business trips and vacations. When Pete pushes a swing with one hand and types notes to Jade with the other, something is familiar: a man finding that a relationship outside his marriage gives him something he wants. But something is unfamiliar: the simultaneity of lives, the interleaving of romance with a shout-out to a six-year-old. Pete says that his online marriage is an essential part of his "life mix." I ask him about this expression. I have never heard it before. Pete explains that the life mix is the mash-up of what you have on- and offline. Now, we ask not of our satisfactions in life but in our life mix. We have moved from multitasking to multi-lifing.

You need mobile communication to get to the notion of the life mix. Until recently, one had to sit in front of a computer screen to enter virtual space. This meant that the passage through the looking glass was deliberate and bounded by the time you could spend in front of a computer. Now, with a mobile device as portal, one moves into the virtual with fluidity and on the go. This makes it easier to use our lives as avatars to manage the tensions of everyday existence. We use social networking to be "ourselves," but our online performances take on lives of their own. Our online selves develop distinct personalities. Sometimes we see them as our "better selves." As we invest in them, we want to take credit for them. Recently—although, admittedly, at MIT I live in the land of the techno-sophisticated—I have been given business cards that include people's real-life names, their Facebook handles, and the name of their avatar on Second Life.

In talking about sociable robots, I described an arc that went from seeing simulation as better than nothing to simply better, as offering companions that could meet one's exact emotional requirements. Something similar is happening online. We may begin by thinking that e-mails, texts, and Facebook messaging are thin gruel but useful if the alternative is sparse communication with the people we care about. Then, we become accustomed to their special pleasures—we can have connection when and where we want or need it, and we can easily make it go away. In only a few more steps, you have people describing life on Facebook as better than anything they have ever known. They use the site to share their thoughts, their music, and their photos. They expand their reach in a continually growing community of acquaintance. No matter how esoteric their interests, they are surrounded by enthusiasts, potentially drawn from all over the world. No matter how parochial the culture around them, they are cosmopolitan. In this spirit, when Pete talks about Second Life, he extols its international flavor and his "in-world" educational opportunities. He makes it clear that he spends time "in physical life" with friends and family. But he says that Second Life "is my preferred way of being with people."[12]

In addition to the time he spends on Second Life, Pete has an avatar on World of Warcraft, and he is a regular on the social-networking sites Facebook, LinkedIn, and Plaxo. Every day he checks one professional and three personal e-mail accounts. I once described this kind of movement among identities with the metaphor of "cycling through."[13] But now, with mobile technology, cycling through has accelerated into the mash-up of a life mix. Rapid cycling stabilizes into a sense of continual copresence. Even a simple cell phone brings us into the world of continual partial attention.[14]

Not that many years ago, one of my graduate students talked to me about the first time he found himself walking across the MIT campus with a friend who took an incoming call on his mobile phone. My student was irritated, almost incredulous. "He put me on 'pause.' Am I supposed to remember where we were and pick up the conversation after he is done with his call?" At the time, his friend's behavior seemed rude and confusing. Only a few years later, it registers as banal. Mobile technology has made each of us "pauseable." Our face-to-face conversations are routinely interrupted by incoming calls and text messages. In the world of paper mail, it was unacceptable for a colleague to read his or her correspondence during a meeting. In the new etiquette, turning away from those in front of you to answer a mobile phone or respond to a text has become close to the norm. When someone holds a phone, it can be hard to know if you have that person's attention. A parent, partner, or child glances down and is lost to another place, often without realizing that they have taken leave. In restaurants, customers are asked to turn their phones to vibrate. But many don't need sound or vibration to know that something has happened on their phones. "When there is an event on my phone, the screen changes," says a twenty-six-year-old lawyer. "There is a brightening of the screen. Even if my phone is in my purse . . . I see it, I sense it. . . . I always know what is happening on my phone."

People are skilled at creating rituals for demarcating the boundaries between the world of work and the world of family, play, and relaxation. There are special times (the Sabbath), special meals (the family dinner), special clothes (the "armor" for a day's labor comes off at home, whether it is the businessman's suit or the laborer's overalls), and special places (the dining room, the parlor, the kitchen, and the bedroom). Now demarcations blur as technology accompanies us everywhere, all the time. We are too quick to celebrate the continual presence of a technology that knows no respect for traditional and helpful lines in the sand.[15]

Sal, sixty-two, a widower, describes one erased line as a "Rip van Winkle experience." When his wife became ill five years before, he dropped out of one world. Now, a year after her death, he wakes up in another. Recently, Sal began to entertain at his home again. At his first small dinner party, he tells me, "I invited a woman, about fifty, who works in Washington. In the middle of a conversation about the Middle East, she takes out her BlackBerry. She wasn't speaking on it. I wondered if she was checking her e-mail. I thought she was being rude, so I asked her what she was doing. She said that she was blogging the conversation. She was *blogging* the conversation." Several months after the event, Sal remains incredulous. He thinks of an evening with friends as private, as if surrounded by an invisible wall. His guest, living the life mix, sees her evening as an occasion to appear on a larger virtual stage.

THE NEW STATE OF THE SELF: MULTITASKING AND THE ALCHEMY OF TIME

In the 1980s, the children I interviewed about their lives with technology often did their homework with television and music in the background and a hand-held video game for distraction. Algebra and Super Mario were part of the same package. Today, such recollections sound almost pastoral. A child doing homework is usually—among other things—attending to Facebook, shopping, music, online games, texts, videos, calls, and instant messages. Absent only is e-mail, considered by most people under twenty-five a technology of the past, or perhaps required to apply to college or to submit a job application.

Subtly, over time, multitasking, once seen as something of a blight, was recast as a virtue. And over time, the conversation about its virtues became extravagant, with young people close to lionized for their ability to do many things at once. Experts went so far as to declare multitasking not just a skill but *the* crucial skill for successful work and learning in digital culture. There was even concern that old-fashioned teachers who could only do one thing at a time would hamper student learning.[16] Now we must wonder at how easily we were smitten.

When psychologists study multitasking, they do not find a story of new efficiencies. Rather, multitaskers don't perform as well on any of the tasks they are attempting.[17] But multitasking feels good because the body rewards it with neurochemicals that induce a multitasking "high." The high deceives multitaskers into thinking they are being especially productive. In search of the high, they want to do even more. In the years ahead, there will be a lot to sort out. We fell in love with what technology made easy. Our bodies colluded.

These days, even as some educators try to integrate smartphones into classrooms, others experiment with media fasts to get students down to business. At my university, professors are divided about whether they should meddle at all. Our students, some say, are grown-ups. It is not for us to dictate how they take notes or to get involved if they let their attention wander from class-related materials. But when I stand in back of our Wi-Fi enabled lecture halls, students are on Facebook and YouTube, and they are shopping, mostly for music. I want to engage my students in conversation. I don't think they should use class time for any other purpose. One year, I raised the topic for general discussion and suggested using notebooks (the paper kind) for note taking. Some of my students claimed to be relieved. "Now I won't be tempted by Facebook messages," said one sophomore. Others were annoyed, almost surly. They were not in a position to defend their right to shop and download music in class, so they insisted that they liked taking notes on their computers. I was forcing them to take notes by hand and then type them into computer documents later. While they were complaining about this two-step process, I was secretly thinking what a good learning strategy this might be. I maintained my resolve, but the following year, I bowed to common practice and allowed students to do what they wished. But I notice, along with several of my colleagues, that the students whose laptops are open in class do not do as well as the others.[18]

When media are always there, waiting to be wanted, people lose a sense of choosing to communicate. Those who use BlackBerry smartphones talk about the fascination of watching their lives "scroll by." They watch their lives as though watching a movie. One says, "I glance at my watch to sense the time; I glance at my BlackBerry to get a sense of my life."[19] Adults admit that interrupting their work for e-mail and messages is distracting but say they would never give it up. When I ask teenagers specifically about being interrupted during homework time, for example, by Facebook messages or new texts, many seem not to understand the question. They say things like, "That's just how it is. That's just my life." When the BlackBerry movie of one's life becomes one's life, there is a problem: the BlackBerry version is the unedited version of one's life. It contains more than one has time to live. Although we can't keep up with it, we feel responsible for it. It is, after all, our life. We strive to be a self that can keep up with its e-mail.

Our networked devices encourage a new notion of time because they promise that one can layer more activities onto it. Because you can text while doing something else, texting does not seem to take time but to give you time. This is more than welcome; it is magical. We have managed to squeeze in that extra little bit, but the fastest living among us encourage us to read books with titles such as *In Praise of Slowness*.[20] And we have found ways of spending more time with friends and family in which we hardly give them any attention at all.

We are overwhelmed across the generations. Teenagers complain that parents don't look up from their phones at dinner and that they bring their phones to school sporting events. Hannah, sixteen, is a solemn, quiet high school junior. She tells me that for years she has tried to get her mother's attention when her mother comes to fetch her after school or after dance lessons. Hannah says, "The car will start; she'll be driving still looking down, looking at her messages, but still no hello." We will hear others tell similar stories.

Parents say they are ashamed of such behavior but quickly get around to explaining, if not justifying, it. They say they are more stressed than ever as they try to keep up with e-mail and messages. They always feel behind. They cannot take a vacation without bringing the office with them; their office is on their cell phone.[21] They complain that their employers require them to be continually online but then admit that their devotion to their communications devices exceeds all professional expectations.

Teenagers, when pressed for time (a homework assignment is due), may try to escape the demands of the always-on culture. Some will use their parents' accounts so that their friends won't know that they are online. Adults hide out as well. On weekends, mobile devices are left at the office or in locked desk drawers. When employers demand connection, people practice evasive maneuvers. They go on adventure vacations and pursue extreme sports. As I write this, it is still possible to take long plane rides with no cell phone or Internet access. But even this is changing. Wi-Fi has made it to the skies.

In a tethered world, too much is possible, yet few can resist measuring success against a metric of what they could accomplish if they were always available. Diane, thirty-six, a curator at a large Midwestern museum, cannot keep up with the pace set by her technology.

I can hardly remember when there was such a thing as a weekend, or when I had a Filofax and I thought about whose name I would add to my address book. My e-mail program lets me click on the name of the person who wrote me and poof, they are in my address book. Now everyone who writes me gets put in my address book; everybody is a potential contact, a buyer, donor, and fund-raiser. What used to be an address book is more like a database.

I suppose I do my job better, but my job is my whole life. Or my whole life is my job. When I move from calendar, to address book, to e-mail, to text messages, I feel like a master of the universe; everything is so efficient. I am a maximizing machine. I am on my BlackBerry until two in the morning. I don't sleep well, but I still can't keep up with what is sent to me.

Now for work, I'm expected to have a Twitter feed and a Facebook presence about the museum. And do a blog on museum happenings. That means me in all these places. I have a voice condition. I keep losing my voice. It's not from talking too much. All I do is type, but it has hit me at my voice. The doctor says it's a nervous thing.

Diane, in the company of programs, feels herself "a master of the universe." Yet, she is only powerful enough to see herself as a "maximizing machine" that responds to what the network throws at her. She and her husband have decided they should take a vacation. She plans to tell her colleagues that she is going to be "off the grid" for two weeks, but Diane keeps putting off her announcement. She doesn't know how it will be taken. The norm in the museum is that it is fine to take time off for vacations but not to go offline during them. So, a vacation usually means working from someplace picturesque. Indeed, advertisements for wireless networks routinely feature a handsome man or beautiful woman sitting on a beach. Tethered, we are not to deny the body and its pleasures but to put our bodies somewhere beautiful while we work. Once, mobile devices needed to be shown in such advertisements. Now, they are often implied. We know that the successful are always connected. On vacation, one vacates a place, not a set of responsibilities. In a world of constant communication, Diane's symptom seems fitting: she has become a machine for communicating, but she has no voice left for herself.

As Diane plans her "offline vacation," she admits that she really wants to go to Paris, "but I would have no excuse not to be online in Paris. Helping to build houses in the Amazon, well, who would know if they have Wi-Fi? My new non-negotiable for a vacation: I have to be able to at least pretend that there is no reason to bring my computer." But after her vacation in remote Brazil finally takes place, she tells me, "Everybody had their BlackBerries with them. Sitting there in the tent. BlackBerries on. It was as though there was some giant satellite parked in the sky."

Diane says she receives about five hundred e-mails, several hundred texts, and around forty calls a day. She notes that many business messages come in multiples. People send her a text and an e-mail, then place a call and leave a message on her voicemail. "Client anxiety," she explains. "They feel better if they communicate." In her world, Diane is accustomed to receiving a hasty message to which she is expected to give a rapid response. She worries that she does not have the time to take her time on the things that matter. And it is hard to maintain a sense of what matters in the din of constant communication.

The self shaped in a world of rapid response measures success by calls made, e-mails answered, texts replied to, contacts reached. This self is calibrated on the basis of what technology proposes, by what it makes easy. But in the technology-induced pressure for volume and velocity, we confront a paradox. We insist that our world is increasingly complex, yet we have created a communications culture that has decreased the time available for us to sit and think uninterrupted. As we communicate in ways that ask for almost instantaneous responses, we don't allow sufficient space to consider complicated problems.

Trey, a forty-six-year-old lawyer with a large Boston firm, raises this issue explicitly. On e-mail, he says, "I answer questions I can answer right away. And people want me to answer them right away. But it's not only the speed. . . . The questions have changed to ones that I *can* answer right away." Trey describes legal matters that call for time and nuance and says that "people don't have patience for these now. They send an e-mail, and they expect something back fast. They are willing to forgo the nuance; really, the client wants to hear something now, and so I give the answers that can be sent back by return e-mail . . . or maybe answers that will take me a day, max. . . . I feel pressured to think in terms of bright lines." He corrects himself. "It's not the technology that does this, of course, but the technology sets expectations about speed." We are back to a conversation about affordances and vulnerabilities. The technology primes us for speed, and overwhelmed, we are happy to have it help us speed up. Trey reminds me that "we speak in terms of 'shooting off' an e-mail. Nobody 'shoots something off' because they want things to proceed apace."

Trey, like Diane, points out that clients frequently send him a text, an e-mail, and a voicemail. "They are saying, 'Feed me.' They feel they have the right." He sums up his experience of the past decade. Electronic communication has been liberating, but in the end, "it has put me on a speed-up, on a treadmill, but that isn't the same as being productive."

I talk with a group of lawyers who all insist that their work would be impossible without their "cells"—that nearly universal shorthand for the smartphones of today that have pretty much the functionality of desktop computers and more. The lawyers insist that they are more productive and that their mobile devices "liberate" them to work from home and to travel with their families. The women, in particular, stress that the networked life makes it possible for them to keep their jobs and spend time with their children. Yet, they also say that their mobile devices eat away at their time to think. One says, "I don't have enough time alone with my mind." Others say, "I have to struggle to make time to think." "I artificially make time to think." "I block out time to think." These formulations all depend on an "I" imagined as separate from the technology, a self that is able to put the technology aside so that it can function independently of its demands. This formulation contrasts with a growing reality of lives lived in the continuous presence of screens. This reality has us, like the MIT cyborgs, learning to see ourselves as one with our devices. To make more time to think would mean turning off our phones. But this is not a simple proposition since our devices are ever more closely coupled to our sense of our bodies and minds.[22] They provide a social and psychological GPS, a navigation system for tethered selves.

As for Diane, she tries to keep up by communicating during what used to be "downtime"—the time when she might have daydreamed during a cab ride or while waiting in line or walking to work. This may be time that we need (physiologically and emotionally) to maintain our ability to focus.[23] But Diane does not permit it to herself. And, of course, she uses our new kind of time: the time of attention sharing.

Diane shies away from the telephone because its real-time demands make too much of a claim on her attention. But like the face-to-face interactions for which it substitutes, the telephone can deliver in ways that texts and e-mails cannot. All parties are present. If there are questions, they can be answered. People can express mixed feelings. In contrast, e-mail tends to go back and forth without resolution. Misunderstandings are frequent. Feelings get hurt. And the greater the misunderstanding, the greater the number of e-mails, far more than necessary. We come to experience the column of unopened messages in our inboxes as a burden. Then, we project our feelings and worry that our messages are a burden to others.

We have reason to worry. One of my friends posted on Facebook, "The problem with handling your e-mail backlog is that when you answer mail, people answer back! So for each 10 you handle, you get 5 more! Heading down towards my goal of 300 left tonight, and 100 tomorrow." This is becoming a common sentiment. Yet it is sad to hear ourselves refer to letters from friends as "to be handled" or "gotten rid of," the language we use when talking about garbage. But this is the language in use.

An e-mail or text seems to have been always on its way to the trash. These days, as a continuous stream of texts becomes a way of life, we may say less to each other because we imagine that what we say is almost already a throwaway. Texts, by nature telegraphic, can certainly be emotional, insightful, and sexy. They can lift us up. They can make us feel understood, desired, and supported. But they are not a place to deeply understand a problem or to explain a complicated situation. They are momentum. They fill a moment.

FEARFUL SYMMETRIES

When I speak of a new state of the self, itself, I use the word "itself" with purpose. It captures, although with some hyperbole, my concern that the connected life encourages us to treat those we meet online in something of the same way we treat objects—with dispatch. It happens naturally: when you are besieged by thousands of e-mails, texts, and messages—more than you can respond to—demands become depersonalized. Similarly, when we Tweet or write to hundreds or thousands of Facebook friends as a group, we treat individuals as a unit. Friends become fans. A college junior contemplating the multitudes he can contact on the Net says, "I feel that I am part of a larger thing, the Net, the Web. The world. It becomes a thing to me, a thing I am part of. And the people, too, I stop seeing them as individuals, really. They are part of this larger thing."

With sociable robots, we imagine objects as people. Online, we invent ways of being with people that turn them into something close to objects. The self that treats a person as a thing is vulnerable to seeing itself as one. It is important to remember that when we see robots as "alive enough" for us, we give them a promotion. If when on the net, people feel just "alive enough" to be "maximizing machines" for e-mails and messages, they have been demoted. These are fearful symmetries.

In Part One, we saw new connections with the robotic turn into a desire for communion that is no communion at all. Part Two also traces an arc that ends in broken communion. In online intimacies, we hope for compassion but often get the cruelty of strangers. As I explore the networked life and its effects on intimacy and solitude, on identity and privacy, I will describe the experience of many adults. Certain chapters focus on them almost exclusively. But I return again and again to the world of adolescents. Today's teenagers grew up with sociable robots as playroom toys. And they grew up networked, sometimes receiving a first cell phone as early as eight. Their stories offer a clear view of how technology reshapes identity because identity is at the center of adolescent life. Through their eyes, we see a new sensibility unfolding.

These days, cultural norms are rapidly shifting. We used to equate growing up with the ability to function independently. These days always-on connection leads us to reconsider the virtues of a more collaborative self. All questions about autonomy look different if, on a daily basis, we are together even when we are alone.

The network's effects on today's young people are paradoxical. Networking makes it easier to play with identity (for example, by experimenting with an avatar that is interestingly different from you) but harder to leave the past behind, because the Internet is forever. The network facilitates separation (a cell phone allows children greater freedoms) but also inhibits it (a parent is always on tap). Teenagers turn away from the "real-time" demands of the telephone and disappear into role-playing games they describe as "communities" and worlds." And yet, even as they are committed to a new life in the ether, many exhibit an unexpected nostalgia. They start to resent the devices that force them into performing their profiles; they long for a world in which personal information is not taken from them automatically, just as the cost of doing business. Often it is children who tell their parents to put away the cell phone at dinner. It is the young who begin to speak about problems that, to their eyes, their elders have given up on.

I interview Sanjay, sixteen. We will talk for an hour between two of his class periods. At the beginning of our conversation, he takes his mobile phone out of his pocket and turns it off.[24] At the end of our conversation, he turns the phone back on. He looks at me ruefully, almost embarrassed. He has received over a hundred text messages as we were speaking. Some are from his girlfriend who, he says, "is having a meltdown." Some are from a group of close friends trying to organize a small concert. He feels a lot of pressure to reply and begins to pick up his books and laptop so he can find a quiet place to set himself to the task. As he says good-bye, he adds, not speaking particularly to me but more to himself as an afterthought to the conversation we have just had, "I can't imagine doing this when I get older." And then, more quietly, "How long do I have to continue doing this?"

CHAPTER 9

growing up tethered

roman, eighteen, admits that he texts while driving and he is not going to stop. "I know I should, but it's not going to happen. If I get a Facebook message or something posted on my wall . . . I have to see it. I have to." I am speaking with him and ten of his senior classmates at the Cranston School, a private urban coeducational high school in Connecticut. His friends admonish him, but then several admit to the same behavior. Why do they text while driving? Their reasons are not reasons; they simply express a need to connect. "I interrupt a call even if the new call says 'unknown' as an identifier—I just have to know who it is. So I'll cut off a friend for an 'unknown,'" says Maury. "I need to know who wanted to connect. . . . And if I hear my phone, I have to answer it. I don't have a choice. I have to know who it is, what they are calling for." Marilyn adds, "I keep the sound on when I drive. When a text comes in, I have to look. No matter what. Fortunately, my phone shows me the text as a pop up right up front . . . so I don't have to do too much looking while I'm driving." These young people live in a state of waiting for connection. And they are willing to take risks, to put themselves on the line. Several admit that tethered to their phones, they get into accidents when walking. One chipped a front tooth. Another shows a recent bruise on his arm. "I went right into the handle of the refrigerator."

I ask the group a question: "When was the last time you felt that you didn't want to be interrupted?" I expect to hear many stories. There are none. Silence.

"I'm waiting to be interrupted right now," one says. For him, what I would term "interruption" is the beginning of a connection.

Today's young people have grown up with robot pets and on the network in a fully tethered life. In their views of robots, they are pioneers, the first generation that does not necessarily take simulation to be second best. As for online life, they see its power—they are, after all risking their lives to check their messages—but they also view it as one might the weather: to be taken for granted, enjoyed, and sometimes endured. They've gotten used to this weather but there are signs of weather fatigue. There are so many performances; it takes energy to keep things up; and it takes time, a lot of time. "Sometimes you don't have time for your friends except if they're online," is a common complaint. And then there are the compulsions of the networked life—the ones that lead to dangerous driving and chipped teeth.

Today's adolescents have no less need than those of previous generations to learn empathic skills, to think about their values and identity, and to manage and express feelings. They need time to discover themselves, time to think. But technology, put in the service of always-on communication and telegraphic speed and brevity, has changed the rules of engagement with all of this. When is downtime, when is stillness? The text-driven world of rapid response does not make self-reflection impossible but does little to cultivate it. When interchanges are reformatted for the small screen and reduced to the emotional shorthand of emoticons, there are necessary simplifications. And what of adolescents' need for secrets, for marking out what is theirs alone?

I wonder about this as I watch cell phones passed around high school cafeterias. Photos and messages are being shared and compared. I cannot help but identify with the people who sent the messages to these wandering phones. Do they all assume that their words and photographs are on public display? Perhaps. Traditionally, the development of intimacy required privacy. Intimacy without privacy reinvents what intimacy means. Separation, too, is being reinvented. Tethered children know they have a parent on tap—a text or a call away.

DEGREES OF SEPARATION

Mark Twain mythologized the adolescent's search for identity in the Huck Finn story, the on-the-Mississippi moment, a time of escape from an adult world. Of course, the time on the river is emblematic not of a moment but of an ongoing process through which children separate from their parents. That rite of passage is now transformed by technology. In the traditional variant, the child internalizes the adults in his or her world before crossing the threshold of independence. In the modern, technologically tethered variant, parents can be brought along in an intermediate space, such as that created by the cell phone, where everyone important is on speed dial. In this sense, the generations sail down the river together, and adolescents don't face the same pressure to develop the independence we have associated with moving forward into young adulthood.

When parents give children cell phones—most of the teenagers I spoke with were given a phone between the ages of nine and thirteen—the gift typically comes with a contract: children are expected to answer their parents' calls. This arrangement makes it possible for the child to engage in activities—see friends, attend movies, go shopping, spend time at the beach—that would not be permitted without the phone. Yet, the tethered child does not have the experience of being alone with only him- or herself to count on. For example, there used to be a point for an urban child, an important moment, when there was a first time to navigate the city alone. It was a rite of passage that communicated to children that they were on their own and responsible. If they were frightened, they had to experience those feelings. The cell phone buffers this moment.

Parents want their children to answer their phones, but adolescents need to separate. With a group of seniors at Fillmore, a boys' preparatory school in New York City, the topic of parents and cell phones elicits strong emotions. The young men consider, "If it is always possible to be in touch, when does one have the right to be alone?"

Some of the boys are defiant. For one, "It should be my decision about whether I pick up the phone. People can call me, but I don't have to talk to them." For another, "To stay free from parents, I don't take my cell. Then they can't reach me. My mother tells me to take my cell, but I just don't." Some appeal to history to justify ignoring parents' calls. Harlan, a distinguished student and athlete, thinks he has earned the right to greater independence. He talks about older siblings who grew up before cell phones and enjoyed greater freedom: "My mother makes me take my phone, but I never answer it when my parents call, and they get mad at me. I don't feel I should have to. Cell phones are recent. In the last ten years, everyone started getting them. Before, you couldn't just call someone whenever. I don't see why I have to answer when my mom calls me. My older sisters didn't have to do that." Harlan's mother, unmoved by this argument from precedent, checks that he has his phone when he leaves for school in the morning; Harlan does not answer her calls. Things are at an unhappy stalemate.

Several boys refer to the "mistake" of having taught their parents how to text and send instant messages (IMs), which they now equate with letting the genie out of the bottle. For one, "I made the mistake of teaching my parents how to text-message recently, so now if I don't call them when they ask me to call, I get an urgent text message." For another, "I taught my parents to IM. They didn't know how. It was the stupidest thing I could do. Now my parents IM me all the time. It is really annoying. My parents are upsetting me. I feel trapped and less independent."

Teenagers argue that they should be allowed time when they are not "on call." Parents say that they, too, feel trapped. For if you know your child is carrying a cell phone, it is frightening to call or text and get no response. "I didn't ask for this new worry," says the mother of two high school girls. Another, a mother of three teenagers, "tries not to call them if it's not important." But if she calls and gets no response, she panics:

> I've sent a text. Nothing back. And I know they have their phones. Intellectually, I know there is little reason to worry. But there is something about this unanswered text. Sometimes, it made me a bit nutty. One time, I kept sending texts, over and over. I envy my mother. We left for school in the morning. We came home. She worked. She came back, say at six. She didn't worry. I end up imploring my children to answer my every message. Not because I feel I have a right to their instant response. Just out of compassion.

Adolescent autonomy is not just about separation from parents. Adolescents also need to separate from each other. They experience their friendships as both sustaining and constraining. Connectivity brings complications. Online life provides plenty of room for individual experimentation, but it can be hard to escape from new group demands. It is common for friends to expect that their friends will stay available—a technology-enabled social contract demands continual peer presence. And the tethered self becomes accustomed to its support.

Traditional views of adolescent development take autonomy and strong personal boundaries as reliable signs of a successfully maturing self. In this view of development, we work toward an independent self capable of having a feeling, considering it, and deciding whether to share it. Sharing a feeling is a deliberate act, a movement toward intimacy. This description was always a fiction in several ways. For one thing, the "gold standard" of autonomy validated a style that was culturally "male." Women (and indeed, many men) have an emotional style that defines itself not by boundaries but through relationships.[1] Furthermore, adolescent conversations are by nature exploratory, and this in healthy ways. Just as some writers learn what they think by looking at what they write, the years of identity formation can be a time of learning what you think by hearing what you say to others. But given these caveats, when we think about maturation, the notion of a bounded self has its virtues, if only as a metaphor. It suggests, sensibly, that before we forge successful life partnerships, it is helpful to have a sense of who we are.[2]

But the gold standard tarnishes if a phone is always in hand. You touch a screen and reach someone presumed ready to respond, someone who also has a phone in hand. Now, technology makes it easy to express emotions while they are being formed. It supports an emotional style in which feelings are not fully experienced until they are communicated. Put otherwise, there is every opportunity to form a thought by sending out for comments.

THE COLLABORATIVE SELF

Julia, sixteen, a sophomore at Branscomb, an urban public high school in New Jersey, turns texting into a kind of polling. Julia has an outgoing and warm presence, with smiling, always-alert eyes. When a feeling bubbles up, Julia texts it. Where things go next is guided by what she hears next. Julia says,

> If I'm upset, right as I feel upset, I text a couple of my friends . . . just because I know that they'll be there and they can comfort me. If something exciting happens, I know that they'll be there to be excited with me, and stuff like that. So I definitely feel emotions when I'm texting, as I'm texting. . . . Even before I get upset and I know that I have that feeling that I'm gonna start crying, yeah, I'll pull up my friend . . . uh, my phone . . . and say like . . . I'll tell them what I'm feeling, and, like, I need to talk to them, or see them.

"I'll pull up my friend . . . uh, my phone." Julia's language slips tellingly. When Julia thinks about strong feelings, her thoughts go both to her phone and her friends. She mixes together "pulling up" a friend's name on her phone and "pulling out" her phone, but she does not really correct herself so much as imply that the phone is her friend and that friends take on identities through her phone.

After Julia sends out a text, she is uncomfortable until she gets one back: "I am always looking for a text that says, 'Oh, I'm sorry,' or 'Oh, that's great.'" Without this feedback, she says, "It's hard to calm down." Julia describes how painful it is to text about "feelings" and get no response: "I get mad. Even if I e-mail someone, I want the response, like, right away.[3] I want them to be, like, right there answering me. And sometimes I'm like, 'Uh! Why can't you just answer me?' . . . I wait, like, depending on what it is, I wait like an hour if they don't answer me, and I'll text them again. 'Are you mad? Are you there? Is everything okay?'" Her anxiety is palpable. Julia must have a response. She says of those she texts, "You want them there, because you need them." When they are not there, she moves on with her nascent feelings, but she does not move on alone: "I go to another friend and tell them."

Claudia, seventeen, a junior at Cranston, describes a similar progression. "I start to have some happy feelings as soon as I start to text." As with Julia, things move from "I have a feeling, I want to make a call" to "I want to have a feeling, I need to make a call," or in her case, send a text. What is not being cultivated here is the ability to be alone and reflect on one's emotions in private. On the contrary, teenagers report discomfort when they are without their cell phones.[4] They need to be connected in order to feel like themselves. Put in a more positive way, both Claudia and Julia share feelings as part of discovering them. They cultivate a collaborative self.

Estranged from her father, Julia has lost her close attachments to his relatives and was traumatized by being unable to reach her mother during the day of the September 11 attacks on the Twin Towers. Her story illustrates how digital connectivity—particularly texting—can be used to manage specific anxieties about loss and separation. But what Julia does—her continual texting, her way of feeling her feelings only as she shares them—is not unusual. The particularities of every individual case express personal history, but Julia's individual "symptom" comes close to being a generational style.[5]

Sociologist David Riesman, writing in the mid-1950s, remarked on the American turn from an inner- to an other-directed sense of self.[6] Without a firm inner sense of purpose, people looked to their neighbors for validation. Today, cell phone in hand, other-directedness is raised to a higher power. At the moment of beginning to have a thought or feeling, we can have it validated, almost prevalidated. Exchanges may be brief, but more is not necessarily desired. The necessity is to have someone be there.

Ricki, fifteen, a freshman at Richelieu, a private high school for girls in New York City, describes that necessity: "I have a lot of people on my contact list. If one friend doesn't 'get it,' I call another." This marks a turn to a hyper-other-directedness. This young woman's contact or buddy list has become something like a list of "spare parts" for her fragile adolescent self. When she uses the expression "get it," I think she means "pick up the phone." I check with her if I have gotten this right. She says, "'Get it,' yeah, 'pick up,' but also 'get it,' 'get *me*.'" Ricki counts on her friends to finish her thoughts. Technology does not cause but encourages a sensibility in which the validation of a feeling becomes part of establishing it, even part of the feeling itself.

I have said that in the psychoanalytic tradition, one speaks about narcissism not to indicate people who love themselves, but a personality so fragile that it needs constant support.[7] It cannot tolerate the complex demands of other people but tries to relate to them by distorting who they are and splitting off what it needs, what it can use. So, the narcissistic self gets on with others by dealing only with their made-to-measure representations. These representations (some analytic traditions refer to them as "part objects," others as "self-objects") are all that the fragile self can handle. We can easily imagine the utility of inanimate companions to such a self because a robot or a computational agent can be sculpted to meet one's needs. But a fragile person can also be supported by selected and limited contact with people (say, the people on a cell phone "favorites" list). In a life of texting and messaging, those on that contact list can be made to appear almost on demand. You can take what you need and move on. And, if not gratified, you can try someone else.

Again, technology, on its own, does not cause this new way of relating to our emotions and other people. But it does make it easy. Over time, a new style of being with each other becomes socially sanctioned. In every era, certain ways of relating come to feel natural. In our time, if we can be continually in touch, needing to be continually in touch does not seem a problem or a pathology but an accommodation to what technology affords. It becomes the norm.

The history of what we think of as psychopathology is dynamic. If in a particular time and place, certain behaviors seem disruptive, they are labeled pathological. In the nineteenth century, for example, sexual repression was considered a good and moral thing, but when women lost sensation or the ability to speak, these troubling symptoms were considered a disease, hysteria. With more outlets for women's sexuality, hysterical symptoms declined, and others took their place. So, the much-prescribed tranquilizers of the 1950s spoke to women's new anxieties when marginalized in the home after a fuller civic participation during World War II.

Now, we have symptoms born of fears of isolation and abandonment. In my study of growing up in the networked culture, I meet many children and teenagers who feel cast off. Some have parents with good intentions who simply work several jobs and have little time for their children. Some have endured divorce—sometimes multiple divorces—and float from one parent to another, not confident of their true home. Those lucky children who have intact families with stable incomes can experience other forms of abandonment. Busy parents are preoccupied, often by what is on their cell phones. When children come home, it is often to a house that is empty until a parent returns from work.

For young people in all of these circumstances, computers and mobile devices offer communities when families are absent. In this context, it is not surprising to find troubling patterns of connection and disconnection: teenagers who will only "speak" online, who rigorously avoid face-to-face encounters, who are in text contact with their parents fifteen or twenty times a day, who deem even a telephone call "too much" exposure and say that they will "text, not talk." But are we to think of these as pathologies? For as social mores change, what once seemed "ill" can come to seem normal. Twenty years ago, as a practicing clinical psychologist, if I had met a college junior who called her mother fifteen times a day, checking in about what shoes to buy and what dress to wear, extolling a new kind of decaffeinated tea, and complaining about the difficulty of a physics problem set, I would have thought her behavior problematic. I would have encouraged her to explore difficulties with separation. I would have assumed that these had to be addressed for her to proceed to successful adulthood. But these days, a college student who texts home fifteen times a day is not unusual.

High school and college students are always texting—while waiting in line at the cafeteria, while eating, while waiting for the campus shuttle. Not surprisingly, many of these texts are to parents. What once we might have seen as a problem becomes how we do things. But a behavior that has become typical may still express the problems that once caused us to see it as pathological. Even a typical behavior may not be in an adolescent's developmental interest.

Consider Leo, a college sophomore far from home, who feels crippling loneliness. He tells me that he "handles" this problem by texting and calling his mother up to twenty times a day. He remarks that this behavior does not make him stand out; everyone he knows is on a phone all day. But even if invisible, he considers his behavior a symptom all the same.

These days, our relationship to the idea of psychological autonomy is evolving. I have said that central to Erik Erikson's thinking about adolescents is the idea that they need a moratorium, a "time out," a relatively consequence-free space for experimentation. But in Erikson's thinking, the self, once mature, is relatively stable. Though embedded in relationships, in the end it is bounded and autonomous.[8] One of Erikson's students, psychiatrist Robert Jay Lifton, has an alternative vision of the mature self. He calls it *protean* and emphasizes its multiple aspects.[9] Thinking of the self as protean accents connection and reinvention. This self, as Lifton puts it, "fluid and many-sided," can embrace and modify ideas and ideologies. It flourishes when provided with things diverse, disconnected, and global.

Publicly, Erikson expressed approval for Lifton's work, but after Erikson's death in 1994, Lifton asked the Erikson family if he might have the books he had personally inscribed and presented to his teacher. The family agreed; the books were returned. In his personal copy of Lifton's *The Protean Self*, Erikson had written extensive marginal notes. When he came to the phrase "protean man," Erikson had scrawled "protean boy?"[10] Erikson could not accept that successful maturation would not result in something solid. By Erikson's standards, the selves formed in the cacophony of online spaces are not protean but juvenile. Now I suggest that the culture in which they develop tempts them into narcissistic ways of relating to the world.

THE AVATAR OF ME

Erikson said that identity play is the work of adolescence. And these days adolescents use the rich materials of online life to do that work. For example, in a game such as The Sims Online (think of this as a very junior version of Second Life), you can create an avatar that expresses aspects of yourself, build a house, and furnish it to your taste. Thus provisioned, you can set about reworking in the virtual aspects of life that may not have gone so well in the real.

Trish, a timid and anxious thirteen-year-old, has been harshly beaten by her alcoholic father. She creates an abusive family on The Sims Online, but in the game her character, also thirteen, is physically and emotionally strong. In simulation, she plays and replays the experience of fighting off her aggressor. A sexually experienced girl of sixteen, Katherine, creates an online innocent. "I want to have a rest," she says. Beyond rest, Katherine tells me she can get "practice at being a different kind of person. That's what Sims is for me. Practice."

Katherine "practices" on the game at breakfast, during school recess, and after dinner. She says she feels comforted by her virtual life. I ask her if her activities in the game have led her to do anything differently in her life away from it. She replies, "Not really," but then goes on to describe how her life is in fact beginning to change: "I'm thinking about breaking up with my boyfriend. I don't want to have sex anymore, but I would like to have a boyfriend. My character on Sims has boyfriends but doesn't have sex. They [the boyfriends of her Sims avatar] help her with her job. I think to start fresh I would have to break up with my boyfriend." Katherine does not completely identify with her online character and refers to her avatar in the third person. Yet, The Sims Online is a place where she can see her life anew.

This kind of identity work can take place wherever you create an avatar. And it can take place on social-networking sites as well, where one's profile becomes an avatar of sorts, a statement not only about who you are but who you want to be. Teenagers make it clear that games, worlds, and social networking (on the surface, rather different) have much in common. They all ask you to compose and project an identity. Audrey, sixteen, a junior at Roosevelt, a suburban public high school near New York City, is explicit about the connection between avatars and profiles. She calls her Facebook profile "my Internet twin" and "the avatar of me."

Mona, a freshman at Roosevelt, has recently joined Facebook. Her parents made her wait until her fourteenth birthday, and I meet her shortly after this long-awaited day. Mona tells me that as soon as she got on the site, "Immediately, I felt power." I ask her what she means. She says, "The first thing I thought was, 'I am going to broadcast the real me.'" But when Mona sat down to write her profile, things were not so straightforward. Whenever one has time to write, edit, and delete, there is room for performance. The "real me" turns out to be elusive. Mona wrote and rewrote her profile. She put it away for two days and tweaked it again. Which pictures to add? Which facts to include? How much of her personal life to reveal? Should she give any sign that things at home were troubled? Or was this a place to look good?

Mona worries that she does not have enough of a social life to make herself sound interesting: "What kind of personal life should I *say* I have?" Similar questions plague other young women in her class. They are starting to have boyfriends. Should they list themselves as single if they are just starting to date someone new? What if they consider themselves in a relationship, but their boyfriends do not? Mona tells me that "it's common sense" to check with a boy before listing yourself as connected to him, but "that could be a very awkward conversation." So there are misunderstandings and recriminations. Facebook at fourteen can be a tearful place. For many, it remains tearful well through college and graduate school. Much that might seem straightforward is fraught. For example, when asked by Facebook to confirm someone as a friend or ignore the request, Helen, a Roosevelt senior, says, "I always feel a bit of panic. . . . Who should I friend? . . . I really want to only have my cool friends listed, but I'm nice to a lot of other kids at school. So I include the more unpopular ones, but then I'm unhappy." It is not how she wants to be seen.

In the Victorian era, one controlled whom one saw and to whom one was connected through the ritual of calling cards. Visitors came to call and, not necessarily expecting to be received, left a card. A card left at your home in return meant that the relationship might grow. In its own way, friending on Facebook is reminiscent of this tradition. On Facebook, you send a request to be a friend. The recipient of the request has the option to ignore or friend you. As was the case in the Victorian era, there is an intent to screen. But the Victorians followed socially accepted rules. For example, it was understood that one was most open to people of similar social standing. Facebook is more democratic—which leaves members to make up their own rules, not necessarily understood by those who contact them. Some people make a request to be a Facebook friend in the spirit of "I'm a fan" and are accepted on that basis. Other people friend only people they know. Others friend any friend of a friend, using Facebook as a tool to expand their acquaintanceships. All of this can be exciting or stressful—often both at the same time, because friending has consequences. It means that someone can see what you say about yourself on your profile, the pictures you post, and your friends' postings on your "wall," the shared communication space for you and your friends. Friending someone gives that person implicit permission to try to friend your friends. In fact, the system constantly proposes that they do so.

Early in this project, I was at a conference dinner, sitting next to an author whose publisher insisted that she use Facebook as a way to promote her new book. The idea was to use the site to tell people where she would be speaking and to share the themes of her book with an ever-expanding potential readership.

Her publisher hoped this strategy would make her book "go viral." She had expected the Facebook project to feel like business, but instead she described complicated anxieties about not having enough friends, and about envy of her husband, also a writer, who had more friends than she. It also felt wrong to use the word "friends" for all of those she had "friended," since so many of the friended were there for professional reasons alone. She left me with this thought: "This thing took me right back to high school."

I promised her that when I joined Facebook I would record my first feelings, while the site was still new to me. My very first feelings now seem banal: I had to decide between "friending" plan A (this will be a place for people I actually know) and plan B (I will include people who contact me because they say they appreciate my work). I tried several weeks on plan A and then switched to the more inclusive Plan B, flattered by the attention of strangers, justifying my decision in professional terms.

But now that I had invited strangers into my life, would I invite myself into the lives of strangers? I would have anticipated not, until I did that very thing. I saw that one of my favorite authors was a Facebook friend of a friend. Seized by the idea that I might be this writer's friend, I made my request, and he accepted me. The image of a cafeteria came to mind, and I had a seat at his virtual table. But I felt like a gatecrasher. I decided realistically that I was taking this way too seriously. Facebook is a world in which fans are "friends." But of course, they are not friends. They have been "friended." That makes all the difference in the world, and I couldn't get high school out of my mind.

PRESENTATION ANXIETY

What are the truth claims in a Facebook profile? How much can you lie? And what is at stake if you do? Nancy, an eighteen-year-old senior at Roosevelt, answers this question. "On the one hand, low stakes, because no one is really checking." Then, with a grimace, she says, "No, high stakes. Everyone is checking." A few minutes later, Nancy comes back to the question: "Only my best friends will know if I lie a little bit, and they will totally understand." Then she laughs. "All of this, it is, I guess, a bit of stress."[11]

At Cranston, a group of seniors describe that stress. One says, "Thirteen to eighteen are the years of profile writing." The years of identity construction are recast in terms of profile production. These private school students had to write one profile for their applications to middle school, another to get into high school, and then another for Facebook. Now they are beginning to construct personae for college applications. And here, says Tom, "You have to have a slightly different persona for the different colleges to which you are applying: one for Dartmouth, a different one, say, for Wesleyan." For this aficionado of profile writing, every application needs a different approach. "By the time you get to the questions for the college application, you are a professional profile writer," he says. His classmate Stan describes his online profiles in great detail. Each serves a different purpose, but they must overlap, or questions of authenticity will arise. Creating the illusion of authenticity demands virtuosity. Presenting a self in these circumstances, with multiple media and multiple goals, is not easy work. The trick, says Stan, is in "weaving profiles together . . . so that people can see you are not too crazy. . . . What I learned in high school was profiles, profiles, profiles, how to make a me."

Early in my study, a college senior warned me not to be fooled by "anyone you interview who tells you that his Facebook page is 'the real me.' It's like being in a play. You make a character." Eric, a college-bound senior at Hadley, a boys' preparatory school in rural New Jersey, describes himself as savvy about how you can "mold a Facebook page." Yet, even he is shocked when he finds evidence of girls using "shrinking" software to appear thinner on their profile photographs. "You can't see that they do it when you look at the little version of the picture, but when you look at a big picture, you can see how the background is distorted." By eighteen, he has become an identity detective. The Facebook profile is a particular source of stress because it is so important to high school social life. Some students feel so in its thrall that they drop out of Facebook, if only for a while, to collect themselves.

Brad, eighteen, a senior at Hadley, is about to take a gap year to do community service before attending a small liberal arts college in the Midwest. His parents are architects; his passion is biology and swimming. Brad wants to be part of the social scene at Hadley, but he doesn't like texting or instant messaging. He is careful to make sure I know he is "no Luddite." He has plenty of good things to say about the Net. He is sure that it makes it easier for insecure people to function. Sometimes the ability to compose his thoughts online "can be reassuring," he says, because there is a chance to "think through, calculate, edit, and make sure you're as clear and concise as possible." But as our conversation continues, Brad switches gears. Even as some are able to better function because they feel in control, online communication also offers an opportunity to ignore other people's feelings. You can avoid eye contact. You can elect not to hear how "hurt or angry they sound in their voice." He says, "Online, people miss your body language, tone of voice. You are not really you." And worst of all, online life has led him to mistrust his friends. He has had his instant messages "recorded" without his knowledge and forwarded on "in a cut-and-paste world."

In fact, when I meet Brad in the spring of his senior year, he tells me he has "dropped out" of online life. "I'm off the Net," he says, "at least for the summer, maybe for my year off until I go to college." He explains that it is hard to drop out because all his friends are on Facebook. A few weeks before our conversation, he had made a step toward rejoining but immediately he felt that he was not doing enough to satisfy its demands. He says that within a day he felt "rude" and couldn't keep up. He felt guilty because he didn't have the time to answer all the people who wrote to him. He says that he couldn't find a way to be "a little bit" on Facebook—it does not easily tolerate a partial buy-in. Just doing the minimum was "pure exhaustion."

In the world of Facebook, Brad says, "your minute movie preferences matter. And what groups you join. Are they the right ones?" Everything is a token, a marker for who you are:

> When you have to represent yourself on Facebook to convey to anyone who doesn't know you what and who you are, it leads to a kind of obsession about minute details about yourself. Like, "Oh, if I like the band State Radio and the band Spoon, what does it mean if I put State Radio first or Spoon first on my list of favorite musical artists? What will people think about me?" I know for girls, trying to figure out, "Oh, is this picture too revealing to put? Is it prudish if I don't put it?" You have to think carefully for good reason, given how much people will look at your profile and obsess over it. You have to know that everything you put up will be perused very carefully. And that makes it necessary for you to obsess over what you do put up and how you portray yourself. . . . And when you have to think that much about what you come across as, that's just another way that . . . you're thinking of yourself in a bad way.

For Brad, "thinking of yourself in a bad way" means thinking of yourself in reduced terms, in "short smoke signals" that are easy to read. To me, the smoke signals suggest a kind of reduction and betrayal. Social media ask us to represent ourselves in simplified ways. And then, faced with an audience, we feel pressure to conform to these simplifications. On Facebook, Brad represents himself as cool and in the know—both qualities are certainly part of who he is. But he hesitates to show people online other parts of himself (like how much he likes Harry Potter). He spends more and more time perfecting his online Mr. Cool. And he feels pressure to perform him all the time because that is who he is on Facebook.

At first Brad thought that both his Facebook profile and his college essays had gotten him into this "bad way" of thinking, in which he reduces himself to fit a stereotype. Writing his Facebook profile felt to him like assembling cultural references to shape how others would see him. The college essay demanded a victory narrative and seemed equally unhelpful: he had to brag, and he wasn't happy. But Brad had a change of heart about the value of writing his college essays. "In the end I learned a lot about how I write and think—what I know how to think about and some things, you know, I really can't think about them well at all." I ask him if Facebook might offer these kinds of opportunities. He is adamant that it does not: "You get reduced to a list of favorite things. 'List your favorite music'—that gives you no liberty at all about how to say it." Brad says that "in a conversation, it might be interesting that on a trip to Europe with my parents, I got interested in the political mural art in Belfast. But on a Facebook page, this is too much information. It would be the kiss of death. Too much, too soon, too weird. And yet . . . it is part of who I am, isn't it? . . . You are asked to make a lot of lists. You have to worry that you put down the 'right' band or that you *don't* put down some Polish novel that nobody's read." And in the end, for Brad, it is too easy to lose track of what is important:

> What does it matter to anyone that I prefer the band Spoon over State Radio? Or State Radio over Cake? But things like Facebook . . . make you think that it really does matter. . . . I look at someone's profile and I say, "Oh, they like these bands." I'm like, "Oh, they're a poser," or "they're really deep, and they're into good music." We all do that, I think. And then I think it doesn't matter, but . . . the thing is, in the world of Facebook it *does* matter. Those minute details *do* matter.

Brad, like many of his peers, worries that if he is modest and doesn't put down all of his interests and accomplishments, he will be passed over. But he also fears that to talk about his strengths will be unseemly. None of these conflicts about self presentation are new to adolescence or to Facebook. What is new is living them out in public, sharing every mistake and false step. Brad, attractive and accomplished, sums it up with the same word Nancy uses: "Stress. That's what it comes down to for me. It's just worry and stressing out about it." Now Brad only wants to see friends in person or talk to them on the telephone. "I can just act how I want to act, and it's a much freer way." But who will answer the phone?

CHAPTER 10

no need to call

"So many people hate the telephone," says Elaine, seventeen. Among her friends at Roosevelt High School, "it's all texting and messaging." She herself writes each of her six closest friends roughly twenty texts a day. In addition, she says, "there are about forty instant messages out, forty in, when I'm at home on the computer." Elaine has strong ideas about how electronic media "levels the playing field" between people like her—outgoing, on the soccer team, and in drama club—and the shy: "It's only on the screen that shy people open up." She explains why: "When you can think about what you're going to say, you can talk to someone you'd have trouble talking to. And it doesn't seem weird that you pause for two minutes to think about what you're going to say before you say it, like it would be if you were actually talking to someone."

Elaine gets specific about the technical designs that help shy people express themselves in electronic messaging. The person to whom you are writing shouldn't be able to see your process of revision or how long you have been working on the message. "That could be humiliating." The best communication programs shield the writer from the view of the reader. The advantage of screen communication is that it is a place to reflect, retype, and edit. "It is a place to hide," says Elaine.

The notion that hiding makes it easier to open up is not new. In the psychoanalytic tradition, it inspired technique. Classical analysis shielded the patient from the analyst's gaze in order to facilitate free association, the golden rule of saying whatever comes to mind. Likewise, at a screen, you feel protected and less burdened by expectations. And, although you are alone, the potential for almost instantaneous contact gives an encouraging feeling of already being together. In this curious relational space, even sophisticated users who know that electronic communications can be saved, shared, and show up in court, succumb to its illusion of privacy. Alone with your thoughts, yet in contact with an almost tangible fantasy of the other, you feel free to play. At the screen, you have a chance to write yourself into the person you want to be and to imagine others as you wish to them to be, constructing them for your purposes.[1] It is a seductive but dangerous habit of mind. When you cultivate this sensibility, a telephone call can seem fearsome because it reveals too much.

Elaine is right in her analysis: teenagers flee the telephone. Perhaps more surprisingly, so do adults. They claim exhaustion and lack of time; always on call, with their time highly leveraged through multitasking, they avoid voice communication outside of a small circle because it demands their full attention when they don't want to give it.

Technologies live in complex ecologies. The meaning of any one depends on what others are available. The telephone was once a way to touch base or ask a simple question. But once you have access to e-mail, instant messaging, and texting, things change. Although we still use the phone to keep up with those closest to us, we use it less outside this circle.[2] Not only do people say that a phone call asks too much, they worry it will be received as demanding too much. Randolph, a forty-six-year-old architect with two jobs, two young children, and a twelve-year-old son from a former marriage, makes both points. He avoids the telephone because he feels "tapped out. . . . It promises more than I'm willing to deliver." If he keeps his communications to text and e-mail, he believes he can "keep it together." He explains, "Now that there is e-mail, people expect that a call will be more complicated. Not about facts. A fuller thing. People expect it to take time—or else you wouldn't have called."

Tara, a fifty-five-year-old lawyer who juggles children, a job, and a new marriage, makes a similar point: "When you ask for a call, the expectation is that you have pumped it up a level. People say to themselves: 'It's urgent or she would have sent an e-mail.'" So Tara avoids the telephone. She wants to meet with friends in person; e-mail is for setting up these meetings. "That is what is most efficient," she says. But efficiency has its downside. Business meetings have agendas, but friends have unscheduled needs. In friendship, things can't always wait.

Tara knows this; she feels guilty and she experiences a loss: "I'm at the point where I'm processing my friends as though they were items of inventory . . . or clients."

Leonora, fifty-seven, a professor of chemistry, reflects on her similar practice: "I use e-mail to make appointments to see friends, but I'm so busy that I'm often making an appointment one or two months in the future. After we set things up by e-mail, we do not call. Really. I don't call. They don't call. They feel that they have their appointment. What do I feel? I feel I have 'taken care of that person.'" Leonora's pained tone makes it clear that by "taken care of" she means that she has crossed someone off a to-do list. Tara and Leonora are discontent but do not feel they have a choice. This is where technology has brought them. They subscribe to a new etiquette, claiming the need for efficiency in a realm where efficiency is costly.

AUDREY: A LIFE ON THE SCREEN

We met Audrey, sixteen, a Roosevelt junior who talked about her Facebook profile as "the avatar of me." She is one of Elaine's shy friends who prefers texting to talking. She is never without her phone, sometimes using it to text even as she instant-messages at an open computer screen. Audrey feels lonely in her family. She has an older brother in medical school and a second, younger brother, just two years old. Her parents are divorced, and she lives half time with each of them. Their homes are about a forty-five-minute drive apart. This means that Audrey spends a lot of time on the road. "On the road," she says. "That's daily life." She sees her phone as the glue that ties her life together. Her mother calls her to pass on a message to her father. Her father does the same. Audrey says, "They call me to say, 'Tell your mom this. . . . Make sure your dad knows that.' I use the cell to pull it together." Audrey sums up the situation: "My parents use me and my cell like instant messenger. I am their IM."

Like so many other children who tell me similar stories, Audrey complains of her mother's inattention when she picks her up at school or after sports practice. At these times, Audrey says, her mother is usually focused on her cell phone, either texting or talking to her friends. Audrey describes the scene: she comes out of the gym exhausted, carrying heavy gear. Her mother sits in her beaten-up SUV, immersed in her cell, and doesn't even look up until Audrey opens the car door. Sometimes her mother will make eye contact but remain engrossed with the phone as they begin the drive home. Audrey says, "It gets between us, but it's hopeless. She's not going to give it up. Like, it could have been four days since I last spoke to her, then I sit in the car and wait in silence until she's done."[3]

Audrey has a fantasy of her mother, waiting for her, expectant, without a phone. But Audrey is resigned that this is not to be and feels she must temper her criticism of her mother because of her own habit of texting when she is with her friends. Audrey does everything she can to avoid a call.[4] "The phone, it's awkward. I don't see the point. Too much just a recap and sharing feelings. With a text . . . I can answer on my own time. I can respond. I can ignore it. So it really works with my mood. I'm not bound to anything, no commitment. . . . I have control over the conversation and also more control over what I say."

Texting offers protection:

> Nothing will get spat at you. You have time to think and prepare what you're going to say, to make you appear like that's just the way you are. There's planning involved, so you can control how you're portrayed to this person, because you're choosing these words, editing it before you send it. . . . When you instant-message you can cross things out, edit what you say, block a person, or sign off. A phone conversation is a lot of pressure. You're always expected to uphold it, to keep it going, and that's too much pressure. . . . You have to just keep going . . . "Oh, how was your day?" You're trying to think of something else to say real fast so the conversation doesn't die out.

Then Audrey makes up a new word. A text, she argues, is better than a call because in a call "there is a lot less *boundness* to the person." By this she means that in a call, she could learn too much or say too much, and things could get "out of control." A call has insufficient boundaries. She admits that "later in life I'm going to need to talk to people on the phone. *But not now.*" When texting, she feels at a reassuring distance. If things start to go in a direction she doesn't like, she can easily redirect the conversation—or cut it off: "In texting, you get your main points off; you can really control when you want the conversation to start and end. You say, 'Got to go, bye.' You just do it . . . much better than the long drawn-out good-byes, when you have no real reason to leave, but you want to end the conversation." This last is what Audrey likes least—the end of conversations. A phone call, she explains, requires the skill to end a conversation "when you have no real reason to leave. . . . It's not like there is a reason. You just want to. I don't know how to do that. *I don't want to learn.*"

Ending a call is hard for Audrey because she experiences separation as rejection; she projects onto others the pang of abandonment she feels when someone ends a conversation with her. Feeling unthreatened when someone wants to end a conversation may seem a small thing, but it is not. It calls upon a sense of self-worth; one needs to be at a place where Audrey has not arrived. It is easier to avoid the phone; its beginnings and endings are too rough on her.

Audrey is not alone in this. Among her friends, phone calls are infrequent, and she says, "Face-to-face conversations happen way less than they did before. It's always, 'Oh, talk to you online.'" This means, she explains, that things happen online that "should happen in person . . . Friendships get broken. I've had someone ask me out in a text message. I've had someone break up with me online." But Audrey is resigned to such costs and focuses on the bounties of online life.

One of Audrey's current enthusiasms is playing a more social, even flirtatious version of herself in online worlds. "I'd like to be more like I am online," she says. As we've seen, for Audrey, building an online avatar is not so different from writing a social-networking profile. An avatar, she explains, "is a Facebook profile come to life." And avatars and profiles have a lot in common with the everyday experiences of texting and instant messaging. In all of these, as she sees it, the point is to do "a performance of you."

> Making an avatar and texting. Pretty much the same. You're creating your own person; you don't have to think of things on the spot really, which a lot of people can't really do. You're creating your own little ideal person and sending it out. Also on the Internet, with sites like MySpace and Facebook, you put up the things you like about yourself, and you're not going to advertise the bad aspects of you.

> You're not going to post pictures of how you look every day. You're going to get your makeup on, put on your cute little outfit, you're going to take your picture and post it up as your default, and that's what people are going to expect that you are every day, when really you're making it up for all these people. . . . You can write anything about yourself; these people don't know. You can create who you want to be. You can say what kind of stereotype mold you want to fit in without . . . maybe in real life it won't work for you, you can't pull it off. But you can pull it off on the Internet.

Audrey has her cell phone and its camera with her all day; all day she takes pictures and posts them to Facebook. She boasts that she has far more Facebook photo albums than any of her friends. "I like to feel," she says, "that my life is up there." But, of course, what is up on Facebook is her edited life. Audrey is pre-occupied about which photographs to post. Which put her in the best light? Which show her as a "bad" girl in potentially appealing ways? If identity play is the work of adolescence, Audrey is at work all day: "If Facebook were deleted, I'd be deleted.... All my memories would probably go along with it. And other people have posted pictures of me. All of that would be lost. If Facebook were undone, I might actually freak out.... That is where I am. It's part of your life. It's a second you." It is at this point that Audrey says of a Facebook avatar: "It's your little twin on the Internet."

Since Audrey is constantly reshaping this "twin," she wonders what happens to the elements of her twin that she edits away. "What does Facebook do with pictures you put on and then take off?" She suspects that they stay on the Internet forever, an idea she finds both troubling and comforting. If everything is archived, Audrey worries that she will never be able to escape the Internet twin. That thought is not so nice. But if everything is archived, at least in fantasy, she will never have to give her up. That thought is kind of nice.

On Facebook, Audrey works on the twin, and the twin works on her. She describes her relationship to the site as a "give-and-take." Here's how it works: Audrey tries out a "flirty" style. She receives a good response from Facebook friends, and so she ramps up the flirtatious tone. She tries out "an ironic, witty" tone in her wall posts. The response is not so good, and she retreats. Audrey uses the same kind of tinkering as she experiments with her avatars in virtual worlds. She builds a first version to "put something out there." Then comes months of adjusting, of "seeing the new kinds of people I can hang with" by changing how she represents herself. Change your avatar, change your world.

Audrey says that her online avatars boost her real-life confidence. Like many other young women on Second Life, Audrey makes her avatar more conventionally attractive than she is in the real. Audrey is a pretty girl, with long red hair, styled in a single braid down her back. Her braid and her preference for floral prints give her an old-fashioned look. On Second Life, Audrey's hair is modern and blunt cut, her body more developed, her makeup heavier, her clothes more suggestive. There are no floral prints. A promotional video for the game asserts that this is a place to "connect, shop, work, love, explore, be different, free yourself, free your mind, change your looks, love your looks, love your life."[5] But is loving your life as an avatar the same as loving your life in the real? For Audrey, as for many of her peers, the answer is unequivocally yes. Online life is practice to make the rest of life better, but it is also a pleasure in itself. Teenagers spend hours depleting allowances, shopping for clothes and shoes for their online selves. These virtual goods have real utility; they are required for avatars with full social lives.

Despite her enthusiasm for Second Life, Audrey's most emotional online experience has taken place on MySpace—or more precisely, on Italian MySpace. During her sophomore year at Roosevelt, Audrey met a group of Italian exchange students. They introduced her to the site. At that point, Audrey had taken one year of high school Italian, just enough to build a profile with some help from her friends. She admits that this profile bears only a glancing relationship to the truth. On Italian MySpace, Audrey is older and more experienced. When her profile went up, a lot of men sent her messages in Italian. She found this thrilling and responded enthusiastically. The game was on. Now, a year later, it continues: "I message back in the little Italian that I know. I don't usually respond to those things, but since I figure my real information isn't on there, and they're in Italy and I'm in America, why not? It's fun to step outside yourself. You can't really do this with your friends in real life." For Audrey, Italian MySpace is like chat rooms: "You do it with people you're never going to speak to or assume you're never going to speak to."

Audrey's focus on "people you're never going to speak to" brings to mind once again how Erik Erikson thought about the moratorium necessary for adolescent development. Writing in the 1950s and early 1960s, he could think of the American "high school years" as offering this relatively consequence-free environment.[6] These days, high school is presented to its students and their parents as anything but consequence free. Audrey is in a highly competitive college preparatory program—the fast track in her high school—and is continually reminded of the consequences of every grade, every SAT score, every extracurricular choice. She thinks of her high school experience as time in a professional school where she trains to get into college. Real life provides little space for consequence-free identity play, but Italian MySpace provides a great deal.

Long after the Italian exchange students are gone, Audrey keeps her page on Italian MySpace. As she talks about its pleasures, I think of my first trip to Europe in the summer after my sophomore year in college. In its spirit, my behavior in the real was not so different from Audrey's in the virtual. I hitchhiked from Paris to Rome, against my parents' clear instructions. I left everything about my identity behind except for being a nineteen-year-old American. I saw no reason for anyone to know me as a serious, academically disciplined student. I preferred to simply be nineteen. I never lied, but I never told any of the young Romans I hung around with that I wasn't simply a lighthearted coed. Indeed, during that summer of not quite being me, it was not so clear that I was not a lighthearted coed. My Roman holiday only worked if I didn't bring my new Italian friends into the rest of my life. Audrey, too, needs to compartmentalize. On Italian MySpace she cultivates friendships that she keeps separate from her "real" American Facebook account.

When I tell Audrey about my month in Rome, she gives me the smile of a coconspirator. She offers that she has done "that kind of thing as well." The previous summer she went on a school trip to Puerto Rico. "I wore kinds of shorts and tops that I would never wear at home. There, my reputation isn't on the line; there's no one I care about judging me or anything, so why not?" Audrey and I talk about the difference between our transgressive real-world travels—mine to Italy, hers to Puerto Rico—and what she can do online. Once our respective trips were over, we were back at home with our vigilant families and everyday identities. But Audrey can go online and dress her avatars in sexy outfits whenever she wants. Her racier self is always a few clicks away on Italian MySpace. She can keep her parallel lives open as windows on her screen.

WHAT HAPPENS ON FACEBOOK, STAYS ON . . . ?

Every day Audrey expresses herself through a group of virtual personae. There are Facebook and Italian MySpace profiles; there are avatars in virtual worlds, some chat rooms, and a handful of online games. Identity involves negotiating all of these and the physical Audrey. When identity is multiple in this way, people feel "whole" not because they are *one* but because the relationships among aspects of self are fluid and undefensive. We feel "ourselves" if we can move easily among our many aspects of self.[7]

I once worried that teenagers would experience this virtual nomadism as exhausting or confusing. But my concerns didn't take into account that in online life, the *site supports the self.* Each site remembers the choices you've have made there, what you've said about yourself, and the history of your relationships. Audrey says it can be hard to decide where to go online, because where she goes means stepping into who she is in any given place, and in different places, she has different pastimes and different friends. What Pete called the "life mix" refers to more than combining a virtual life with a physically embodied one. Even for sixteen-year-old Audrey, many virtual lives are in play.

Not surprisingly, there are moments when life in the life mix gets tense. Audrey tells a story about a boy from school who was online with her and several of her girlfriends in the game World of Warcraft. They were all present as avatars, but each knew the real-life identity of the other players in their group. The online setting emboldened the normally shy young man, who, Audrey says, "became aggressive. He started talking tough." Audrey says that online, she and her friends began to laugh at him, to tease him a bit, "because knowing who he is in person, we were like, 'Are you kidding me?'" But the girls were also upset. They had never seen their friend behave like this. The next day, when they saw him at school, he just walked away. He could not own what had happened online. Shame about his virtual self changed his life in the real. Audrey calls this kind of thing the "spillover effect." It happens frequently, she says, but "it is not a good thing."

Audrey has developed a strategy to avoid such spillovers. If she is online in any setting where she knows the real identity of those with her, she treats what happens there as if it were shared under attorney-client privilege. Put otherwise, she takes an online space such as Facebook, where her identity is "known," and reconstructs it as a place that will be more useful as a context for the much-needed moratorium. For Audrey, what happens on the Internet should stay on the Internet, at least most of the time. Audrey compares the Internet to Alcoholics Anonymous:

> If you went to an AA group and you said, "I'm an alcoholic," and your friend was there . . . you don't talk about it outside of there even if you two are in the same group. It's that kind of understanding. So, on Facebook, I'm not anonymous. But not many people will bring up Internet stuff in real life.
>
> Unless there's a scandal, no one will call you on what you write on Facebook. It's kind of a general consensus that you created your profile for a reason. No one's going to question why you choose to put this or that in your "About Me" [a section of the Facebook profile]. People are just going to leave that alone. Especially if they actually know who you are, they don't really care what you write on the Internet.

Audrey's friends see her bend reality on Facebook but are willing to take her online self on its own terms. She offers them the same courtesy. The result is more leeway to experiment with emotions and ideas in digital life. Audrey says, "Even on AIM [the free instant messaging service offered by America Online], I could have long conversations with someone and the next day [in person] just be like, 'Hey.'" You split the real and virtual to give the virtual the breathing space it needs.

Sometimes, says Audrey, "people take what they show online and try to bring it back to the rest of their lives," but this to sorry effect. As an example, Audrey describes her "worst Internet fight." It began in a chat room where she quarreled with Logan, a classmate. Feeling that she had been in the wrong, the next day Audrey told Logan she was sorry, face-to-face. This real-world apology did not quiet things down. Instead, Logan brought the quarrel back into the online world. He posted his side of the story to Audrey's Facebook wall. Now, all of her friends could read about it. Audrey felt compelled to retaliate in kind. Now, his Facebook wall related her angry version of things. At school, Audrey and Logan shared many friends, who felt they had to take sides. Day after day, hours were spent in angry exchanges, with an expanding group of players.

What strikes Audrey most about this Internet fight is that, in the end, it had been "about close to nothing." She explains, "I said something I shouldn't have. I apologized. If it had happened at a party, it would have ended in five minutes." But she had said it on the Internet, its own peculiar echo chamber. For Audrey, the hurt from this incident, six months in the past, is still raw: "We were really good friends, and now we don't even look at each other in the hall."

Audrey is comforted by the belief that she had done her best. Even though she had broken her rule about keeping the virtual and real separate, she insists that trying to make things "right" in person had given her friendship with Logan its best chance: "An online apology. It's cheap. It's easy. All you have to do is type 'I'm sorry.' You don't have to have any emotion, any believability in your voice or anything. It takes a lot for someone to go up to a person and say, 'I'm sorry,' and that's when you can really take it to heart. If someone's going to take the easy way out and rely on text to portray all these forgiving emotions, it's not going to work." Eventually Logan did apologize, but only online. Accordingly, the apology failed: "It might have been different if he said it in person, but he didn't. With an online apology, there are still unanswered questions: 'Is he going to act weird to me now? Are we just going to be normal?' You don't know how the two worlds are going to cross." An online apology is only one of the easy "shortcuts" that the Net provides. It is a world of many such temptations.

Audrey says that she took her worst shortcut a year before when she broke up with a boyfriend online. Teenage girls often refer to television's *Sex and the City* to make a point about when not to text. In a much-discussed episode, the heroine's boyfriend breaks up with her by leaving a Post-it note. You shouldn't break up by Post-it note and you shouldn't break up by text. Audrey says she knew this rule; her break up on instant messenger had been a lapse. She still has not entirely forgiven herself:

> I was afraid. I couldn't do it on the phone, and I couldn't do it in person. It was the kind of thing where I knew it had to end because I didn't feel the same way, one of those things. I felt so bad, because I really did care for him, and I couldn't get myself to say it. It was one of those. . . . I wasn't trying to chicken out, I just couldn't form the words, so I had to do it on-line, and I wish I hadn't. He deserved to have me do it in person. . . . I'm very sorry for it. I just think it's a really cold move, and kind of lame.

Audrey was still so upset by the online breakup that in our conversation, she comes to her own defense. She tells me about a time when she behaved better: "I was in an argument with a friend and I began to write a Facebook message about it but I stopped myself." She explains that breaking up with a boyfriend online is very bad, but "well, at least you can just cut ties. With a friend you actually have to work it out. It's not as easy as 'I don't want to be friends with you anymore.'" And now that friendships span the physical and virtual, you have to "work it out" across worlds.

Audrey's etiquette for how to work things out across worlds is complicated. She finds face-to-face conversation difficult and avoids the telephone at all cost. Yet, as we've seen, she also thinks there are things that should only be done face-to-face, like breaking up with a boy and the "whole heartfelt baring of souls." When her parents separated, she had to move and change school districts. She was disappointed when one of her friends at her former school sent her an instant message to tell her she would be missed. Audrey's comment: "It was really sweet, but I just wished that—it would have meant so much more if we could've done that face-to-face. And I understood. We don't see each other every day, and if you feel it right now, on the Internet, you can tell them right now; you don't have to wait or anything. I really appreciated it, but it was different reading it than hearing it in her voice."

As Audrey tells me this story, she becomes aware that she is suggesting a confusing set of rules. She tries to impose some order: "I try to avoid the telephone, I like texting and instant messaging, and I am so often on Facebook that I probably give the impression that I want everything to happen online." But some things, such as a friend's good-bye to a friend, she wants to have happen in person. Like Tara and Leona, Audrey makes no suggestion that "talking" on a telephone could ever be of much help. Telephones are for logistical arrangements, if complicated (often overlapping) text messages have confused a situation.

When Audrey considers whether her school friend said good-bye in a text because she didn't care or wasn't "brave enough to say something nice face-to-face," Audrey admits that the latter is more likely and that she can identify with this.[8] If you send fond feelings or appreciation digitally, you protect yourself from a cool reception. One of the emotional affordances of digital communication is that one can always hide behind deliberated nonchalance.

FINER DISTINCTIONS

"Whassup?" Reynold, a sixteen-year-old at Silver Academy, a small urban Catholic high school in Pennsylvania, savors the phrase. "With instant messaging, 'Whassup?' is all you need to say." Reynold makes it clear that IM does not require "content." You just need to be there; your presence says you are open to chat. A text message is more demanding: "You need more of a purpose. Texting is for 'Where are you, where am I, let's do this, let's do that.'" Among friends, however, "texting can be just as random as IM." Reynold likes this: "Among close friends, you can text to just say 'Whassup?'"

I discuss online communications with eight junior and senior boys at Silver who eagerly take up Reynold's question: When should one use texting, IM, Facebook wall posts, or Facebook and MySpace messaging? (Messaging on social networks is the closest these students get to e-mailing except to deal with teachers and college and job applications.) One senior is critical of those who don't know the rules: "Some people try to have conversations on texts, and I don't like that." In this group, there is near consensus that one of the pleasures of digital communication is that it does not need a message. It can be there to trigger a feeling rather than transmit a thought. Indeed, for many teenagers who discover their feelings by texting them, communication is the place where feelings are born.

Not far into this conversation, the emphasis on nonchalance runs into the complication that Audrey signaled: the composition of any message (even the most seemingly casual) is often studied. And never more so than when dealing with members of the opposite sex. John, sixteen, is an insecure young man with a crush who turns to a Cyrano, digital style. When he wants to get in touch with a girl he really likes, John hands his phone over to a friend he knows to be skilled at flirting by text. In fact, he has several stand-ins. When one of these friends does his texting, John is confident that he sounds good to his Roxanne. In matters of the heart, the quality of one's texts is as crucial as the choice of communications medium.

High school students have a lot to say about what kinds of messages "fit" with what kinds of media. This, one might say, is their generational expertise. Having grown up with new media that had no rules, they wrote some out of necessity. At Richelieu, Vera, a sophomore, says that texting brings "social pressure" because when she texts someone and the person does not get back to her, she takes it hard. With instant messaging, she feels less pressure because "if somebody doesn't get back to you, well, you can just assume they stepped away from their computer." Her classmate Mandy disagrees: "When I am ignored on IM, I get very upset." Two other classmates join the conversation. One tells Mandy that her reaction is "silly" and betrays a misunderstanding of "how the system works." A gentler girl tries to reason Mandy out of her hurt feelings: "Everyone knows that on IM, it is assumed you are busy, talking with other people, doing your homework, you don't have to answer." Mandy is not appeased: "I don't care. When I send a message out, it is hurtful if I don't get anything back."

Mandy presses her point. For her, the hurt of no response follows from what she calls the "formality" of instant messenging. In her circle, instant messages are sent in the evening, when one is working on homework on a laptop or desktop. This presumed social and technical setting compels a certain gravitas. Mandy's case rests on an argument in the spirit of Marshall McLuhan. The medium is the message: if you are at your computer, the medium is formal, and so is the message. If you are running around, shopping, or having a coffee, and you swipe a few keys on your phone to send a text, the medium is informal, and so is the message, no matter how much you may have edited the content.

The defenders of the "nonchalance" of instant messaging stand their ground: when you send an IM, it is going to a person "who has maybe ten things going on." Even though sitting at a computer, the recipient could well be doing homework, playing games on Facebook, or watching a movie. In all of this noise, your instant message can easily get lost. And sometimes, people stay signed on to instant messenger even though they have left the computer. All of this means, Vera sums up, "that IM can be a lower risk way to test the waters, especially with a boy, than sending a text. You can just send out something without the clear expectation that you will get something back." Though designed for conversation, IM is also perfect for the noncommittal, for "Whassup."

All the Richelieu sophomores agree that the thing to avoid is the telephone. Mandy presents a downbeat account of a telephone call: "You wouldn't want to call because then you would have to get into a conversation." And conversation, "Well, that's something where you only want to have them when you want to have them." For Mandy, this would be "almost never. . . . It is almost always too prying, it takes too long, and it is impossible to say 'good-bye.'" She shares Audrey's problem. Awkward good-byes feel too much like rejection. With texting, she says, "you just ask a question and then it's over."

This distaste for the phone crosses genders. A sixteen-year-old boy at Fillmore will not speak on the telephone except when his mother makes him call a relative. "When you text, you have more time to think about what you're writing. When you talk on the phone, you don't really think about what you're saying as much as in a text. On the telephone, too much might show." He prefers a deliberate performance that can be made to seem spontaneous. This offhand, seeming-not-to-care style has always been an emotional staple of adolescence, but now it is facilitated by digital communication: you send out a feeler; you look like you don't much care; things happen.

A text message might give the impression of spontaneity to its recipient, but teenagers admit they might spend ten minutes editing its opening line to get it just right. Spencer, a senior at Fillmore, says, "You forget the time you put into it when you get a text message back. You never think that anyone else put thought into theirs. So you sort of forget that you put time into yours." I ask him if he ever has sent a hastily composed text, and he assures me that this sometimes happens. "But not the ones that really count. . . . Before I send an important one, I switch it around, a lot." Deval, one of his classmates, says he is a very fast "thumb typist" and refers to his text messages as "conversations." One day we meet at noon. By that time, he says, he has "already sent out perhaps a hundred texts," most of them in two conversational threads. One conversation, Deval explains, "was with my buddy about his game last night. I wasn't able to go. Another was with my cousin who lives in Montreal, and she was asking about this summer and stuff. I'm going to be going to Canada for college. Since I'm going to be near them next year, she was asking whether I was going to come visit this summer."

I ask Deval how this conversation by text differs from placing a call to his Montreal cousin. He has spent the better part of the morning texting back and forth to her. Avoiding the phone cannot be about efficient time management. His answer is immediate: "She has an annoying voice." And besides, he says, "Texting is more direct. You don't have to use conversation filler." Their interaction on text "was just information." Deval says, "She was asking me direct questions; I was giving her direct answers. A long phone conversation with somebody you don't want to talk to that badly can be a waste of time."

Texting makes it possible for Deval to have a "conversation" in which he does not have to hear the sound of a voice he finds irritating. He has a way to make plans to live with his cousin during the summer without sharing any pleasantries or showing any interest in her. Both parties are willing to reduce their interchange to a transaction that scheduling software could perform. The software would certainly be comfortable with "no conversation filler" and "just information."

And yet, Deval does not know if texting is for life. He says that he might, not now, but sometime soon, "force himself" to talk on the phone. "It might be a way to teach yourself to have a conversation . . . For later in life, I'll need to learn how to have a conversation, learn how to find common ground so I can have something to talk about, rather than spending my life in awkward silence. I feel like phone conversations nowadays will help me in the long run because I'll be able to have a conversation." These days, of course, even those who are "later in life" have come to avoid telephone conversations. If you feel that you're always on call, you start to hide from the rigors of things that unfold in real time.

OVERWHELMED ACROSS THE GENERATIONS

The teenagers I studied were born in the late 1980s and early 1990s. Many were introduced to the Internet through America Online when they were only a little past being toddlers. Their parents, however, came to online life as grown-ups. In this domain, they are a generation that, from the beginning, has been playing catch-up with their children. This pattern continues: the fastest-growing demographic on Facebook is adults from thirty-five to forty-four.[9] Conventional wisdom stresses how different these adults are from their children—laying out fundamental divides between those who migrated to digital worlds and those who are its "natives." But the migrants and natives share a lot: perhaps above all, the feeling of being overwhelmed. If teenagers, overwhelmed with demands for academic and sexual performance, have come to treat online life as a place to hide and draw some lines, then their parents, claiming exhaustion, strive to exert greater control over what reaches them. And the only way to filter effectively is to keep most communications online and text based.

So, they are always on, always at work, and always on call. I remember the time, not many years ago, when I celebrated Thanksgiving with a friend and her son, a young lawyer, who had just been given a beeper by his firm. At the time, everyone at the table, including him, joked about the idea of his "legal emergencies." By the following year, he couldn't imagine not being in continual contact with the office. There was a time when only physicians had beepers, a "burden" shared in rotation. Now, we have all taken up the burden, reframed as an asset—or as just the way it is.

We are on call for our families as well as our colleagues. On a morning hike in the Berkshires, I fall into step with Hope, forty-seven, a real estate broker from Manhattan. She carries her BlackBerry. Her husband, she says, will probably want to be in touch. And indeed, he calls at thirty-minute intervals. Hope admits, somewhat apologetically, that she is "not fond" of the calls, but she loves her husband, and this is what he needs. She answers her phone religiously until finally a call comes in with spotty reception. "We're out of range, thank goodness," she says, as she disables her phone. "I need a rest."

Increasingly, people feel as though they must have a reason for taking time alone, a reason not to be available for calls. It is poignant that people's thoughts turn to technology when they imagine ways to deal with stresses that they see as having been brought on by technology. They talk of filters and intelligent agents that will handle the messages they don't want to see. Hope and Audrey, though thirty years apart in age, both see texting as the solution to the "problem" of the telephone. And both redefine "stress" in the same way—as pressure that happens in real time. With this in mind, my hiking partner explains that she is trying to "convert" her husband to texting. There will be more messages; he will be able to send more texts than he can place calls. But she will not have to deal with them "as they happen."

Mixed feelings about the drumbeat of electronic communication do not suggest any lack of affection toward those with whom we are in touch. But a stream of messages makes it impossible to find moments of solitude, time when other people are showing us neither dependency nor affection. In solitude we don't reject the world but have the space to think our own thoughts. But if your phone is always with you, seeking solitude can look suspiciously like hiding.

We fill our days with ongoing connection, denying ourselves time to think and dream. Busy to the point of depletion, we make a new Faustian bargain. It goes something like this: if we are left alone when we make contact, we can handle being together.

A thirty-six-year-old nurse at a large Boston hospital begins her day with a visit to her mother. Then she shops for food, cleans the house, and gets ready for work. After an eight-hour shift and dinner, it is after 9 p.m. "I am in no state to socialize," she says. "I don't even have the energy to try to track people down by phone. My friends from nursing school are all over the country. I send some e-mails. I log onto Facebook and feel less alone. Even when people are not there, like, exactly when I'm there, it seems like they are there. I have their new pictures, the last thing they were doing. I feel caught up." A widow of fifty-two grew up on volunteer work and people stopping by for afternoon tea. Now she works full-time as an office manager. Unaccustomed to her new routine, she says she is "somewhat surprised" to find that she has stopped calling friends. She is content to send e-mails and Facebook messages. She says, "A call feels like an intrusion, as though I would be intruding on my friends. But also, if they call me, I feel they are intruding . . . After work—I want to go home, look at some photos from the grandchildren on Facebook, send some e-mails and feel in touch. I'm tired. I'm not ready for people—I mean people in person." Both women feel put upon by what used to be sustaining, a telephone call. Its design flaw: it can only happen in real time. The flight to e-mail begins as a "solution" to fatigue. It ends with people having a hard time summoning themselves for a telephone call, and certainly not for "people in person."

Dan, a law professor in his mid-fifties, explains that he never "interrupts" his colleagues at work. He does not call; he does not ask to see them. He says, "They might be working, doing something. It might be a bad time." I ask him if this behavior is new. He says, "Oh, yes, we used to hang out. It was nice." He reconciles his view that once collegial behavior now constitutes interruption by saying, "People are busier now." But then he pauses and corrects himself. "I'm not being completely honest here: it's also that *I* don't want to talk to people now. *I* don't want to be interrupted. I think I should want to, it would be nice, but it is easier to deal with people on my BlackBerry."[10]

This widespread attitude makes things hard for Hugh, twenty-five, who says that he "needs more than e-mails and Facebook can provide." If his friends don't have time to see him, he wants them to talk to him on the phone so that he can have "the full attention of the whole person." But when he texts his friends to arrange a call, Hugh says that he has to make his intentions clear: he wants "private cell time." He explains, "This is time when the person you are calling makes a commitment that they will not take calls from other people. They are not doing anything else." He says he feels most rejected when, while speaking on the phone with a friend, he becomes aware that his friend is also texting or on Facebook, something that happens frequently. "I don't even want them to be walking. I can't have a serious conversation with someone while they are on their way from one sales meeting to another. Private cell time is the hardest thing to get. People don't want to make the commitment."

Some young people—aficionados of the text message and the call to "touch base"—echo Hugh's sentiments about the difficulty of getting "full attention." One sixteen-year-old boy says, "I say to people, talk to *me*. Now is my time." Another tries to get his friends to call him from landlines because it means they are in one place as they speak to him, and the reception will be clear. He says, "The best is when you can get someone to call you back on a landline.... That is the best." Talking on a landline with no interruptions used to be an everyday thing. Now it is exotic, the jewel in the crown.

Hugh says that recently, when he does get private cell time, he comes to regret it. By demanding that people be sitting down, with nothing to do but chat with him, he has raised the bar too high: "They're disappointed if I'm, like, not talking about being depressed, about contemplating a divorce, about being fired." Hugh laughs. "You ask for private cell time, you better come up with the goods."

The barrier to making a call is so high that even when people have something important to share, they hold back. Tara, the lawyer who admits to "processing" her friends by dealing with them on e-mail, tells me a story about a friendship undermined. About four times a year, Tara has dinner with Alice, a classmate from law school. Recently, the two women exchanged multiple e-mails trying to set a date. Finally, after many false starts, they settled on a time and a restaurant. Alice did not come to the dinner with good news. Her sister had died. Though they lived thousands of miles apart, the sisters had spoken once a day. Without her sister, without these calls, Alice feels ungrounded.

At dinner, when Alice told Tara her about her sister's death, Tara became upset, close to distraught. She and Alice had been e-mailing for months. Why hadn't Alice told her about this? Alice explained that she had been taken up with her family, with arrangements. And she said, simply, "I didn't think it was something to discuss over e-mail." Herself in need of support, Alice ended up comforting Tara.

As Tara tells me this story, she says that she was ashamed of her reaction. Her focus should have been—and should now be—on Alice's loss, not on her own ranking as a confidant. But she feels defensive as well. She had, after all, "been in touch." She'd e-mailed; she'd made sure that their dinner got arranged. Tara keeps coming back to the thought that if she and Alice had spoken on the telephone to set up their dinner date, she would have learned about her friend's loss. She says, "I would have heard something in her voice. I would have suspected. I could have drawn her out." But for Tara, as for so many, the telephone call is for family. For friends, even dear friends, it is close to being off the menu.

Tara avoids the voice but knows she has lost something. For the young, this is less clear. I talk with Meredith, a junior at Silver Academy who several months before had learned of a friend's death via instant message and had been glad that she didn't have to see or speak to anyone. She says, "It was a day off, so I was at home, and I hadn't seen anyone who lives around me, and then my friend Rosie IM'ed me and told me my friend died. I was shocked and everything, but I was more okay than I would've been if I saw people. I went through the whole thing not seeing anyone and just talking to people online about it, and I was fine. I think it would've been much worse if they'd told me in person."

I ask Meredith to say more. She explains that when bad news came in an instant message, she was able to compose herself. It would have been "terrible," she says, to have received a call. "I didn't have to be upset in front of someone else." Indeed, for a day after hearing the news, Meredith only communicated with friends by instant message. She describes the IMs as frequent but brief: "Just about the fact of it. Conversations like, 'Oh, have you heard?' 'Yeah, I heard.' And that's it." The IMs let her put her emotions at a distance. When she had to face other people at school, she could barely tolerate the rush of feeling: "The second I saw my friends, it got so much worse." Karen and Beatrice, two of Meredith's friends, tell similar stories. Karen learned about the death of her best friend's father in an instant message. She says, "It was easier to learn about it on the computer. It made it easier to hear. I could take it in pieces. I didn't have to look all upset to anyone." Beatrice reflects, "I don't want to hear bad things, but if it is just texted to me, I can stay calm."

These young women prefer to deal with strong feelings from the safe haven of the Net. It gives them an alternative to processing emotions in real time. Under stress, they seek composure above all. But they do not find equanimity. When they meet and lose composure, they find a new way to flee: often they take their phones out to text each other and friends not in the room. I see a vulnerability in this generation, so quick to say, "Please don't call." They keep themselves at a distance from their feelings. They keep themselves from people who could help.

VOICES

When I first read how it is through our faces that we call each other up as human beings, I remember thinking I have always felt that way about the human voice. But like many of those I study, I have been complicit with technology in removing many voices from my life.

I had plans for dinner with a colleague, Joyce. On the day before we were to meet, my daughter got admitted to college. I e-mailed Joyce that we would have much to celebrate. She e-mailed back a note of congratulations. She had been through the college admissions process with her children and understood my relief. At dinner, Joyce said that she had thought of calling to congratulate me, but a call had seemed "intrusive." I admitted that I hadn't called her to share my good news for the same reason. Joyce and I both felt constrained by a new etiquette but were also content to follow it. "I feel more in control of my time if I'm not disturbed by calls," Joyce admitted.

Both Joyce and I have gained something we are not happy about wanting. License to feel together when alone, comforted by e-mails, excused from having to attend to people in real time. We did not set out to avoid the voice but end up denying ourselves its pleasures. For the voice can only be experienced in real time, and both of us are so busy that we don't feel we have it to spare.

Apple's visual voicemail for the iPhone was welcomed because it saves you the trouble of having to listen to a message to know who sent it. And now there are applications that automatically transcribe voicemail into text. I interview Maureen, a college freshman, who is thrilled to have discovered one of these programs. She says that only her parents send her voicemail: "I love my parents, but they don't know how to use the phone. It's not the place to leave long voice messages. Too long to listen to. Now, I can scroll through the voicemail as text messages. Great."

Here, in the domain of connectivity, we meet the narrative of better than nothing becoming simply better. People have long wanted to connect with those at a distance. We sent letters, then telegrams, and then the telephone gave us a way to hear their voices. All of these were better than nothing when you couldn't meet face-to-face. Then, short of time, people began to use the phone instead of getting together. By the 1970s, when I first noticed that I was living in a new regime of connectivity, you were never really "away" from your phone because answering machines made you responsible for any call that came in. Then, this machine, originally designed as a way to leave a message if someone was not at home, became a screening device, our end-of-millennium Victorian calling card. Over time, voicemail became an end in itself, not the result of a frustrated telephone call. People began to call purposely when they knew that no one would be home. People learned to let the phone ring and "let the voicemail pick it up."

In a next step, the voice was taken out of voicemail because communicating with text is faster. E-mail gives you more control over your time and emotional exposure. But then, it, too, was not fast enough. With mobile connectivity (think text and Twitter), we can communicate our lives pretty much at the rate we live them. But the system backfires. We express ourselves in staccato texts, but we send out a lot and often to large groups. So we get even more back—so many that the idea of communicating with anything but texts seems too exhausting. Shakespeare might have said, we are "consumed with that which we are nourished by."[11]

I sketched out this narrative to a friend for whom it rang true as a description but seemed incredible all the same. A professor of poetry and a voracious reader, she said, "We cannot all write like Lincoln or Shakespeare, but even the least gifted among of us has this incredible instrument, our voice, to communicate the range of human emotion. Why would we deprive ourselves of that?"

The beginning of an answer has become clear: in text, messaging, and e-mail, you hide as much as you show. You can present yourself as you wish to be "seen." And you can "process" people as quickly as you want to. Listening can only slow you down. A voice recording can be sped up a bit, but it has to unfold in real time. Better to have it transcribed or avoid it altogether. We work so hard to give expressive voices to our robots but are content not to use our own.

Like the letters they replace, e-mail, messaging, texting, and, more recently, Tweeting carry a trace of the voice. When Tara regretted that she had not called her friend Alice—on the phone she would have heard her friend's grief—she expressed the point of view of someone who grew up with the voice and is sorry to have lost touch with it. Hers is a story of trying to rebalance things in a traditional framework. We have met Trey, her law partner. He confronts something different, something he cannot rebalance.

> My brother found out that his wife is pregnant and he put it on his *blog*. He didn't call me first. I called him when I saw the blog entry. I was mad at him. He didn't see why I was making a big deal. He writes his blog every day, as things happen, that's how he lives. So when they got home from the doctor—bam, right onto the blog. Actually, he said it was part of how he celebrated the news with his wife—to put it on the blog together with a picture of him raising a glass of champagne and she raising a glass of orange juice. Their idea was to celebrate on the blog, almost in real time, with the photos and everything. When I complained they made me feel like such a *girl*. Do you think I'm old-school?[12]

Trey's story is very different from Tara's. Trey's brother was not trying to save time by avoiding the telephone. His brother did not avoid or forget him or show preference to other family members. Blogging is part of his brother's intimate life. It is how he and his wife celebrated the most important milestone in their life as a family. In a very different example of our new genres of online intimacy, a friend of mine underwent a stem cell transplant. I felt honored when invited to join her family's blog. It is set up as a news feed that appears on my computer desktop. Every day, and often several times a day, the family posts medical reports, poems, reflections, and photographs. There are messages from the patient, her husband, her children, and her brother, who donated his stem cells. There is progress and there are setbacks. On the blog, one can follow this family as it lives, suffers, and rejoices for a year of treatment. Inhibitions lift. Family members tell stories that would be harder to share face-to-face. I read every post. I send e-mails. But the presence of the blog changes something in my behavior. I am grateful for every piece of information but feel strangely shy about calling. Would it be an intrusion? I think of Trey. Like him, I am trying to get my bearings in a world where the Net has become a place of intimate enclosure.

The Net provides many new kinds of space. On one end of the spectrum, I interview couples who tell me that they text or e-mail each other while in bed. Some say they want to leave a record of a request or a feeling "on the system."

And there are family blogs—places to announce a wedding or the progress of an illness or share photographs with the grandparents. These are all places to be yourself. At the other end of the spectrum, there are places where one constructs an avatar—from games to virtual communities—where people go to find themselves, or to lose themselves, or to explore aspects of themselves. On this spectrum, as we've seen, things are never clear-cut. As Audrey put it, a Facebook profile is "an avatar of me." And when you play Ringo Starr on a simulation of the Beatles, your avatar may feel like a second self. In simulation culture we become cyborg, and it can be hard to return to anything less.

CHAPTER 11

reduction and betrayal

In the mid-1990s, computer scientist and technological utopian Raymond Kurzweil created an avatar, Ramona, which he put into a virtual world. At that time, most players of online role-playing games had text-based avatars, complete with long descriptions of their histories and relationships, as well as the clothes they were wearing. Kurzweil looked forward to a new era. He didn't want to describe himself as Grace Slick. He wanted to *be* Grace Slick. Kurzweil created a virtual world and made a beautiful, sexy avatar who sang before the psychedelic backdrops of his choosing. This was Ramona. In the real, Kurzweil wore high-tech gear that captured his every gesture and turned them into Ramona's movements. His own voice was transformed into Ramona's female voice. Watching Kurzweil perform as Ramona was mesmerizing. And Kurzweil himself was mesmerized. It was an occasion, he said, for him to reflect on the difficulties of inhabiting another body and on how he had to retrain his movements—the way he held his head, the shape of his gestures—to become an avatar of another gender. These days, certain aspects of that experience, once so revolutionary, have become banal. We have turned them into games.

One such game, The Beatles: Rock Band, was released in September 2009 and hailed by the *New York Times* as a "transformative entertainment experience."[1] As in its older cousin, Rock Band, players hold game controllers in the shape of musical instruments and microphones that will transform the sounds they make into the sounds produced by screen avatars. Here the goal of play is to simulate the playing and singing of the Beatles. Such games are said to open music up to those who have no talent or no guitar. It is hoped that if children practice on such games, they will end up wanting to play a real instrument.

Like Kurzweil with Ramona, you have an avatar that you drive toward competency, and you have all that goes on in your head. The game sets you up not just to perform as a rock star but to feel like one, with all the attendant dreams and fantasies.

In online worlds and massively multiplayer online role-playing games, you have virtuosity and fantasy—and something more: your performances put you at the center of a new community with virtual best friends and a sense of belonging. It is not unusual for people to feel more comfortable in an unreal place than a real one because they feel that in simulation they show their better and perhaps truer self. With all of this going on, who will hold a brief for the real?

SERIOUS PLAY: A SECOND LIFE

When I joined Second Life, I was asked to choose a name for my avatar. I have often imagined having a name other than Sherry. It has never seemed quite right. Is it the Four Seasons song of the early 1960s that keeps it stuck in the world of junior high? But when I finally had the chance to be known as something else, I was confused. It was easy to dislike the name Sherry but not so easy to know what name I wanted. Fortunately, the system offered me choices. Once I chose, I felt relieved. Rachel. Something about this new name appealed. What was it? And with a question that simple, life on the screen became an identity workshop.[2]

Online worlds and role-playing games ask you to construct, edit, and perform a self. Yet, in these performances, like the performances we saw with sociable robots, something else breaks through. When we perform a life through our avatars, we express our hopes, strengths, and vulnerabilities.[3] They are a kind of natural Rorschach.[4] We have an opportunity to see what we wish for and what we might be missing. But more than this, we may work through blocks and address insecurities. People can use an avatar as "practice" for real life. As I've said, our lives on the screen may be play, but they are serious play.

Of course, people don't forge online identities with the idea that they are embarking on a potentially "therapeutic" exercise. Experimentation and self-reflection sneak up on you. You begin the process of building an avatar to play a game or join an online community; you imagine that it will be a simple matter, but then, suddenly, it is not. You can't, for example, decide on a name.

Joel, twenty-six, has given much thought to such questions of identity and online representation. For him, Second Life is quite literally his second life. In person, Joel appears far younger than his years. He is slender, casually dressed, with a slash of dark, tousled hair. Only a few years ago, his youthful appearance bothered Joel. He felt it was hard for people to take him seriously. Now, happily engaged to be married and settled down in a job he enjoys, Joel has made peace with his appearance. He still wishes he looked older but admits, "In the end, I suppose it can be helpful. Underestimation has its uses." Joel grew up hoping to be an artist, but practical considerations led him to study computer science. He is a programmer, talented and sought after.

Joel runs a software-design team at an elite biotechnology firm. He is challenged by the work, but his search for more creative outlets in programming brought him to Second Life. This is where Pete, whom we met earlier, had his virtual love affair with the beautiful avatar Jade. Joel has no interest in a Second Life romance. He wants a place to explore his potential as an artist and a leader. In real life, he does not feel confirmed in either. But both are integral to who he wants to be. In the safety of the online world, Joel performs them to become them.

Anthropologist Victor Turner writes that we are most free to explore identity in places outside of our normal life routines, places that are in some way "betwixt and between." Turner calls them *liminal*, from the Latin word for "threshold." They are literally on the boundaries of things.[5] Thomas Mann's imagined world in *The Magic Mountain* is a place out of time and place; this is what Second Life is for Joel, a place on the border between reality and fantasy. While many in Second Life build an avatar that is sexy, chic, and buff—a physical embodiment of a certain kind of ideal self—Joel goes in a different direction. He builds a fantasy version of how he sees himself, warts and all. He makes his avatar a pint-sized elephant named Rashi, a mix of floppy-eared sweetness and down-to-earth practicality. On Second Life, Rashi has a winsome side but is respected as an artist and programmer. That is, Joel creates beautiful buildings and virtual sculptures by programming at his keyboard; his avatar Rashi gets the credit in Second Life. More than being an artist, Joel (as Rashi) also takes charge of things. He organizes virtual building projects and gallery installations. Rashi is the kind of manager

Joel wants to be: strict but always calm and nonthreatening. Although an elephant, Rashi offers many possibilities for identity exploration to a man trying to bring together his artistic and managerial talents.

On Second Life, Joel could have built a tall and commanding avatar. He could have given his avatar a military bearing, or an Einsteinian "genius" allure. Instead, he crafted an avatar that faces the same challenges he does in the physical real. The avatar, like the man behind him, often has to prove his talent and self-discipline. For although he can be formal in manner, Rashi does, after all, resemble Dumbo more than the man in the gray flannel suit. So, like Joel, the elephant Rashi often works on teams whose members expect a lack of seriousness when they first meet him and then are taken aback by his dedication and technical virtuosity.

From the earliest days of online role-playing games, there were those who saw virtual places as essential to their life off the screen because online experiences were helping them to grow. One young man told me how he had "come out" online and saw this as practice for coming out to his friends and then to his family. A young woman who had lost a leg in a car crash and now wore a prosthetic limb felt ready to resume a sexual life after the accident but was still awkward and anxious. She created an online avatar with a prosthetic leg and had virtual relationships. Online, she practiced talking about her prosthetic limb and taking it off before being intimate with her virtual lovers. She grew more comfortable with her physical body through the experience of her virtual body. Another dedicated player described himself as a too-timid man. Online, he practiced greater assertiveness by playing a woman he called "a Katherine Hepburn type." Over time, he was able to bring assertiveness into his day-to-day life as a man. This is the kind of crossover effect that Joel is trying to effect. In the virtual, he cultivates skills he wants to use in the real.

In thinking about online life, it helps to distinguish between what psychologists call acting out and working through. In acting out, you take the conflicts you have in the physical real and express them again and again in the virtual. There is much repetition and little growth. In working through, you use the materials of online life to confront the conflicts of the real and search for new resolutions. This is how Joel uses Rashi. He has made a space for learning how to combine whimsy and gravitas.

Ever since high school, Joel has earned money building websites. He takes pleasure in beating deadlines and saving clients' money through clever design. Joel credits this to teenage experiences in what he calls the "hacker" culture. Then,

Joel felt part of a community of technical virtuosos who worked within a strict ethical code. Using the computer, hackers would play tricks on each other—these were the "hacks"—but they never played tricks on people outside the group, who could not defend themselves. (A classic hack might be to make a computer seem to crash, only to have it revive when a hacker in the know touched it with a particular keystroke.) If a young hacker did not play by these rules, senior hackers would step in and make things right. Joel mourns the passing of the hacker ethic. In today's virtual worlds, he says, "there is more mischief." Clever people who don't feel a commitment to the community are in a position to do real damage. On Second Life, through Rashi, Joel has become an enforcer of "old-school" hacker standards. His elephant is there to keep people in line. Property is to be respected. People's work is not to be destroyed. Rashi, with his elephant ears and mournful eyes, is a disheveled superhero, but he gets the job done.

Joel joined Second Life as soon as it was announced. He became a beta tester, meaning that he worked in the world before it was released to the public. His job was to help remove programming bugs, to make the environment as good as it could be. Joel's first impression of Second Life was negative. "I didn't like it. It was silly. Predictable. Good for techies." He dropped out for a while, but then came back in search of a creative space. He had heard about a group of "builders," artistic people who used the Second Life programming language to construct extraordinary and irreverent virtual architecture and art installations. In Second Life, these builders have status; they have made Second Life a significant destination for artists. Over time, Joel found a more welcoming community of artists in Second Life than he could in the real. Joel threw himself into the work of the group. He says, "If I was going to do it, I was going to do it well."

LIFE ON THE SCREEN

In Second Life, Rashi is a master builder who adds a subtle design vision to any project. He is also very kind. This means that through Rashi, Joel has a rich virtual social life. It brings him into contact with a range of people—artists, intellectuals, writers, businesspeople—he would not ordinarily meet. Rashi is often invited to parties where avatars eat, drink, dance, and chat. Whenever he attends a formal function, Rashi makes an elegant (online) scrapbook of the event and sends it as a gift to his avatar host or hostess.

The week before Joel and I meet, Rashi attended a Second Life wedding. Two avatars got married, and Rashi was asked to be ring bearer. Joel accepted with pleasure and designed an elaborate elephant tuxedo for the occasion. Since the dress code listed on the wedding invitation was "creative formal," Joel rendered the tuxedo in an iridescent multicolor fabric. He shows me the screenshot album he created after the event, the one that Rashi sent as a gift to the bride and groom. Rashi's generosity draws people to him, as does his emotional composure. In real life, Joel is a contented man, and this state of mind projects into the game. Perhaps it is this calm that attracted Noelle, a Second Life avatar who presents as a depressed Frenchwoman. Noelle has most recently been talking to Rashi about suicide, that is, suicide in the real. Joel and I sit at his computer on a day after he, as Rashi, has spent many hours "talking her down."

Noelle tells Rashi that their talks help her, and this makes Joel very happy. He also worries about her. Sometimes he thinks of himself as her father, sometimes as her brother. But since their entire relationship takes place in Second Life, the question of Noelle's authenticity is unclear. Recently, however, it is very much on Joel's mind. Who is she really? Is he talking with a depressed woman who has taken on the avatar Noelle, also depressed? Or is the person behind Noelle someone very different who is simply "playing" a depressed person online? Joel says that he would be "okay" if Noelle turns out not to be French. That would not seem a betrayal. But to have spent hours offering counsel to a woman who says she is contemplating suicide, only to find out it was "just a game"— that would feel wrong. Although delivered from Rashi to Noelle, the advice he gives, as Joel sees it, is from him as a human being to the purportedly depressed woman who is Noelle's puppeteer.

On the game, Joel makes it a rule to take people "at interface value." That is, he relates to what an avatar presents in the online world. And this is how he wants to be taken by other people. He wants to be treated as a whimsical elephant who is a good friend and a virtuoso programmer. Yet, Joel has been talking to Noelle about the possible death of the *real* person behind the avatar. And even though he doesn't think Noelle is exactly as she presents—for one thing her name is surely not Noelle, any more than his is Rashi—he counts on her being enough like her avatar that their relationship is worth the time he puts into it. He certainly is "for real" in his hours of counseling her. He believes that their relationship means something, is worth something, but not if she is "performing" depression. Or, for that matter, if she is a he.

Joel is aware of how delicate a line he walks in his virtual relationship with Noelle. Yet, he admits that the ground rules are not clear. There is no contract stipulating that an avatar will be "truthful" to the reality of the person playing it. Some people create three or four avatars to have the experience of playing different aspects of self, genders other than their own, ages different from their own. Joel knows all of this. But he is moving in another direction. Most recently, Joel's real-world business cards include his avatar name on Second Life.

We can guess why Joel doesn't like the telephone. When he makes or receives a call, he feels impatient and fidgety. He says that a call is "too much interruption"; he prefers to text or instant-message. Second Life avatars are able to communicate with each other in real time with text and speech, but because players are so often in and out of the world, this is a place of asynchronous messaging. As I watch Joel on Second Life, he moves through hundreds of messages as though gliding in a layered space. For him these messages, even those sent hours or days before, seem "of the moment." He experiences the asynchronous as synchronous. He has mastered a kind of information choreography. He speeds through pop-up messages and complex exchanges, surfing waves of information, graceful and in control. He only has to read one or two sentences of a message before he begins his response. Working without interruption, he feels both connected and pleasurably isolated.

Joel is in the same zone between connection and disconnection when he "parks" his avatar and flies without a body through Second Life. When he does this, Joel's "self" in the game is no longer Rashi. Joel explains that when he flies this way, he becomes a camera; his "I" becomes a disembodied "eye." Joel jokingly refers to his ability to fly "bodiless" through Second Life as an "out-of-avatar experience." He brings up an ethical issue: only some people can fly as he does, people who are experts. And when he flies this way, other people can't see him or know he is looking at them. Joel acknowledges the problem but is not troubled by it. He is comfortable with his privilege because he knows he does not abuse it. He sees himself as a benign caretaker. His "eye" belongs to a superhero surveying his city on the hill. And besides, says Joel, this isn't life. This is a game with a skill set that anyone is free to learn. Flying as an invisible eye is one such skill. He has paid his dues and this gives him the right to an activity that in another context might be thought of as spying.

Maria, a thirty-three-year-old financial analyst, can also fly as an "eye" through Second Life, but what she most enjoys in this virtual world is that life there is writ large. "The joy of Second Life is the heightened experience," she says. Time and relationships speed up. Emotions ramp up: "The time from meeting to falling in love to marrying to passionate breakups . . . that all can happen in very short order. . . . It is easy to get people on Second Life to talk about the boredom of the everyday. But on Second Life there is overstimulation." Maria explains that "the world leads people to emphasize big emotional markers. There is love, marriage, divorce—a lot of emotional culminating points are compressed into an hour in the world. . . . You are always attending to something big." What you hear from people is "I want to [virtually] kill myself, I want to get married, I am in love, I want to go to an orgy." Joel and Maria both say that after they leave the game, they need time to "decompress." From Maria's point of view, Second Life is not like life, but perhaps like life on speed. Yet, one of the things that Maria describes as most exhausting, "cycling through people," others on Second Life describe as most sustaining. For them, the joy of this online world is that it is a place where "new friendships come from."

Second Life gives Nora, thirty-seven, a happy feeling of continual renewal: "I never know who I'll meet 'in world.'" She contrasts this with the routine of her life at home with two toddlers. "At home I always know who I will meet. No one if I stay in with the kids. Or a bunch of nannies if I take the kids to the park. Or a bunch of bored rich-lady moms—I guess they're like me—if I take them shopping at Formaggio [a well-known purveyor of gourmet foods] or for snacks at the Hi-Rise [a well-known coffee shop/bakery]." Nora is bored with her life but not with her Second Life. She says of her online connections, "They are always about something, always about a real interest." But connections all about shared "interests" mean that Nora discards people when her "interests" change. She admits that there is a very rapid turnover in her Second Life friendships: "I toss people. . . . I make friends and then move on. . . . I know it gives me something of a reputation, but I like that there are always new people." Alexa, a thirty-one-year-old architecture student, has a similar experience. She says of Second Life, "There is always someone else to talk to, someone else to meet. I don't feel a commitment."

A Second Life avatar offers the possibility of virtual youth and beauty and, with these, sexual encounters and romantic companionship not always available in the physical real. These may be engaged in to build confidence for real-life encounters, but sometimes practice seems perfect. Some citizens of Second Life claim that they have found, among other things, sex, art, education, and acceptance. We hear the familiar story: life on the screen moves from being better than nothing to simply being better. Here, the self is reassuringly protean. You can experiment with different kinds of people, but you don't assume the risks of real relationships. Should you get bored or into trouble, you can, as Nora puts it, "move on." Or you can "retire" your avatar and start again.

Does loving your Second Life resign you to your disappointments in the real? These days, if you can't find a good job, you can reimagine yourself as successful in the virtual. You can escape a depressing apartment to entertain guests in a simulated mansion. But while for some the virtual may subdue discontents, for others it seems just a way to escape the doldrums. "During graduate school I spent four years on World of Warcraft [often referred to as WoW]," says Rennie, a thirty-two-year-old economist. "I loved the adventure, the puzzles, the mystery. I loved how I worked with so many different kinds of people. Once I was on a quest with a dancer from New York, a sixteen-year-old math prodigy from Arizona, and a London banker. Their perspectives were so interesting. The collaboration was awesome. It was the best thing in my life." Now, married with children, Rennie still slips away to World of Warcraft whenever he can. "It's better," he says, "than any vacation." What made it great in graduate school still obtains: it is his fastest, surest way to meet new people and find some thrills and challenges. "A vacation, well, it can work out or not. WoW always delivers."

ADAM

Simulation engages Adam, forty-three, to the point where everything else disappears and he just has to stay in the zone. His simulations of choice are the games Quake and Civilization. The first he plays in a group; the other he plays with online "bots," the artificial intelligences that take the roles of people. Adam likes who he is in these games—a warrior and a world ruler—more than who he is outside of them. His handicaps are in the real; in the games he is a star.

Adam is single, an aspiring singer and songwriter. Beyond this, he dreams of writing a screenplay. To make ends meet, he provides technical support for an insurance company and takes care of an elderly man on the weekends. Neither of these "real jobs" engages him. He is barely holding on to them. He says, "They are slipping away," under pressure from his game worlds, into which he disappears for up to fifteen-hour stretches. Adam gets little sleep, but he does not consider cutting back on his games. They are essential to his self-esteem, for it is inside these worlds that he feels most relaxed and happy. Adam describes a moment in Quake. "You're walking through shadow, you can see—there's snow on the ground, you're walking through a shadow landscape, and then you're walking out to the light, and you can see the sunlight!"

In one of the narratives on Quake, the greatest warriors of all time fight for the amusement of a race called the Vadrigar. It is a first-person shooter game.

You, as player, are the gun. Adam describes it as a "testosterone-laced thing, where you blow up other guys with various weapons that you find on a little map." Adam explains that when he plays Quake on a computer network, he can have one-on-one duels or join a team for tournament play. If he plays Quake alone, he duels against bots.

Now Adam plays alone. But in the past, he enjoyed playing Quake with groups of people. These game friends, he says, were the people who had "counted most" in his life. And he had played an online version of Scrabble with a woman named Erin, who became his closest friend. He doesn't have contact with Erin any more. She moved on to another game.

Adam thinks back to his earlier days on games with nostalgia. He recalls that the group sessions began at the office. "Five or six guys were hooked up to the server. We would play in our cubicles when management had a long meeting. . . . As long as the Notes server didn't explode, we would be able to blast away at each other and have a grand old time. And that got me hooked." After a while, the group moved to playing tournaments at people's homes. There was food and drink. And an easy way to be with people. Normally shy, Adam says that the game gave him things to talk about. "It didn't have to be really personal. It could all be about the game."

> Somebody would have a decent-enough network at their home, and we would take our computers there, hook them up, pizza would be ordered, onion dip, lots of crappy food, piles of Coca-Cola. There's actually a specialized drink for this sort of thing, called "Balz"—B-A-L-Z. Have you heard of it? I think it's spelled B-A-L-Z. But the point is, it's hypercaffeinated, something akin to Red Bull. For gaming, we'd set up the thing in the guy's basement, and we'd do it for four or five hours and blast away. . . . We'd be screaming at each other. . . . We'd all be able to hang out during the game and shoot the shit during the game or after the game, and that was a lot of fun.

From gathering in people's homes, the group went on to rent conference rooms at a hotel, with each participant contributing $50. Meetings now included food, dim lights, and marathon sessions of Quake, played for nine or ten hours at a stretch. Adam says that no one in the group wanted to leave: "And you keep going, you know, 'Gotta keep doing it again. Let's do it again! Blast away, you know.'" But the games in homes and hotel rooms have not happened in a long time. Now, Adam is most often on Quake as a single player, teaming up with the computer, with bots for companions. Adam says that the bots "do a great job." It is easy to forget that they are not people. Although he says it was "more of an ego trip to play with people as the competition, the bots are fine." Different bots have distinct personalities. They hold forth with scripted lines that simulate real player chat—usually irreverent and wise guy. In fact, Adam finds that "conversations with the [human] players . . . are about things that bots can talk about as well." He explains that the bots are competent conversationalists because conversations on Quake tend to follow predictable patterns. There is talk about "the maps . . . the places to hide, places to get certain bombs, places to get certain forms of invincibility." The bots can handle this.

Adam reminisces about moments of mastery on Quake; for him, mastery over the game world is a source of joy. "Over time," Adam says, "you learn where things are. . . . You get really good." In one play session, Adam ran around, as a cockroach, in a setting called "the Bathroom." He admits that "it might not sound like much," but it had engaged him, mind and body: "There are little tricks, you know, there are little slides, you can slide around, and you can leap up, and you're going down the sink, you slide down the sink, you end up in the cabinet, you run up a little ramp then find another place . . . then you get to this other spot where you can grab this pair of wings and fly around the room, just blasting away."

When Adam played Quake with his office mates, his favorite game had been a virtual version of Capture the Flag. Teams of players raid an opponent's base to take its flag while holding on to their own. Capture the Flag had everything Adam likes best: competition, flying, and losing himself in the person—that agile and masterful person—he becomes in the game.

> You want to beat your buddies. You want to make sure that you've out-done them. You capture one flag, and there's this series of jets you can grab, and you can fly over to the other end and you grab the flag and fly back. And you're flying, and all of a sudden, you hear [makes loud explosion noise, claps hands] and then "Boom." . . . Red team scores [in a dramatic voice, then "ding ding ding ding," indicating music]. And it's like "DAMMIT!" [loudly] You get a whole idea of what the hell's going on with the intensity of it. [laughs] You sorry you asked me for this? [laughs]

The game of Quake, played with his office friends and now played in single-player mode, makes Adam feel better about who he is in the game than who he is outside it. Adam says that he shows more skill at Capture the Flag than he does at his technical job, which he considers rote and beneath him. Beyond mastery, games offer the opportunity to perform roles he finds ennobling. Adam wants to be a generous person, but power is a prerequisite for benevolence. In life Adam feels he has none. In games he has a great deal. Indeed, in Civilization, which he now plays alone, Adam is in charge of nothing less than building the world.

> These games take so long, you can literally play it for days. One time when I played it, I had just got the game, and I got so addicted, I stayed home the next day and I played. . . . I think it was like noon the next day, or like nine o'clock the next day, I played all night long. And I ended up winning. You get so advanced. You get superadvanced technology. The first wave of technology is like a warrior, and the next big advance is you got, like, a spear and a shield, and then later on you get these things like Aegis. . . . It's a ship. It's a modern-day ship, or like nuclear weapons. . . . And you can actually build a spaceship and can leave the planet. . . . That would be a way of winning the game. . . .

To succeed in Civilization, Adam has to juggle exploration, conquest, economics, and diplomacy. He needs to exploit culture and technology—there is in-game research to produce an alphabet, build the pyramids, and discover gunpowder. He gets to choose the nature of his government; he feels good when he changes over from despotism to monarchy. "When you change the game to monarchy, [and you want to speed the production of something in a city] then you don't lose citizens, you lose gold. *So it gives you this feeling that you're humane.*"

But those toward whom Adam feels humane are not human. His benevolence is toward artificial intelligences. Adam has not forgotten that the bots are programs, but in the game he sees them as programs *and* as people. He exhibits the simultaneous vision we saw when people approached sociable robots. Adam enjoys the gratitude of his (AI) subjects. The fact that he takes good care of them makes him feel good about himself. This, in turn, makes him feel indebted to them. His sense of attachment grows. These are his bots, his people who aren't. He speaks of them in terms of endearment.

Adam talks about how good it feels when "up steps some little guy" (a bot of course) who comes out of battle ready to go over to his side. "Once that one guy comes over," he says, "there will be more and more of them." Unlike in real life, allegiance within the game comes with its own soundtrack. Adam says, "There's this little sound effect of a bunch of tribesmen going [grunt noise], and it echoes. It's fucking great."

The dictionary says that "humane" implies compassion and benevolence. Adam's story has taken us to the domain of compassion and benevolence toward the inanimate. There are echoes here of the first rule of the Tamagotchi primer: we nurture what we love, and we love what we nurture. Adam has beings to care for and the resources to do so. They "appreciate" what he does for them. He feels that this brings out the best in him. He wants to keep playing Civilization so that he can continue to feel good. On Civilization, Adam plays at gratifications he does not believe will come to him any other way.

Laboratory research suggests that how we look and act in the virtual affects our behavior in the real.[6] I found this to be the case in some of my clinical studies of role-playing games. Experimenting with behavior in online worlds—for example, a shy man standing up for himself—can sometimes help people develop a wider repertoire of real-world possibilities.[7] On this subject, I have also said that virtual experience has the greatest chance of being therapeutic if it becomes grist for the mill in a therapeutic relationship. In Adam's case, there is no evidence that online accomplishment is making him feel better about himself in the real. He says he is letting other things "slip away"—Erin, the girl he liked on the word game; his job; his hopes of singing and writing songs and screenplays. None of these can compete with his favorite simulations, which he describes as familiar and comforting, as places where he feels "special," both masterful and good.

Success in simulation tempers Adam's sense of disappointment with himself. He says that it calms him because, in games, he feels that he is "creating something new." But this is creation where someone has already been. Like playing the guitar in The Beatles: Rock Band, it is not creation but the *feeling* of creation. It suits Adam's purposes. He says he is feeling "less energetic than ever before." The games make him feel that he is living a better life. He can be adventurous and playful because the games present "a format that has already been established, that you don't have to create. You're creating something as you go along with it, but it's a format that provides you with all the grunt work already, it's already there, it's set up, and you just got this little area—it's a fantasy, it's a form of wish fulfillment. And you can go and do that." And yet, in gaming he finds something exhilarating and *his*.

Adam describes his creativity in Civilization as "just the right amount of creating. It's not like you really have to do something new. But it feels new. . . . It's a very comforting kind of thing, this repetitive sort of thing, it's like, 'I'm building a city—oh, yes, I built a city.'" These are feelings of accomplishment on a time scale and with a certainty that the real cannot provide.

This is the sweet spot of simulation: the exhilaration of creativity without its pressures, the excitement of exploration without its risks. And so Adam plays on, escaping to a place where he does not have to think beyond the game. A jumble of words comes out when he describes how he feels when he puts the game aside: "gravity, weight, movement away, bathroom, food, television." And then, without the game, there comes a flurry of unwelcome questions: "What am I going to do next? What are the things I really ought to be doing? . . . Off the game, I feel the weight of depression because I have to write my resume."

Although Adam fears he will soon be out of work, he has not been writing songs or a screenplay. He has not finished his resume or filed his taxes. These things feel overwhelming. The games are reassuring, their payoff guaranteed. Real life takes too many steps and can always disappoint.

Adam gets what he wants from the games, but he no longer feels himself— or at least a self he admires—without them. Outside the games, he is soon to be jobless. Outside the games, he is unable to act on goals, even for so small a thing as a trip to the accountant. The woman he considers his most intimate friend has moved on to a different game. Adam's thoughts turn back to the people with whom he had once played Quake. Their conversations had been mostly about game strategy, but Adam says, "That doesn't matter. There's something about the electronic glow that makes people connected in some weird way." Adam feels down. His real life is falling apart. And so he moves back, toward the glow.

TEMPTATION

We are tempted, summoned by robots and bots, objects that address us as if they were people. And just as we imagine things as people, we invent ways of being with people that turn them into something close to things.

In a program called Chatroulette, you sit in front of your computer screen and are presented with an audio and video feed of a randomly chosen person, also logged into the game. You can see, talk to, and write each other in real time. The program, written by a Russian high school student, was launched in November 2009. By the following February, it had 1.5 million users. This translates into about thirty-five thousand people logged onto Chatroulette at any one time. Some are in their kitchens, cooking, and want some company; some are masturbating; some are looking for conversation. Some are simply curious about who else is out there. In only a few months, Chatroulette had contributed a new word to the international lexicon: "nexting." This is the act of moving from one online contact to another by hitting the "next" button on your screen. On average, a Chatroulette user hits "next" every few seconds.

My own first session on Chatroulette took place in March 2010, during a class I teach at MIT. A student suggested it as a possible paper topic, and in our wired classroom, it took only a few seconds for me to meet my first connection. It was a penis. I hit next, and we parted company. Now my screen filled with giggling teenage girls. They nexted me. My third connection was another penis, this one being masturbated. Next. My fourth was a group of young Spanish men in a dimly lit room. They seemed to be having dinner by candlelight. They smiled and waved. Encouraged, I said, "Hi!" and was mortified by their friendly response, typed out: "Hello, old woman!" My class, protective, provided moral support and moved into the frame. I felt, of course, compelled to engage the Spaniards in lively conversation—old woman indeed! No one wanted to "next" on. But I needed to get back to other class business, so the Spaniards were made to disappear.

Chatroulette takes things to an extreme: faces and bodies become objects. But the mundane business of online life has its own reductions. The emoticon emotions of texting signal rather than express feelings. When we talk to artificial intelligences in our game worlds, we speak a language that the computer will be able to parse. Online, it becomes more difficult to tell which messages come from programs because we have taught ourselves to sound like them.[8] At the extreme—and the extreme is in sight—when we sound like programs, we are perhaps less shocked when they propose themselves as interlocutors. In science fiction terms, as a friend put it to me, "We can't identify the replicants because the people, inexplicably, took to acting like them."

As I have been writing this book, many people who enjoy computer games have asked me, "What's my problem? What's wrong with Scrabble or chess played online or against a computer? What's wrong with the new and artistic world of computer games?" Nothing is wrong with them. But looking to games for amusement is one thing. Looking to them for a life is another. As I have said, with robots, we are alone and imagine ourselves together. On networks, including game worlds, we are together but so lessen our expectations of other people that we can feel utterly alone. In both cases, our devices keep us distracted. They provide the sense of safety in a place of excluding concentration. Some call it the "zone."[9]

Psychologist Mihaly Csíkszentmihalyi examines the idea of "zone" through the prism of what he calls "flow," the mental state in which a person is fully immersed in an activity with focus and involvement.[10] In the flow state, you have clear expectations and attainable goals. You concentrate on a limited field so that anxiety dissipates and you feel fully present. Flow would capture how Rudy, eighteen, describes the pleasure of computer games: "I like the game best if you get sucked in. That's why I like playing single-player, not online, because you can get sucked into a character. There's this whole different world you can pretend to be in, pretty much. That's why it's different from a movie. When you're watching a movie, you're watching all the things happening, but when you're playing a video game, you're inside of it, and you can become the character you're playing as. It feels like you're there."

In the flow state, you are able to act without self-consciousness. Overstimulated, we seek out constrained worlds. You can have this experience at a Las Vegas gambling machine or on a ski slope. And now, you can have it during a game of Civilization or World of Warcraft. You can have it playing The Beatles: Rock Band. You can have it on Second Life. And it turns out, you can have it when texting or e-mailing or during an evening on Facebook. All of these are worlds that compel through their constraints, creating a pure space where there is nothing but its demands. It is flow that brings so many of us the experience of sitting down to do a few e-mails in the morning, just to "clear the decks" for a day of work, and then finding ourselves, five hours later, shocked that the day is half gone and no work has been done at all.

"I have to do my e-mail," says Clara, a thirty-seven-year-old accountant, looking down at her BlackBerry during a lunch break. "It's very tense," she says, "but it's also relaxing. Because when I'm doing it, that's all there is."[11] In her study of slot machine gambling in Las Vegas, anthropologist Natasha Schüll argues that Americans face too many choices, but they are not real choices.[12] They provide the illusion of choice—just enough to give a sense of overload, but not enough to enable a purposeful life. To escape, gamblers flee to a machine zone where the goal is not to win but to be. Gambling addicts simply want to stay in the game, comfortable in a pattern where other things are shut out. To make her point, Schüll cites my work on the psychology of video games.[13] From the earliest days, video game players were less invested in winning than in going to a new psychic place where things were always a bit different, but always the same. The gambler and video game player share a life of contradiction: you are overwhelmed, and so you disappear into the game. But then the game so occupies you that you don't have room for anything else.

When online life becomes your game, there are new complications. If lonely, you can find continual connection. But this may leave you more isolated, without real people around you. So you may return to the Internet for another hit of what feels like connection. Again, the Shakespeare paraphrase comes to mind: we are "consumed with that which we were nourished by."

"I'm trying to write," says a professor of economics. "My article is due. But I'm checking my e-mail every two minutes. And then, the worst is when I change the setting so that I don't have to check the e-mail. It just comes in with a 'ping.' So now I'm like Pavlov's dog. I'm sitting around, waiting for that ping. I should ignore it. But I go right to it." An art critic with a book deadline took drastic measures: "I went away to a cabin. And I left my cell phone in the car. In the trunk. My idea was that maybe I would check it once a day. I kept walking out of the house to open the trunk and check the phone. I felt like an addict, like the people at work who huddle around the outdoor smoking places they keep on campus, the outdoor ashtray places. I kept going to that trunk." It is not unusual for people to estimate that when at work, but taken up by search, surfing, e-mail, photos, and Facebook, they put in double the amount of hours to accommodate the siren of the Web.

Our neurochemical response to every ping and ring tone seems to be the one elicited by the "seeking" drive, a deep motivation of the human psyche.[14] Connectivity becomes a craving; when we receive a text or an e-mail, our nervous system responds by giving us a shot of dopamine. We are stimulated by connectivity itself. We learn to require it, even as it depletes us. A new generation already suspects this is the case. I think of a sixteen-year-old girl who tells me, "Technology is bad because people are not as strong as its pull."

Her remark reminds me of Robin, twenty-six, a young woman in advertising who complains that her life has been swallowed by the demands of e-mail. When I first meet her, she has what she describes as a "nervous rash" and says she is going on a retreat in western Canada to "detox from my e-mail." When I run into her three months later, there has been no retreat. She has found a doctor who diagnosed her rash as eczema. She explains that it can be brought on by stress, so surely e-mail had its role to play. But there is a pill she can take and a cream she can apply. And if she does all of this, she can stay online. It is easier to fix the eczema than to disconnect.

For many people, the metaphor of addiction feels like the only possible way to describe what is going on. I will have more to say about this later. For now, it must be given its due. Adam, whose only current passion is playing Civilization, says, "I've never taken opiates, but I imagine it's an electronic version of that. I guess television's that way too, but this is an opiate, or a numbing kind of thing. And you can find yourself satisfied in doing that."

At first Adam describes Civilization as enhancing. "There are diplomatic wins, conquests, victories." But he moves quickly to a language of compulsion. His achievements in the game—from instituting universal suffrage to building cultural wonders—seem dosed, dispensed like a drug designed to keep him hooked. Game success is fed to him in a way that "makes it hard to stop playing." He says,

> You just gotta keep having more popcorn, more potato chips. So what keeps the taste going? Well, I gotta achieve these little various things. . . . One city is building another rifleman . . . or you want universal suffrage. But, once you get universal suffrage, there's like . . . [makes a booming noise] "Universal suffrage has been built in Washington," and they show this great bronzed image. . . . You get this reward of this image that you normally don't see. It's a very comforting kind of thing, this repetitive sort of thing.

In Adam's story we see the comfort of retreat that Schüll describes, where one feels a sense of adventure in a zone of predictable action. Simulation offers the warmth of a technological cocoon. And once we feel humane because we are good friends to bots, perhaps it is not so surprising that we confide in online strangers, even about the most personal matters. On confessional sites our expectations of each other are reduced, but people are warmed by their electronic hearth. Just as simulation makes it possible to do things you can't accomplish in the real—become a guitar virtuoso or live like a benevolent prince—online confession gives you permission *not* to do things you *should* do in the real, like apologize and make amends.

CHAPTER 12

true confessions

i regularly read online confessional sites for six months. One afternoon of reading brings me to "The only reason I haven't killed myself yet is because my mother would kill herself. . . . I'm in love with a boy I've never met but we IM each other every day and talk about what we'll do or where we'll live when we're married. . . . My bulimia has made me better at giving blowjobs."

On most confessional sites, people log on anonymously and post a confession, sometimes referred to as a secret. On some sites, the transaction ends there. On most, the world is asked to respond. The world may be kind or ignore you. Or the world may be harsh. On PostSecret, a site where confessions are sent as illustrated postcards before being scanned and put on the site, a woman creates an image depicting a reed-thin model and writes, "If, in order to look like this, I would have to have my foot amputated . . . I would cut it off in a second."[1] A year later I come back to the PostSecret site and its troubled minds: "My mother had an affair with the first boy I slept with." "Divorcing you was a mistake." "I used to be dependent on him. Now I'm dependent on the drugs he sold me when we broke up."

On PostSecret there are exchanges between postcard writers and those who respond to them with an e-mail. The message "I wonder if white people know how lucky they are to be white" evokes "I wonder if straight people know how lucky they are to be straight" and "I wonder if any white/non-white, straight/not straight people know how lucky they are not to be autistic." The postcard that says, "I am having neck surgery tomorrow and I hope I die," brings forth "I hope that feeling dies and your surgery gives you another reason to live. You're in my prayers."[2]

These writers hold a mirror up to our complex times. There are important things to learn or be reminded of: Relationships we complain about nevertheless keep us connected to life. Advertising exerts a deadly tyranny. People reach out to strangers in kindness. Loneliness is so great that marriage to someone we have only met on a website can seem our best hope. On the electronic frontier, we forge connections that bring us back to earlier times and earlier technologies. We fall in love with twenty-first-century pen pals. Often their appeal is that we don't know who they "really" are. So they might be perfect.

In the world of PostSecret, the ability to be tentative, to speak in half-thoughts, gives permission to speak. Nancy, twenty-two, sends cards to PostSecret nearly once a month.[3] She says, "I don't have enough discipline to keep a diary. I don't think I'm important enough to do that. But I'm able to send my postcard." For a postcard, her simplest formulation is formulation enough. It is nice to think that the cards could be her start toward feeling worthy of more.

That the Internet is a place to simplify and heighten experience is very much on my mind as I read confessional sites. Market incentives are, after all, at work; each story competes with others. Exaggeration might increase readership. And since all confessions are anonymous, who will ever know? But if people are not truthful here, these confessions are fiction. Or perhaps, online confessions are a new genre altogether. When people create avatars, they are not themselves but express important truths about themselves. Online confession, another Internet performance zone, also occupies an intermediate space. Here, statements may not be true, but true enough for writers to feel unburdened and for readers to feel part of a community.

PostSecret holds annual picnics at which people can meet each other and see the actual paper postcards that were mailed to the site. At the first picnic, a young man explains how the site consoles him. He clearly means to say that it "offers the assumption of acceptance." But he makes a slip and says that the site "offers the pretense of acceptance." Both are true. His slip captures the site for me. Sometimes acceptance is there. Sometimes it is not. But it all works as a new fantasy—someone is listening.

Some people dash off their postcards, but others use the making of the postcard as an opportunity to take stock. Crafting a postcard demands a pause. That pause is PostSecret's great strength. Louisa, thirty-two, a mother of two, says, "You know what's on your mind, but here, you get to see what is *most* on your mind." On other sites, posting seems more impulsive. But on all of them, a confession that once might have been made within the bounds of friendship, family, or church now takes place with no bounds or bonds at all. It goes out to whoever is on the site. When confessions happen in real physical space, there is talk and negotiation.

Confessing to a friend might bring disapproval. But disapproval, while hard to take, can be part of an ongoing and sustaining relationship. It can mean that someone cares enough to consider your actions and talk to you about their feelings. And if a face-to-face confession meets criticism, we have some basis for evaluating its source. None of this happens in an online confession to strangers. One says one's piece, and the opinions of others come as a barrage of anonymous reactions. It is hard, say those who post, to pay attention only to the kind ones.

VENTING

When I talk to people about online confession, I hear many of the same comments that come up during conversations about robot companionship: "It can do no harm." "People are lonely. This gives them someplace to turn." "It helps get things off your chest." On the face of it, there are crucial differences between talking to human readers on a confessional site and to a machine that can have no idea of what a confession is. That the two contexts provoke similar reactions points to their similarities. Confessing to a website and talking to a robot deemed "therapeutic" both emphasize getting something "out." Each act makes the same claim: bad feelings become less toxic when released. Each takes as its premise the notion that you can deal with feelings without dealing directly with a person.[4] In each, something that is less than conversation begins to seem like conversation. Venting feelings comes to feel like sharing them.

There is a danger that we will come to see this reduction in our expectations as a new norm. There is the possibility that chatting with anonymous humans can make online bots and agents look like good company. And there is the possibility that the company of online bots makes anonymous humans look good. We ask less of people and more of technology.

Older people—say over thirty-five—talk about online confession as a substitute for things they want and don't have (like a trusted pastor or friend). Younger people are more likely to take online confession on its own terms. It's new; it's interesting. Some read confessional sites simply to see what's there. Some say they take comfort in learning that others have the same troubles that they do. Some say they do it for fun. And, of course, some use the sites for their own confessions, describing them with no intended irony, as a way to speak in private. Most Internet sites keep track of who has visited them. Online confessional sites make a point of saying that they do not. Sixteen-year-old Darren says, "Confession sites offer anonymity if you just want to get a secret out there."

Darren's family is from Vietnam. They are Catholic, very strict and religious. His father checks his homework every night and personally supervises extra lessons if he sees things slipping. His parents make his significant decisions for him, using what he calls their "rational rule." He says they will choose his college by "measuring its cost relative to what different options will mean for my future career." Darren adds with some edge to his voice, "I will be surprised if the 'rational' choice for my career is not engineering." In all of this, Darren acquiesces. He does not express displeasure with his family culture, but he has looked for a place outside its bounds where "I can just shout my own feelings."

Several of Darren's Vietnamese friends use confessional sites; that is where he learned about them. Darren explains that when he and his friends confess, they all make up false screen names. He says, "We put our secrets up, and we just want to show it to a stranger, not a friend but a stranger. You want to express your emotion. You write it down and write it on the website and you just want a stranger who doesn't know you to look at it. Not your friends." Darren also thinks that a robotic confidant sounds like a good idea. That the robot would lack emotion does not bother him at all. In fact, he sees its lack of emotion as potentially "a good thing." Unlike his family, the robot would be "nonjudgmental." Darren's reaction to the idea of talking to a computer program: "I could get out some pure feelings."

In Darren's community, he has no place to take what he calls his "irrational positions." He says that it would be shaming to share them, even with his friends. This is where a future robot would be helpful and the Internet is helpful now. I never find out what Darren's "irrational positions" are, but Sheryl, thirty-two, a nurse in western Massachusetts, is willing to say what she has shared online. There have been "inappropriate" romantic encounters with coworkers and she has taken two vacations with some of the money set aside for her parents' retirement. She says that regarding both situations—the money and the men— online confession was a solace: "The most important thing is that after you make your confession, you read the others. You know you are not alone. A lot of other people have done almost the same things you did."

Sheryl's online confessions do not lead her to talk to those she has wronged or to try to make amends. She goes online to feel better, not to make things right. She thinks that most people use confessional sites this way. She says, "Many posts begin with 'I could never tell my husband, but . . . I could never tell my mother, but . . . ' " I ask her if online confession makes it easier *not* to apologize. Her answer is immediate: "Oh, I definitely think so. This is my way to make my peace . . . and move on." I am taken aback because I did not expect such a ready response. But Sheryl has already given this thought. She refers to the Twelve Steps, a program to combat addiction. She explains steps eight and nine: "Step eight says to make a 'list of all persons we had harmed, and become willing to make amends to them all.' Step nine says to make 'direct amends to such people.'" Sheryl then points out that step nine exempts you from taking these actions if amends "would injure those or others." Sheryl is going with the exemption. She is ready to confess, not apologize.

The distinction between confession and apology comes up regularly in conversations about online communication and social-networking sites. There is a lot of apologizing on Facebook, for example, but I am often told that these apologies don't count. They are more like confessions because a real apology has to deal more directly with the person you have wronged. Maria, the thirty-three-year-old financial analyst who said that the intensity of Second Life could be exhausting, does not like it when people try "to make things right" by e-mail. She thinks apologies must be made in person. "But," she continues, "people don't do that any more. . . . When people confess on the computer, they think they have done their job and now it is up to others to respond. But I think if you have hurt me, why should it be my job to come tell you that it is all right?" Recall sixteen-year-old Audrey's derisive account of an online apology: "It's cheap. It's easy. All you have to do is type 'I'm sorry.'" That pretty much describes how eighteen-year-old Larry handles things: "I don't apologize to people any more. I just put my excuses on as my status [referring to Facebook]. The people who are affected know who I mean." Sydney, twenty-three, a first-year law student, takes exception: "Saying you are sorry as your status . . . that is not an apology. That is saying 'I'm sorry' to Facebook."

The elements of an apology are meant to lay the psychological groundwork for healing—and this means healing both for the person who has been offended and for the person who has offended. First, you have to know you have offended, you have to acknowledge the offense to the injured party, and you have to ask what you can do to make things right.

Technology makes it easy to blur the line between confession and apology, easy to lose sight of what an apology is, not only because online spaces offer themselves as "cheap" alternatives to confronting other people but because we may come to the challenge of an apology already feeling disconnected from other people. In that state, we forget that what we do affects others.

Young people, bruised by online skirmishes, can be the most articulate about looking back to the best practices of the past in the pursuit of a classic apology. Two sophomore girls at Silver Academy make the point that there is too much online apology going around. For one, "Texting an apology is really impersonal. You can't hear their voice. They could be sarcastic, and you wouldn't know." The other agrees: "It's harder to say 'Sorry' than text it, and if you're the one receiving the apology, you know it's hard for the person to say 'Sorry.' But that is what helps you forgive the person—that they're saying it in person, that they actually have the guts to actually want to apologize." In essence, both young women are saying that forgiveness follows from the experience of empathy. You see someone is unhappy for having hurt you. You feel sure that you are standing together with them. When we live a large part of our personal lives online, these complex empathetic transactions become more elusive. We get used to getting less.

THE CRUELTY OF STRANGERS

Harriet, thirty-two, posts to online confessional sites when she feels depressed, maybe two or three times a month. She prefers sites on which her readers can leave comments. She says, "It makes me feel in contact." Otherwise, she says, "it's like putting a post in a glass bottle and putting it in the ocean." At first she claims that "critical comments" about her posts don't bother her. But only a few minutes later, when we talk about specifics, Harriet admits that, somewhat to her surprise, they can hurt a lot. Her worst experience came after confessing that she had been seduced by her uncle as a teenager. "My aunt never found out. She recently died. He's dead too. There is no one I can tell now who it would matter to. So I went online, just to tell. People were really critical, and it hurt. I thought there would be some, like, religious people who wouldn't like it. But really I got a lot of disapproval." Harriet begins by saying, "Who cares what strangers think?" She ends up describing a human vulnerability: if you share something intimate with a stranger, you invest in that person's opinion. Anonymity does not protect us from emotional investment. In talking about online confessions, people say they are satisfied if they get their feelings out, but they still imagine an ideal narrative: they are telling their stories to people who care. Some online confessions reach sympathetic ears, but the ideal narrative is just that, an ideal.

When Roberta, thirty-eight, types her online confessions, she describes being in a state of mind that is close to dissociative. When reality is too painful (for example, the reality of abuse), people may feel that they have left their bodies and are watching themselves from above. Leaving the self is a way not to feel something intolerable. So, Roberta types her confessions but sometimes doesn't remember the details of doing so. Then she leaves the site and returns to read comments. They are not always supportive, and the dissociative state returns. She says,

> When I was about fourteen, I began an affair with my mother's boyfriend. He lived with us since I was ten. . . . When I confessed online, I found that I didn't even know I was typing. . . . Later that day, I checked back and there were some very positive comments but there were some that said I had completely betrayed my mom. . . . I should tell her. Others said I shouldn't tell her but that I was a bitch. I didn't faint or anything. But I . . . found myself in the kitchen and I don't remember how I got there.

We build technologies that leave us vulnerable in new ways. In this case, we share our burdens with unseen readers who may use us for their own purposes. Are those who respond standing with us, or are they our judges, "grading" each confession before moving on to the next? With some exceptions, when we make ourselves vulnerable, we expect to be nurtured.[5] This is why people will sometimes, often prematurely, tell their "sad stories" to others they hardly know. They hope to be repaid in intimacy. The online setting increases the number of people to whom one applies for a caring response. But it also opens one up to the cruelty of strangers. And by detaching words from the person uttering them, it can encourage a coarsening of response. Ever since e-mail first became popular, people have complained about online "flaming." People say outrageous things, even when they are not anonymous. These days, on social networks, we see fights that escalate for no apparent reason except that there is no physical presence to exert a modulating force.

When Audrey described an Internet fight in her school, we saw how flaming works: "Someone says a cross word. Someone calls someone else a name. Large numbers of people take sides. . . . They had a fight for a weekend. Twenty or thirty interchanges a day." In her opinion, by the end of the weekend, nothing had been resolved. Nothing had been learned about how to deal with other people. "No one could even say, really, what the fight was about." But people who were friends no longer spoke to each other. Freed from the face-to-face, some people develop an Internet-specific road rage. Online, Audrey knows, it is easier to be a bully.

Yet teenagers, knowing this, are frequent visitors to online confessional sites. Brandi, eighteen, compares them to Facebook and MySpace, her other online places. Through her eyes, it becomes clear that what they have in common is that people form a relationship to the site as well as to those on it. "Online," says Brandi, "I get the private out of my system. . . . I put my unhappiness onto the site."

With such displacement of feeling, it is not surprising that the online world becomes fraught with emotion. On confessional sites, people who disagree about a particular confession begin to "scream" at each other. They displace their strong investments in some issue—abortion, child abuse, euthanasia—in fights with strangers. They put their "unhappiness onto the site" because, often, they are most angry at others for what they dislike in themselves.[6]

Jonas, forty-two, admits to being "addicted" to a range of confessional sites, some religious, most of them not. He interrupts his work by "dipping in" to one or another of them during the day. Divorced, Jonas is preoccupied by the idea that he is becoming estranged from his son, who is choosing to spend more time with his mother. Jonas doesn't think there is anything in particular to blame; he and his son haven't had a fight. "I'm just seeing him less and less." But with this issue on his mind, he tells me that he became enraged by a particular online confession by a woman named Lesley, who is concerned about her nineteen-year-old son. Lesley and her son had a falling-out during his junior year of high school, and it was never repaired. Shortly after graduation, the son joined the army and was sent to Iraq. Lesley worries that she drove her son away. Jonas says, "I attacked Lesley for being a bad mother. . . . I said she was close to re-sponsible if her son dies."

It seems apparent that instead of exploring his feelings about his own son, Jonas had lashed out at Lesley. Of course, this kind of thing happens between friends. It happens in families. But it is endemic on the Internet. There is no barrier to displacement, no barrier to rage. Online confessionals, with their ethic of "getting the private out," as Brandi put it, reassure users with the promise that they do not need to talk to someone in person—expression alone is helpful. And, of course, it sometimes may be. I think of the authors on PostSecret who might feel better when they make postcards that say "Divorcing you was a mistake" and "Celebrating the last year you abused me. They don't make cards like that." But confessional sites are often taken as therapy and they are not. For beyond self-expression, therapy seeks new ways to approach old conflicts. And thinking of Jonas and Leslie, therapy works because it helps us see when we project feelings onto others that we might understand as our own.

It is useful to think of a symptom as something you "love to hate" because it offers relief even as it takes you away from addressing an underlying problem. To me, online confessional sites can be like symptoms—a shot of feeling good that can divert attention from what a person really needs. One high school senior tells me that she visits online confessional sites at least twice a week. Most recently, she has been writing descriptions of sleeping with her best friend's boyfriend. When I ask her what she does after she writes her confessions, she says that she stays alone in her room, smoking. She thinks that she has unburdened herself and now wants to be alone. Or perhaps the confession has left her depleted.

Like a conversation with a robot, online confession appeals because someone silent wants to speak. But if we use these sites to relieve our anxieties by getting them "out there," we are not necessarily closer to understanding what stands behind them. And we have not used our emotional resources to build sustaining relationships that might help. We cannot blame technology for this state of affairs. It is people who are disappointing each other. Technology merely enables us to create a mythology in which this does not matter.

SEEKING COMMUNITIES

In what framework does confessing to online strangers make sense? It does not connect us with people who want to know us; rather, it exposes us to those who, like Jonas, may use our troubles to relieve them from looking at their own. It does nothing to improve our practical situations. It may keep us from taking positive action because we already feel we've done "something." I know these things to be true. But people who confess online also tell me that they feel relieved and less alone. This is also true. So, if sites are symptoms, and we need our symptoms, what else do we need? We need trust between congregants and clergy. We need parents who are able to talk with their children. We need children who are given time and protection to experience childhood. We need communities.

Molly, fifty-eight, a retired librarian who lives alone, does not feel part of any community. She doesn't have children; her urban neighborhood, she says, is "not the kind of place people know each other. . . . I don't even recognize the people in the Shaw's [a local supermarket chain]." She says that she has memories of grocery shopping with her father as a girl. Then, she had felt part of a family, a family in a neighborhood. Every visit to Shaw's reminds her of what she doesn't have. She imagines her favorite confessional sites as communities and says that this has been helpful to her, at least to a point. Molly has posted stories of her mother's struggle with alcoholism. She is Catholic, but as both child and adult, she never felt comfortable talking to a priest about her history. "It wasn't something to confess. It just seemed like complaining." Speaking of her "real life," she says, "I don't see the goodness around me. Online I have found some good people." She uses the word "community."

One can only be glad that Molly has found sustenance. But her view of "community" is skewed by what technology affords. Although she claims that on confessional sites she has met "good people," when she gets feedback she doesn't like, Molly leaves the site so that she does not have to look at the criticism again. Communities are places where one feels safe enough to take the good and the bad. In communities, others come through for us in hard times, so we are willing to hear what they have to say, even if we don't like it. What Molly experiences is not community.

Those who run online confessional sites suggest that it is time to "broaden our definition of community" to include these virtual places.[7] But this strips language of its meaning. If we start to call online spaces where we are with other people "communities," it is easy to forget what that word used to mean. From its derivation, it literally means "to give among each other." It is good to have this in mind as a standard for online places. I think it would be fair to say that online confessional sites usually fall below this mark.

Perhaps community should have not a broader but a narrower definition. We used to have a name for a group that got together because its members shared common interests: we called it a club. But in the main, we would not think of confessing our secrets to the members of our clubs. But we have come to a point at which it is near heresy to suggest that MySpace or Facebook or Second Life is not a community. I have used the word myself and argued that these environments correspond to what sociologist Ray Oldenberg called "the great good place."[8] These were the coffee shops, the parks, and the barbershops that used to be points of assembly for acquaintances and neighbors, the people who made up the landscape of life. I think I spoke too quickly. I used the word "community" for worlds of weak ties.[9]

Communities are constituted by physical proximity, shared concerns, real consequences, and common responsibilities. Its members help each other in the most practical ways. On the lower east side of Manhattan, my great grandparents belonged to a block association rife with deep antagonisms. I grew up hearing stories about those times. There was envy, concern that one family was doing better than another; there was suspicion, fear that one family was stealing from another. And yet these families took care of each other, helping each other when money was tight, when there was illness, when someone died. If one family was evicted, it boarded with a neighboring one. They buried each other. What do we owe to each other in simulation? This was Joel's problem as he counseled Noelle in Second Life. What real-life responsibilities do we have for those we meet in games? Am I my avatar's keeper?

AFTER CONFESSION, WHAT?

After a morning immersed in reading online confessions, I suddenly become anxious about my own responsibilities. The sites make it clear that they do not collect IP addresses from those who write in. If they did, they would be responsible for reporting people who confessed to illegal actions. (When people confess to killing someone, the caretakers of these sites do not pursue the question, choosing to interpret these posts as coming from members of the military.) But what of my sense of responsibility? If this is not a game, how do you not get anxious when a woman talks about letting her lover suffocate her until she fears for her life? If this is not a game, how do you not get anxious when a mother talks about nearly uncontrollable desires to shake her baby? My time on confessional sites leaves me jumpy, unable to concentrate. People are in dire straits. And I am there bearing witness.

Yet, my anxiety may be ill placed. Some people tell me that what they post on the Internet bears only a glancing relationship to reality. One young man in his twenties says that the Internet is our new literature. It is an account of our times, not necessarily calling for each individual's truth to be told. A twenty-four-year-old graduate student tells me she goes to confessional sites to say "whatever comes into my mind" in order to get attention. A forty-year-old college professor explains that when he does anything online in an anonymous forum, he takes on the persona of "everyman." For him, anonymity means universality. What he says on the Web does not necessarily follow from his actual experience: if the world is violent, he feels free to write of violence in his own voice. So, when I read online confessions and go cold, am I tuning out the voice of a woman who was raped at nine, or have I ceased to believe that the confessional Internet can connect me to real people and their true stories?

Trained psychoanalytically, I am primed not to ask what is true but what things mean. That doesn't suggest that truth is unimportant, but it does say that fantasies and wishes carry their own significant messages. But this perspective depends on listening to a person, in person. It depends on getting to know that person's life history, his or her struggles with family, friendship, sexuality, and loss. On the Internet, I feel an unaccustomed desire to know if someone is telling "the truth."

A good therapy helps you develop a sense of irony about your life so that when you start to repeat old and unhelpful patterns, something within you says, "There you go again; let's call this to a halt. You can do something different." Often the first step toward doing something different is developing the capacity to not act, to stay still and reflect. Online confession keeps you moving. You've done your job. You've gotten your story out. You're ready for your responses. We did not need the invention of online confessional sites to keep us busy with ways to externalize our problems instead of looking at them. But among all of its bounties, here the Internet has given us a new way not to think.

I grant that confessional sites leave some people feeling better for "venting" and knowing that, in their misery, they are not alone. But here is how they leave me: I am anxious about my inability to help. I feel connected to these people and their stories, but I realize that to keep reading, I must inure myself to what is before my eyes. Certain kinds of confessions (and, unfortunately, some of the most brutal ones) start to read like formulaic writing in well-known genres. When this happens, I start to tune out and then feel terribly upset. I think of Joel on Second Life and his doubts about Noelle's really being suicidal. Am I watching a performance? Or, more probably, how much performance am I watching? Am I becoming coarsened, or am I being realistic?

CHAPTER 13

anxiety

M arcia, sixteen, a sophomore at Silver Academy, has her own problems. "Right now," she says, on-screen life "is too much to bear." She doesn't like what the Internet brings out in her—certainly not her better angels. Online, she gives herself "permission to say mean things." She says, "You don't have to say it to a person. You don't see their reaction or anything, and it's like you're talking to a computer screen so you don't see how you're hurting them. You can say whatever you want, because you're home and they can't do anything." Drea, a classmate sitting next to her, quips, "Not if they know where you live," but Marcia doesn't want to be taken lightly. She has found herself being cruel, many times. She ends the conversation abruptly: "You don't see the impact that what you say has on anyone else."

Marcia and Drea are part of a group of Silver Academy sophomores with whom I am talking about the etiquette of online life. Zeke says that he had created "fake" identities on MySpace. He scanned in pictures from magazines and wrote profiles for imaginary people. Then, he used their identities to begin conversations about himself, very critical conversations, and he could see who joined in. This is a way, he says, "to find out if people hate you." This practice, not unusual at Silver, creates high anxiety. Zeke's story reminds me of John, also at Silver and also sixteen, who delegated his texting to digitally fluent Cyranos. When John told his story to his classmates, it sparked a fretful conversation about how you never really know who is on the other end when you send or receive a text. Now, after hearing Zeke's story, Carol picks up this theme. "You never know," she says, "who you might be talking to. A kid could start a conversation about your friend, but you have to be careful. It could *be* your friend. On MySpace . . . you can get into a *lot* of trouble."

Others join the discussion of "trouble." One says, "Facebook has taken over my life." She is unable to log off. "So," she says, "I find myself looking at random people's photos, or going to random things. Then I realize after that it was a waste of time." A second says she is afraid she will "miss something" and cannot put down her phone. Also, "it has a camera. It has the time. I can always be with my friends. Not having your phone is a high level of stress." A third sums up all she has heard: "Technology is bad because people are not as strong as its pull."

Anxiety is part of the new connectivity. Yet, it is often the missing term when we talk about the revolution in mobile communications. Our habitual narratives about technology begin with respectful disparagement of what came before and move on to idealize the new. So, for example, online reading, with its links and hypertext possibilities, often receives a heroic, triumphalist narrative, while the book is disparaged as "disconnected." That narrative goes something like this: the old reading was linear and exclusionary; the new reading is democratic as every text opens out to linked pages—chains of new ideas.[1] But this, of course, is only one story, the one technology wants to tell. There is another story. The book is connected to daydreams and personal associations as readers look within themselves. Online reading—at least for the high school and college students I have studied—always invites you elsewhere.[2] And it is only sometimes interrupted by linking to reference works and associated commentaries. More often, it is broken up by messaging, shopping, Facebook, MySpace, and YouTube. This "other story" is complex and human. But it is not part of the triumphalist narrative in which every new technological affordance meets an opportunity, never a vulnerability, never an anxiety.

There were similar idealizations when it became clear that networked computers facilitated human multitasking. Educators were quick to extol the virtues of doing many things at once: it was how the future wanted us to think. Now we know that multitasking degrades performance on everything we try to accomplish. We will surely continue to multitask, deciding to trade optimum performance for the economies of doing many things at once. But online multitasking, like online reading, can be a useful choice without inspiring a heroic narrative.

We have to love our technology enough to describe it accurately. And we have to love ourselves enough to confront technology's true effects on us. These amended narratives are a kind of *realtechnik*. The *realtechnik* of connectivity culture is about possibilities and fulfillment, but it also about the problems and dislocations of the tethered self. Technology helps us manage life stresses but generates anxieties of its own. The two are often closely linked.

So, for example, mobile connections help adolescents deal with the difficulties of separation. When you leave home with a cell phone, you are not as cut off as before, and you can work through separation in smaller steps. But now you may find yourself in text contact with your parents all day. And your friends, too, are always around. You come to enjoy the feeling of never having to be alone. Feeling a bit stranded used to be considered a part of adolescence, and one that developed inner resources. Now it is something that the network makes it possible to bypass. Teenagers say that they want to keep their cell phones close, and once it is with you, you can always "find someone."

Sometimes teenagers use the network to stay in contact with the people they "know for real," but what of online friends? Who are they to you? You may never have met them, yet you walk the halls of your school preoccupied with what you will say to them. You are stalked on Facebook but cannot imagine leaving because you feel that your life is there. And you, too, have become a Facebook stalker. Facebook feels like "home," but you know that it puts you in a public square with a surveillance camera turned on. You struggle to be accepted in an online clique. But it is characterized by its cruel wit, and you need to watch what you say. These adolescent posts will remain online for a lifetime, just as those you "friend" on Facebook will never go away. Anxieties migrate, proliferate.

RISK MANAGEMENT

We have met Julia, sixteen, a sophomore at Branscomb High School for whom texting was a way to acknowledge, even discover, her feelings. An only child, her mother had a heart condition, and Julia spent her early years living with her aunt. When Julia was nine, her mother underwent successful surgery, and Julia was able to move in with her and a new stepfather. When this marriage dissolved, she and her mother set off on their own. Her health restored, Julia's mother put herself through college and now runs a small employment agency.

When she was younger, Julia saw her father once a week. But he wanted more and blamed Julia for the infrequency of their visits. She felt caught between her parents. "So," she now says, "I stopped calling him." Of course, as is often the case in such matters, when Julia stopped calling her father, she wanted desperately for him to call her. "I wanted him to call me, but he didn't want to call me. . . . But if I called him, he would blame me for not talking to him enough." And so their relationship trailed off in bad feelings all around: "It was just like we hadn't been talking to each other as much. And the less we talked, the less we saw of each other, and one day it just stopped completely." For the past four years Julia hasn't seen or spoken to her father.

I meet Julia at a possible turning point. Over winter break, she plans to go on a school-sponsored trip to work at an orphanage in Guatemala. To participate she needs both of her parents' signatures on a permission document. Julia is very nervous that her father won't sign the form ("I just haven't spoken to him in so long"), but in the end, he does and sends her a note that includes his e-mail address. When I meet Julia she is excited and apprehensive: "So, he sent a letter with the signature for my passport, saying he was sorry, and he'd like to keep in touch now. So, I'm gonna start talking to him more. . . . I'm going to try to talk to him through e-mail." Julia is not ready to speak to her father. For her that would be too great a jump—perhaps for her father as well. He did not, after all, send her his telephone number. E-mail was offered as a way to talk without speaking. "E-mail is perfect," Julia says. "We have to build up. If we talk awhile on the computer, then I can call him and then maybe go and see him." As Julia talks about this plan, she nods frequently. It feels right.

Julia knows another way to reach her father: he has a MySpace account. However, she explains that there was "no way" she would ever contact him through it. For one thing, Julia is upset that her father even has the account: "It doesn't seem like something a grown-up should have." What's more, to become her father's MySpace friend would give him too much access to her personal life, and she would have too much information about him. She is not sure she could resist the temptation to "stalk" him—to use the social-networking site to follow his comings and goings without his knowledge. When you're stalking, you follow links, moving from the postings of your prey to those of their friends. You look at photographs of parties and family events at which your prey might be a guest. To whom are they talking? Julia worries that she would try to investigate whether her father was seeing a new woman.

Despite all of this, Julia cannot not help herself from looking up her father's extended family on MySpace—his parents, siblings, cousins, aunts, and uncles.

She says she is not going to contact any of them, at least not until she e-mails her father. She wonders if MySpace might be a way to take small steps to knit together what had been torn apart in her childhood. And how might she talk about any of this with her mother?

As she describes a call to her father as "too much," Julia plays with her new cell phone. She has chosen one with a flip-up keyboard, optimized for texting. "I begged for this," she says. Julia texts her friends many times a day, almost constantly when she is not in class. Julia has to be careful. She explains that if she texts people on Verizon, where she has her account, the texts are free. Texting people on other carriers costs money. "I wish all my friends had Verizon," she says wistfully. Julia has a best friend on Cingular (a rival service), and says, "We don't text together." The solution: "We talk at school." Julia makes it clear that this is not a happy outcome.

I ask Julia about telephone calls. She doesn't like them at all. "I feel weird talking on the phone. My friends call and say, 'What's up?' and I'll say, 'Nothing.' Then I'll say, 'Okay, I gotta go. Bye.' I feel awkward talking on the phone. I just think it's easier to text." Julia's phone is always with her. If she is in class and a text arrives, Julia asks to go to the bathroom to check it out. The texts come in all day, with at least one vibration every five minutes. Knowing she has a message makes her "antsy." She starts to worry. She needs to read the message. Julia tells me that when she goes to the bathroom to check her texts, they are often from people just saying hello. She says, "This makes me feel foolish for having been so scared."

With Julia's permission, one of her teachers has been listening to our conversation about the phone. She asks, sensibly, "Why don't you turn it off?" Julia's answer is immediate: "It might be my mother. There might be an emergency." Her teacher persists gently: "But couldn't your mom call the school?" Julia does not hesitate: "Yeah, but what if it was one of my other friends having the emergency right in school?"

Julia describes the kinds of emergencies that compel her to respond to any signal from her phone. She talks about a hypothetical situation with a "friend" (later Julia will admit that she was describing herself): "Let's say she got into trouble. She knows she didn't do something, but she needs to tell somebody, she needs to tell me. Or, I know this one sounds kind of silly, but if she was having friend or boy trouble, she'd text me or call me. So those are the kind of things." Having a feeling without being able to share it is considered so difficult that it constitutes an "emergency."

Or something might happen to Joe, the father of her best friend Heather. Joe has had multiple heart attacks. "They told him if he has another one he'll probably die. So I'm always like, in my pocket, waiting for a call. Heather would either call me, or her mom would call me. 'Cause I'm really close to their family. Her dad's like my dad. And I love him. . . . So something like that would be an emergency."

Julia shows me her cell phone emergency contact list, which includes Heather, Heather's parents, and all of Heather's siblings. Julia says that she used to have Heather's uncle and aunt on her emergency list as well, "but I got a new phone and I don't have them anymore." She makes a note to get their numbers for her new phone. Along with her mother, these people are her safety net. Her cell phone embodies their presence.

Julia, her life marked by transitions and separations, always fears disconnection. She is poised to mourn. People can leave or be taken from her at any time. Her phone is tied up with a kind of magical thinking that if she can be in touch, the people she loves will not disappear.[3]

Julia's phone, a symbol of connection in a world on the brink, goes some distance toward making her feel safe. She says, "If there was ever an emergency in the school, I could always call 911, or if something happened, if there was a fire, or some strange guy came into the school I could always call my mom to tell her that I was okay, or not okay. So it's good like that too." As Julia talks about her anxieties of disconnection, she begins to talk about the 2001 terrorist attacks on the World Trade Center. When I interview teenagers about cell phones, I often hear stories about 9/11. Remembered through the prism of connectivity, 9/11 was a day when they could not be in touch. Many teachers and school administrators, in that generation that grew up hiding under desks in fear of an atomic attack, reacted to the news of the Twin Towers' collapse by isolating the children under their care. Students were taken out of classrooms and put in basements, the iconic hiding places of the Cold War. On 9/11 Julia spent many hours in such an improvised quarantine. Frightened, she and her classmates had no way to contact their parents. "I was in fourth grade," she said. "I didn't have a cell phone then. I needed to talk to my mother."

For Julia, 9/11 was all the more frightening because one of the girls in her class had an aunt who worked in the World Trade Center. And one of the boys had a relative who was supposed to be flying that day, but he didn't know to where. Only later in the afternoon, with communication restored, did Julia and her friends learn that everyone they had worried about was safe.

It was scary 'cause my teachers didn't know what was going on, and they all brought us into a room, and they didn't really know what to tell us. They just told us that there were bad guys bombing, crashing planes into buildings. We would ask questions, but they didn't know how to answer. We asked if they caught the bad guys. And our teacher said, "Yeah, they are in jail." But none of them could have been in jail because they crashed the plane. So our teachers really didn't know what was going on either.

The trauma of 9/11 is part of the story of connectivity culture. After the bombing of the World Trade Center, Americans accepted an unprecedented level of surveillance of both their persons and their communications. For Julia's generation, 9/11 marked childhood with an experience of being cut off from all comfort. In its shadow, cell phones became a symbol of physical and emotional safety. After the bombing of the World Trade Center, parents who really had not seen the point of giving cell phones to their children discovered a reason: continual contact. Julia took from her experience of 9/11 the conviction that it is "always good" to have your cell phone with you.

At schools all over the country, this is just what teachers try to discourage. They are in a tough spot because students are getting mixed messages. When parents give children phones, there is an implied message: "I love you, and this will make you safe. I give this to you because I care." Then schools want to take the phone away.[4] Some schools demand that phones be silenced and kept out of sight. Others ban them to locker rooms. At Julia's school, teachers try to convince students that they don't need their phones with them at all times. Julia quotes her teachers derisively: "They say, 'Oh, there's a phone in every class-room.'" But Julia makes it clear that she is having none of it: "I feel safer with my own phone. Because we can't all be using the same phone at once." This is a new nonnegotiable: to feel safe, you have to be connected. "If I got in a fight with somebody, I'd call my friend. I'd tell my friend if I got in trouble with the teacher. I'd tell my friend if there was a fight and I was scared. If I was threatened, I'd tell my friends. Or if someone came in and had a knife, I'd text my friends." For all of these imagined possibilities, the phone is comfort. Branscomb High School has metal detectors at its entrance. Uniformed security guards patrol the halls. There have been flare-ups; students have gotten into fights. As she and I speak, Julia's thoughts turn to Columbine and Virginia Tech: "I'm reading a book right now about a school. . . . It's about two kids who brought a gun to a dance and keep everyone hostage, and then killed themselves. And it's a lot like

Columbine. . . . We had an assembly about Columbine just recently. . . . At a time like that, I'd need my cell phone."

We read much about "helicopter parents."[5] They hail from a generation that does not want to repeat the mistakes of its parents (permitting too much independence too soon) and so hover over their children's lives. But today our children hover as well. They avoid disconnection at all cost. Some, like Julia, have divorced parents. Some have families broken twice or three times. Some have parents who support their families by working out of state or out of the country. Some have parents with travel schedules so demanding that their children rarely see them. Some have parents in the military who are stationed abroad. These teenagers live in a culture preoccupied with terrorism. They all experienced 9/11. They have grown up walking through metal detectors at schools and airports. They tend not to assume safe passage. The cell phone as amulet becomes emblematic of safety.

Julia tells her mother where she is at all times. She checks in after school, when she gets on the train, when she arrives home or at a friend's house. If going out, she calls "when we get to the place and when I get home." She says, "It's really hard to think about not having your cell phone. I couldn't picture not having it. . . . I feel like it's attached. Me and my friends say, 'I feel naked without it.'" The naked self feels itself in jeopardy. It is fragile and dependent on connection. Connection can reduce anxieties, but as I have said, it creates problems of its own.

WHAT'S IT ALL ABOUT?

Lisa, seventeen, a junior at the Cranston School, feels disoriented: "I come home from school and go online, and I'm feeling okay, and I talk for two hours on the Web. But then I still have no friends. I'll never know the people I spoke to. They are 'chat people.' Yeah, they could be twelve years old." Her investment in "chat people" leaves her with the question of what her online hours really add up to. It is a question that preoccupies Hannah, sixteen, another Cranston junior. She knows for sure that online connections help her with anxiety about boys. Many of her friends have boyfriends. She has not really started to date. At Cranston, having a boyfriend means pressure for sexual intimacy. She knows she is not ready but does feel left out.

Five years before, when she was eleven, Hannah made an online friend who called himself Ian. She joined an Internet Relay Chat (IRC) channel about 1960s rock bands, a particular passion of hers. Ian, who said he was fourteen at the time, was also on the channel. After a few years of getting to know each other in the group, Hannah says that she and Ian figured out how to create a private chat room. She says, "It felt like magic. All of a sudden we were in this room, by ourselves." Over time, Hannah and Ian got into the habit of talking and playing Scrabble every day, often for hours at a time. Ian said he lived in Liverpool and was about to go off to university. Hannah dreams of meeting him as soon as she goes to college, a year and a half from now, "when it won't seem strange for me to have friends from all over the world and friends who are older." Despite the fact that they have only communicated via typed-out messages, Hannah says, "Ian is the person who knows me best." Hannah doesn't want to add an audio or video channel to their encounters. As things are, Hannah is able to imagine Ian as she wishes him to be. And he can imagine her as he wishes her to be. The idea that we can be exactly what the other desires is a powerful fantasy. Among other things, it seems to promise that the other will never, ever, have reason to leave. Feeling secure as an object of desire (because the other is able to imagine you as the perfect embodiment of his or her desire) is one of the deep pleasures of Internet life.

Online, Hannah practices the kind of flirting that does not yet come easily to her in the real. The safety of the relationship with Ian allows her to explore what it might be like to have a boyfriend and give herself over to a crush. But Hannah also finds the friendship "a bit scary" because, she says, "the person I love most in the world could simply *not show up* on any given day," never to be heard from again. Ian boils down to a probably made-up first name and a history of warm conversations.

Hannah is wistful about Ian. "Even if I feel that I know Ian, I still don't feel that I know him the same way I know someone in real life." Sometimes she feels in a close friendship, and sometimes she sees it all as a house of cards. With the aspect of someone who has discovered that they don't like one of Newton's laws, she says, "I think it's kind of sad, but in order to have a really genuine relationship, there has to be some point where you're with your senses, experiencing some output that person's making with their body, like looking at their face, or hearing their words." Hannah falls silent. When she speaks again, her voice is low. There is something else. Her time on the IRC channel has cost her. The people on the channel are not nice. "I don't like it that I'm friends with mean people," she says. Her online friends "mock and kick and abuse newcomers," and sometimes they even turn against their own. Hannah doesn't think she will become the object of this kind of hostility. But this is not entirely comforting.

For she has become part of the tribe by behaving like its members. She says, "I do sometimes make crueler jokes on IRC than I would in real life. . . . That made me start thinking, 'Do I want to continue being friends with these people?' I started thinking about how vicious they can be. It's a little bit like if someone were in a clique at school, who would viciously reject or mock people they didn't want in their clique, I might say that's a little unfriendly, and why do you want to be friends with people who are so cruel?"

Hannah does not think that her "nicer" school friends are so different from her online friends. Like Marcia, she attributes their cruelty to the Internet because "it can bring out the worst in people. . . . Angers get worse. . . . There are no brakes." And now, she is spending close to twenty hours a week seeking the approval of people whose behavior she finds reprehensible. And whose behavior she emulates to curry favor. It is all exhausting. "Friendship on the Internet," says Hannah, "is much more demanding than in real life." And in the end, with all those demands met, she doesn't know what she really has.

Hannah thought that online friendships would make her feel more in control of her social life. Her "original assumption," she says, had been that she could be online when she felt like it but skip out "and not feel bad" when she was busy. This turned out not to be the case. Her online friends get angry if she doesn't show up in the chat room. And they are high maintenance in other ways. On IRC, they are fast-talking and judgmental. There is pressure to be witty. Hannah says, "I walk around school thinking about what I will talk about with them later." Beyond this, Hannah has recently joined Facebook, which only increases her preoccupations. Most Cranston students agree that the people who know you best in real life—school friends, for example—will be tolerant of an untended Facebook. The harsher judges are more distant acquaintances or people you hope to bring into your circle. These could be more popular students, students from other high schools, or those already in college. The consensus about them: "You know they're looking at all your stuff."

Hannah, sensitive to having all these new eyes on her, becomes drawn into what she describes as "an all-consuming effort to keep up appearances." She says, "On Facebook, things have gotten out of control. . . . You don't have to be on a lot, but you can't be on so little that your profile is totally lame. So, once you are on it, it makes you do enough so that you are not embarrassed." In this construction, tellingly, it is Facebook that has volition.

Other Cranston students describe similar pressures. For one senior boy, "You have to give to Facebook to get from Facebook." He continues, "If you don't use it, people are not going to communicate with you. People are going to see no one's communicating with you, and that, I think, leads to kids spending hours on Facebook every day trying to buff it out." Like a sleek, gym-toned body, an appealing online self requires work to achieve. A sophomore girl says, "I get anxious if my last wall post was from a week ago because it looks like you're a nerd. It really matters. People know it is a way that people are going to judge you." A senior boy painstakingly explains how to keep "your Facebook in shape." First, you have to conserve your energy. "It is a waste of time," he says, "to use Facebook messaging" because these messages are like e-mail, private between the correspondents. "They will do nothing for your image." The essential is "to spend some time every day writing things on other people's walls so that they will respond on your wall." If you do this religiously, you will look popular. If you don't, he says darkly, "there can be a problem." Another senior boy describes the anxieties that attend feeding the beast:

> I go on sometimes and I'm like, "My last wall post was a week ago." I'm thinking, "That's no good, everyone will see this and say, 'He doesn't have any friends.'" So I get really nervous about that and I'm thinking, "I have to write on somebody else's wall so they'll write back to me so it looks like I have friends again." That's my whole mentality on Facebook.

Hannah succumbed to this mentality and her time on Facebook got out of control. She explains how one thing led to another: "You're online. Someone asks you something. You feel like they want to know. It makes you feel good, so you keep on typing. . . . It's like being flattered for hours. But who are they really?" Now, increasingly anxious about that question, even her friendship with Ian seems tenuous. She has started to feel that by investing in Ian, she is becoming more isolated from what she calls "everyday in-person" connections. There are, she observes, "just so many hours in a day." In fact, when I meet her, Hannah is taking a break from the IRC channel. But she misses Ian and does not think it will last long.

HIDE AND STALK

Julia is afraid to "friend" her father on MySpace because she thinks she would not be able to resist the temptation to stalk him. Stalking is a guilty pleasure and a source of anxiety, but Chris, nineteen, a senior at Hadley, explains how it becomes routine. Every phone has a camera, and his friends take photographs all the time. They post their photos on Facebook and label them. This usually includes "tagging" each photograph with the names of all of the people in it. There are a lot of "tagged" photographs of Chris online, "pictures at parties, in the locker room, when I'm messing around with my friends." On Facebook, one can search for all the pictures of any given person. This is often where stalking begins. Chris is handsome and an accomplished athlete. He knows that a lot of girls look at his pictures. "The stalking is a little flattering, but it also makes me feel creeped out. . . . Some of the pictures creep me out, but everybody has all of these kinds of pictures online." And he is not in a position to cast the first stone. For he, too, stalks girls on Facebook who interest him: "I find myself choosing some girl I like and following the trail of her tagged pictures. You can see who she hangs with. Is she popular? Is there a chance she has a boyfriend? I start to do it in a sort of random way, and then, all of a sudden, a couple hours have passed. I'm stalking."

Chris does not judge himself harshly. The public display of locker room photos is "awful" but part of being popular. Also, having people look at you puts you in contact with them. Even when you are alone, you know that people are seeking you out. Teenagers seem to feel that things should be different but are reconciled to a new kind of life: the life they know celebrities live. So, you get used to the idea that if you are drunk or in erotic disarray—things that are likely to happen at some point during high school—someone will take a picture of you, probably by using the camera in their phone. And once on that person's phone, the image will find its way to the Internet, where you will lose control of its further travels.

So, stalking is a transgression that does not transgress. A seventeen-year-old junior at the Fillmore School describes it as "the worst. Normal, but still creepy." Normal because "it's not against the rules to look at people's wall-to-wall conversations [on Facebook]." Creepy because "it's like listening to a conversation that you are not in, and after stalking I feel like I need to take a shower." Just starting college, Dawn, eighteen, says she is "obsessed" with the "interesting people" who are her new classmates: "I spend all night reading people's walls. I track their parties. I check out their girlfriends." She, too, says, "My time on Facebook makes me feel dirty." So stalking may not be breaking any rules, but it has given young people a way to invade each other's privacy that can make them feel like spies and pornographers.

As teenagers turn stalking into part of their lives, they become resigned to incursions into their privacy. Julia says that at Branscomb "you get into trouble if there are MySpace pictures of you at a party where there is beer." She and her friends believe that school officials and the police look at students' MySpace accounts. Julia's response is to police herself and watch over her friends. "I'm, like, always telling them, 'Don't put that picture up there. You'll get into trouble.'" One Branscomb senior says that he has "a regular blog and a secret blog. On my secret blog I have a fake name," but later in our conversation he wonders whether his secret blog can be traced to him through the IP address that tags his computer. He hadn't thought of this before our conversation. He says that thinking about it makes him feel "hopeless."

At Roosevelt High School, sixteen-year-old Angela had her MySpace page "hacked." She explains, "'Hacked' is when people get on your page and change everything. Yeah, that happened to me once. I don't know who did it. But it happened. [voice gets quiet] They changed the whole layout. And they made it as though I was a lesbian. I had to go and erase everything. A lot of people asked me, 'Oh, are you a lesbian now?' I had to explain to everyone, 'No, I got hacked.' It took me a long time to explain. And they'd say, 'Oh, that sucks.'"

When people tamper with your physical mail, they have committed a crime. When people hack your social-networking account, *you* have explaining to do. When Angela first blurted out her story, it was clear that the incident had frightened her. Then, she backtracked and minimized what had happened, saying, "It doesn't really happen every day." This is the defense of those who feel they have no options. Angela is not going to give up MySpace. Anger will serve no purpose. So, instead, she reinterprets what happened to her. She had been inconvenienced. "I was mad because now I had to do everything all over again, but I didn't really care that they did it. It doesn't really happen every day. . . . It doesn't really happen every day."

I hear a similar kind of backpedaling in a discussion of online life at the Silver Academy. When I ask a group of sophomores, "Are any of you worried about your online privacy?" they call out, "Yeah, yes, yeah." Carla and Penny rush to tell their story. They are so excited that they begin to speak together. Then they settle down, and Carla takes over: "I went out to the store with my mother, and I left my phone at home, and Penny here texted me. And I didn't have my phone, but my brother was right near it, and it was buzzing. So my brother decides to text her back as me. And she said something, and my brother was being very rude. And I had to call her up later and tell her that it was my brother texting, not me." At first, the two girls seem to want everyone to know that this is a very upsetting story. But when the group listens with little visible emotion—everyone there has heard of a similar story—the girls retreat. Penny says that Carla's brother was not artful in his impersonation, so maybe she would have figured it out. Carla, now isolated in her anger, backs down. "Yeah, I guess so."

The media has tended to portray today's young adults as a generation that no longer cares about privacy. I have found something else, something equally disquieting. High school and college students don't really understand the rules. Are they being watched? Who is watching? Do you have to do something to provoke surveillance, or is it routine? Is surveillance legal? They don't really understand the terms of service for Facebook or Gmail, the mail service that Google provides. They don't know what protections they are "entitled" to. They don't know what objections are reasonable or possible. If someone impersonates you by getting access to your cell phone, should that behavior be treated as illegal or as a prank? In teenagers' experience, their elders—the generation that gave them this technology—don't have ready answers to such questions.

So Julia, despite worrying that school authorities and the police look over students' online profiles, is quick to admit that she is not really sure that this is the case. But then she adds that that no matter what the truth might be, there is nothing she can do about it. One seventeen-year-old, "scrubbing" her Facebook account under the orders of her high school guidance counselor (concerned about compromising photographs that should be removed before the college-admissions process), is convinced that anyone with enough time and money can find a way onto her Facebook page without her permission. "People keep talking about how colleges look at it and employers too. I guess they just have people signing on, pretending to be friends. I don't really know how it works."

There is an upside to vagueness. What you don't know won't make you angry. Julia says, "Facebook and MySpace are my life." If she learned something too upsetting about what, say, Facebook can do with her information, she would have to justify staying on the site. But Julia admits that whatever she finds out, even if her worst fears of surveillance by high school administrators and local police were true, she would not take action. She cannot imagine her life without Facebook.

Julia ends up a portrait of insecurity and passivity. She wants to hide from the details. She would rather just be careful about what she does than learn too much about who is actually watching. "I put it out of my mind," she says. She tells me that she, personally, feels safe because "I'm kind of boring." That is, it makes no difference if she is watched because there is nothing much to see. A sixteen-year-old girl shrugs off Facebook's privacy policy in similar terms: "Who would care about me and my little life?" Another sixteen-year-old, a boy, says that when he wants to have a private conversation he knows that he has to find a pay phone—"the old fashioned kind" that takes coins. These are disturbing mantras.

Some teenagers say that their privacy concerns are not as bad as they might seem because, in the future, everyone running for office, everyone about to get a judicial appointment or an important corporate job, will have an accessible Internet past with significant indiscretions.[6] In this narrative, implacable digital memory will not be punishing but will create a more tolerant society. Others come up with a generational argument: "Facebook is owned by young people." This idea confuses investors, owners, managers, inventors, spokespeople, and shareholders. It is innocent of any understanding of how corporations work or are governed. But it is not a surprising response. If your life is on Facebook or MySpace or Google, you want to feel that these companies are controlled by good people. Good people are defined as those who share what you feel is your most salient characteristic. For the young, that characteristic is youth.

In fact, from the very beginning, Facebook has been in something of a tug-of-war with its users about how much control it has over their data. The pattern, predictably, is that Facebook declares ownership of all of it and tries to put it to commercial use. Then, there is resistance and Facebook retreats. This is followed by another advance, usually with subtler contours. One sixteen-year-old says, and her comment is typical, "Oh, they [Facebook] keep changing the policy all the time. You can try to change their policy, but usually they just put the policy in fine print." She herself doesn't read the fine print. She assumes that in the end, Facebook will take what it wants. "You can try to get Facebook to change things. Maybe after years they will. Maybe they won't. This is just the way it is." Google's advances and retreats in this arena show a similar pattern.[7] As long as Facebook and Google are seen as necessities, if they demand information, young people know they will supply it. They don't know what else to do.

Some Internet entrepreneurs have made the case that there is not much to do.[8] As early as 1999, Scott McNealy, a cofounder of Sun Microsystems, said,

"You have zero privacy anyway; get over it."[9] A decade later, Eric Schmidt, the CEO of Google, added a new spin: "If you have something you don't want anyone to know, maybe you shouldn't be doing it in the first place." Most recently he is on record predicting that in the near future all young people will be automatically entitled to change their names to escape their online pasts.[10]

PRIVACY AND THE ANXIETIES OF ALWAYS

In the early 1990s, I began studying people experimenting with identity on the Internet. They created avatars and Web pages. They played with romance and revenge. In those early days, it was commonplace for Web sites and virtual locales to disappear because the enthusiasts who ran them lost interest, lost access to a server, or invented something new. When this happened, people migrated to other online places. These migrations could mean "losing" all the work you had put into an avatar and a virtual community. The Internet seemed transient.

The Facebook generation goes online with different expectations. They expect Facebook or its successor company to be there forever. This expectation is incentive to "behave." Of course, people slip up and repent at leisure. Gloria, eighteen, contemplates the things she has posted on Facebook and says, "It's like the Internet could blackmail me." She has grown more careful. She cannot imagine doing something in public that will not end up on Facebook. Any time she goes out to a dance or a party or a coffee shop, friends are taking pictures and posting them. She does not want to misbehave in a way that would cause Facebook to want her off the system.

Hester, eighteen, a college freshman, says that she has started to worry about all the things she has put on the Internet that it is "too late to take away." She says, "That's the one bad thing [about online life]. On a typewriter, you can take the paper out and shred it. But if it's online, it's online. People can copy and paste it; people can e-mail it to each other; people can print it. . . . You need to be careful what you write on the Internet because most of the things . . . if you put it on the Internet, that's it. A lot of people . . . they may or may not have access to it, but still, it's there." This is life in the world of cut and paste. Worse, this is life in the world of cut, edit, and paste. A senior at the Hadley School reviews what can happen to an online conversation: "People can save it, and you don't know they're saving it. Or people can copy and paste it and send it to someone else. You think it is private but its not. . . . And all they have to do is rewrite anything they want. They can send it to a friend, making a person look a lot worse. Nothing you say will necessarily stay the way you said it."

A junior girl at Roosevelt High School is worried: "My SAT tutor told me never to say anything stupid on e-mail because it can always be available to people. Which is a little alarming, because I'm writing every day to my friend in Toronto, and of course I'm mentioning other friends and sometimes summarizing them in ways I wouldn't want them to see, and so I'm kind of hoping no one discovers that." A Roosevelt freshman, already aware that the Internet is a "permanent record," decides to commit her most private thoughts to paper: "I keep my secrets in my diary, not on my computer and not on my website."

We have met eighteen-year-old Brad, wary of the Internet. He knows that his online life is not private. Most of the time, he manages not to think about it. But most recently, he is troubled. What bothers him most are his friends' use of "chat logs." Brad explains: "Anytime you type something, even without your having done anything or agreed to anything, it [the chat log] saves it to a folder." Brad was unaware that there was such a thing until a conversation with a friend brought him up short. At the time, they were both high school juniors, and she mentioned something he had said during freshman year. She had been using chat logs all through high school. Brad says, "I was shocked that this was how she was spending her time . . . going through conversations like that." Now, he is torn between feeling upset that he had been unknowingly "recorded" and feeling angry at himself for being surprised. "After all," he says "I know how IM conversations work. . . . I think I had heard of this but forgot it. I know there's a very good chance . . . that I know certain [people] who have chat logs turned on."

Brad blames himself for being too free in his messaging. The idea that his sophomore year ramblings could find their way onto somebody's Facebook page or blog or "wherever" is intolerable to him. Brad doesn't have a very clear image of what bad things might happen, but his anxiety is real. He says that data capture is "awful." His words could show up anywhere.

Brad says that he no longer sees online life as a place to relax and be himself "because things get recorded. . . . It's just another thing you have to keep in the back of your mind, that you have to do things very carefully." In person, if he loses his temper or is misunderstood in a conversation, he says, "I can be like, 'I'm sorry' or 'Let me repeat myself' . . . or I can crack a joke and laugh it off." Online, even if a person isn't recording you, Facebook is. "I've heard plenty of stories about people leaving messages or posts on people's walls on Facebook that, the next day, they felt bad about, because they felt it was stupid of them. It was spur-of-the-moment, they lost their head or something like that." But there it was, representing you at your worst.

Brad acknowledges that "of course, if you say or do something stupid in person," you can be reminded of it later, but in face-to-face communication, he sees a lot of "wiggle room" for "general human error." Online, it is always possible that people are collecting "visual proof . . . saved written proof" of your mistake. Brad steps back from blaming either what the technology makes possible or the people who record you without permission. He says he is a "realist." By this he means that "anyone who lives in the digital world should know that it is not permissible to lose your temper online or say anything that you would not want to be distributed." And besides, says Brad, "there is never any reason to use online communication for spontaneous feeling. . . . You have no excuse for snapping online because you could have just waited for a couple minutes and not typed anything and cooled down." Here, we see self-policing to the point of trying to achieve a precorrected self.

When Brad talks about "visual proof . . . saved written proof" of damaging exchanges, he sounds like someone hunted. I ask him about people saving his letters. He says that this does not bother him. In a letter, he explains, he thinks before he writes, and sometimes he writes a letter over several times. But to him, even though he "knew better," Internet conversations feel tentative; you get into the habit of thinking as you write. Although everything is "composed," he somehow gets into "an experience of being in a free zone." Audrey, sixteen, described a similar disconnect. She *feels* that online life is a space for experimentation. But she *knows* that electronic messages are forever and that colleges and potential employers have ways of getting onto her Facebook page. What she feels and what she knows do not sync.

Brad and Audrey both experience the paradox of electronic messaging. You stare at a screen on your desk or in your hand. It is passive, and you own the frame; these promise safety and acceptance. In the cocoon of electronic messaging, we imagine the people we write to as we wish them to be; we write to that part of them that makes us feel safe. You feel in a place that is private and ephemeral. But your communications are public and forever. This disconnect between the feeling of digital communication and its reality explains why people continue to send damaging e-mails and texts, messages that document them breaking the law and cheating on their spouses. People try to force themselves to mesh their behavior with what they know rather than how they feel.

But when people want to forget that they do not have privacy on the Internet, the medium colludes.

Recall seventeen-year-old Elaine, who thought that the Internet made it easier for the shy to make friends because they have fewer inhibitions when they can hide behind a screen. Elaine's sense of this "free" space is conflicted. For example, she knows that everything she puts on a site like Facebook will always be there and belong to Facebook. But Elaine has no confidence that, once online, she will be able to remember that she is speaking to posterity. The Internet might be forever, but it takes discipline to keep this in mind. She thinks it is unrealistic to say, "What happens on the Internet, stays on the Internet." She says that this is "just too hard. . . . It's just human nature that things will get out." She is skeptical of those who say they are able to place a wall between their offline and online lives: "Everything that is on the Internet, everyone can copy and paste, or save. . . . If you're having a conversation with someone in speech, and it's not being tape-recorded, you can change your opinion, but on the Internet it's not like that. On the Internet it's almost as if everything you say were being tape-recorded. You can't say, 'I changed my mind.' You can, but at the same time it's already there."

There is truth in a view of the Internet as a place for experimentation and self-expression. Yet, from Elaine's point of view, what she is free to do is to say things that will "be remembered forever." Common sense prevails: "free" combined with "forever" doesn't seem workable. Elaine says, "I feel that my childhood has been stolen by the Internet. I shouldn't have to be thinking about these things." Dawn tried to "scrub" her Facebook page when she got into college. "I wanted a fresh start," she says. But she could only delete so much. Her friends had pictures of her on their pages and messages from her on their walls. All of these would remain. She says, "It's like somebody is about to find a horrible secret that I didn't know I left someplace."

Here, as in Brad's unforgiving self-criticism ("I should have known . . . you have no excuse . . . "), one sees a new regime of self-surveillance at work. As toddlers, these children learned how to type online, and then they discovered it was forever. We see a first generation going through adolescence knowing that their every misstep, all the awkward gestures of their youth, are being frozen in a computer's memory. Some put this out of mind, but some cannot, do not—and I think, should not.

It has taken a generation for people to begin to understand that on the Internet, the words "delete" and "erase" are metaphorical: files, photographs, mail, and search histories are only removed from your sight.[11] The Internet never for-gets. The magnitude of this is hard to believe because one's first instinct is to find it unbelievable. Some teenagers deny what is happening; some respond by finding it "unfair" that they will, like turtles, be carrying themselves on their backs all their lives. Corbin, a senior at Hadley, comments on the idea that noth-ing on the Net will ever go way. He says, "All the things I've written on Facebook will always be around. So you can never escape what you did."

With the persistence of data, there is, too, the persistence of people. If you friend someone as a ten-year-old, it takes positive action to unfriend that person. In principle, everyone wants to stay in touch with the people they grew up with, but social networking makes the idea of "people from one's past" close to an anachronism. Corbin reaches for a way to express his discomfort. He says, "For the first time, people will stay your friends. It makes it harder to let go of your life and move on." Sanjay, sixteen, who wonders if he will be "writing on my friends' walls when I'm a grown-up," sums up his misgivings: "For the first time people can stay in touch with people all of their lives. But it used to be good that people could leave their high school friends behind and take on new identities."

This is the anxiety of always. A decade ago I argued that the fluidity, flexibility, and multiplicity of our lives on the screen encouraged the kind of self that Robert Jay Lifton called "protean."[12] I still think it is a useful metaphor. But the protean self is challenged by the persistence of people and data. The sense of being protean is sustained by an illusion with an uncertain future. The experi-ence of being at one's computer or cell phone feels so private that we easily forget our true circumstance: with every connection we leave an electronic trace.

Similarly, I have argued that the Internet provided spaces for adolescents to experiment with identity relatively free of consequences, as Erik Erikson argued they must have. The persistence of data and people undermines this possibility as well. I talk to teenagers who send and receive six to eight thousand texts a month, spend hours a day on Facebook, and interleave instant messaging and Google searches—all activities that leave a trace. The idea of the moratorium does not easily mesh with a life that generates its own electronic shadow.

Peter Pan, who could not see his shadow, was the boy who never grew up. Most of us are like him. Over time (and I say this with much anxiety), living with an electronic shadow begins to feel so natural that the shadow seems to disappear—that is, until a moment of crisis: a lawsuit, a scandal, an investigation. Then, we are caught short, turn around, and see that we have been the instru-ments of our own surveillance. But most of the time, we behave as if the shadow were not there rather than simply invisible. Indeed, most of the adolescents who worry with me about the persistence of online data try to put it out of their minds. The need for a moratorium space is so compelling that if they must, they are willing to find it in a fiction. This is an understandable and unstable resolution. The idea that you leave a trace because you make a call, send a text, or leave a Facebook message is on some level intolerable. And so, people simply behave as though it were not happening.

Adults, too, live the fiction. Some behave as though e-mail were private, although they know it is not. Others say they never have significant business or personal conversations electronically. They insist that for anything important, they speak on a secure landline. But then, as we talk, they usually admit to the times that they haven't followed their own rules. Most often, there is a shame-faced admission of an indiscretion on e-mail.

Some say this issue is a nonissue; they point out that privacy is a historically new idea. This is true. But although historically new, privacy has well served our modern notions of intimacy and democracy. Without privacy, the borders of intimacy blur. And, of course, when all information is collected, everyone can be turned into an informer.

PRIVACY HAS A POLITICS

It has become commonplace to talk about all the good the Web has done for politics. We have new sources of information, such as news of political events from all over the world that comes to us via photographs and videos taken by the cameras on cell phones. There is organizing and fund-raising; ever since the 2004 primary run of Howard Dean, online connections have been used as a first step in bringing people together physically. The Barack Obama campaign transformed the Dean-era idea of the "meet up" into a tool for bringing supporters out of the virtual and into each other's homes or onto the streets. We diminish none of these very positive developments if we attend to the troubling realities of the Internet when it comes to questions of privacy. Beyond passivity and resignation, there is a chilling effect on political speech.

When they talk about the Internet, young people make a disturbing distinction between embarrassing behavior that will be forgiven and political behavior that might get you into trouble. For high school and college students, stalking and anything else they do *to each other* fall into the first category. Code such antics as embarrassing. They believe that you can apologize for embarrassing behavior and then move on. Celebrity culture, after all, is all about transgression and rehabilitation. (These young people's comfort with "bullying" their peers is part of this pattern—something for which they believe they will be forgiven.) But you can't "take back" political behavior, like signing a petition or being at a demonstration. One eighteen-year-old puts it this way: "It [the Internet] definitely makes you think about going to a protest or something. There would be so many cameras. You can't tell where the pictures could show up."

Privacy has a politics. For many, the idea "we're all being observed all the time anyway, so who needs privacy?" has become a commonplace. But this state of mind has a cost. At a Webby Awards ceremony, an event to recognize the best and most influential websites, I was reminded of just how costly it is. The year I attended the Webbies, the ceremonies took place just as a government wiretapping scandal dominated the press. When the question of illegal eavesdropping arose, a common reaction among the gathered "Weberati" was to turn the issue into a nonissue. There was much talk about "all information being good information," "information wanting to be free," and "if you have nothing to hide, you have nothing to fear." At a pre-awards cocktail party, one Web luminary spoke to me with animation about the wiretapping controversy. To my surprise, he cited Michel Foucault on the panopticon to explain why he was not worried about privacy on the Internet.

For Foucault, the task of the modern state is to reduce its need for actual surveillance by creating a citizenry that will watch itself. A disciplined citizen minds the rules. Foucault wrote about Jeremy Bentham's design for a panopticon because it captured how such a citizenry is shaped.[13] In the panopticon, a wheel-like structure with an observer at its hub, one develops the sense of always being watched, whether or not the observer is actually present. If the structure is a prison, inmates know that a guard can potentially always see them. In the end, the architecture encourages self-surveillance.[14]

The panopticon serves as a metaphor for how, in the modern state, every citizen becomes his or her own policeman. Force becomes unnecessary because the state creates its own obedient citizenry. Always available for scrutiny, all turn their eyes on themselves. By analogy, said my Webby conversation partner, on the Internet, someone might always be watching, so it doesn't matter if, from time to time, someone actually is. As long as you are not doing anything wrong, you are safe. Foucault's critical take on disciplinary society had, in the hands of this technology guru, become a justification for the U.S. government to use the Internet to spy on its citizens. All around us at the cocktail party, there were nods of assent. We have seen that variants of this way of thinking, very common in the technology community, are gaining popularity among high school and college students.

If you relinquish your privacy on MySpace or Facebook about everything from your musical preferences to your sexual hang-ups, you are less likely to be troubled by an anonymous government agency knowing whom you call or what websites you frequent. Some are even gratified by a certain public exposure; it feels like validation, not violation. Being seen means that they are not insignificant or alone. For all the talk of a generation empowered by the Net, any discussion of online privacy generates claims of resignation and impotence. When I talk to teenagers about the certainty that their privacy will be invaded, I think of my very different experience growing up in Brooklyn in the 1950s.

As the McCarthy era swirled about them, my grandparents were frightened. From Eastern European backgrounds, they saw the McCarthy hearings not as a defense of patriotism but as an attack on people's rights. Joseph McCarthy was spying on Americans, and having the government spy on its citizens was familiar from the old world. There, you assumed that the government read your mail, which never led to good. In America, things were different. I lived with my grandparents as a young child in a large apartment building. Every morning, my grandmother took me downstairs to the mailboxes. Looking at the gleaming brass doors, on which, she noted, "people were not afraid to have their names listed, for all to see," my grandmother would tell me, as if it had never come up before, "In America, no one can look at your mail. It's a federal offense. That's the beauty of this country." From the earliest age, my civics lessons at the mailbox linked privacy and civil liberties. I think of how different things are today for children who learn to live with the idea that their e-mail and messages are shareable and unprotected. And I think of the Internet guru at the Webby awards who, citing Foucault with no apparent irony, accepted the idea that the Internet has fulfilled the dream of the panopticon and summed up his political position about the Net as follows: "The way to deal is to just be good."

But sometimes a citizenry should not simply "be good." You have to leave space for dissent, real dissent. There needs to be technical space (a sacrosanct mailbox) and mental space. The two are intertwined. We make our technologies, and they, in turn, make and shape us. My grandmother made me an American citizen, a civil libertarian, a defender of individual rights in an apartment lobby in Brooklyn. I am not sure where to take my eighteen-year-old daughter, who still thinks that Loopt (the application that uses the GPS capability of the iPhone to show her where her friends are) seems "creepy" but notes that it would be hard to keep it off her phone if all her friends had it. "They would think I had something to hide."

In democracy, perhaps we all need to begin with the assumption that everyone has something to hide, a zone of private action and reflection, one that must be protected no matter what our techno-enthusiasms. I am haunted by the sixteen-year-old boy who told me that when he needs to make a private call, he uses a pay phone that takes coins and complains how hard it is to find one in Boston. And I am haunted by the girl who summed up her reaction to losing online privacy by asking, "Who would care about me and my little life?"

I learned to be a citizen at the Brooklyn mailboxes. To me, opening up a conversation about technology, privacy, and civil society is not romantically nostalgic, not Luddite in the least. It seems like part of democracy defining its sacred spaces.

CHAPTER 14

the nostalgia of the young

Cliff, a Silver Academy sophomore, talks about whether it will ever be possible to get back to what came "before texting." Cliff says that he gets so caught up in the back-and-forth of texting that he ends up wasting time in what he thinks are superficial communications "just to get back." I ask him about when, in his view, there might be less pressure for an immediate response. Cliff thinks of two: "Your class has a test. Or you lost your signal." Conspicuously absent—you are doing something else, thinking something else, with someone else.

We have seen young people walk the halls of their schools composing messages to online acquaintances they will never meet. We have seen them feeling more alive when connected, then disoriented and alone when they leave their screens. Some live more than half their waking hours in virtual places. But they also talk wistfully about letters, face-to-face meetings, and the privacy of pay phones. Tethered selves, they try to conjure a future different from the one they see coming by building on a past they never knew. In it, they have time alone, with nature, with each other, and with their families.

Texting is too seductive. It makes a promise that generates its own demand.[1] The promise: the person you text will receive the message within seconds, and whether or not he or she is "free," the recipient will be able to see your text. The demand: when you receive a text, you will attend to it (during class, this might mean a glance down at a silenced phone) and respond as soon as possible. Cliff says that in his circle of friends, that means, "ten minutes, maximum."

> I will tell you how it is at this school. If something comes in on our phone and it's a text, you feel you have to respond. They obviously know you got it. With IM, you can claim you weren't at the computer or you lost your Internet connection and all that. But if it's a text, there's no way you didn't get it. Few people look down at their phone and then walk away from it. Few people do that. It really doesn't happen. . . . Texting is pressure. I don't always feel like communicating. Who says that we always have to be ready to communicate?

Indeed, who says? Listening to what young people miss may teach us what they need. They need attention.

ATTENTION

Teenagers know that when they communicate by instant message, they compete with many other windows on a computer screen. They know how little attention they are getting because they know how little they give to the instant messages they receive. One sophomore girl at Branscomb High School compares instant messaging to being on "cruise control" or "automatic pilot." Your attention is elsewhere. A Branscomb senior says, "Even if I give my full attention to the person I am IMing . . . they are not giving full attention to me." The first thing he does when he makes a call is to gauge whether the person on the other end "is there just for me." This is one advantage of a call. When you text or instant-message, you have no way to tell how much else is going on for the person writing you. He or she could also be on the phone, doing homework, watching TV, or in the midst of other online conversations.

Longed for here is the pleasure of full attention, coveted and rare. These teenagers grew up with parents who talked on their cell phones and scrolled through messages as they walked to the playground. Parents texted with one hand and pushed swings with the other. They glanced up at the jungle gym as they made calls. Teenagers describe childhoods with parents who were on their mobile devices while driving them to school or as the family watched Disney videos. A college freshman jokes that her father read her the Harry Potter novels, periodically interrupted by his BlackBerry. BlackBerries and laptops came on family vacations. Weekends in the country were cut short if there was no Internet service in the hotel. Lon, eighteen, says when that happened, his father "called it a day." He packed up the family and went home, back to a world of connections.

From the youngest ages, these teenagers have associated technology with shared attention. Phones, before they become an essential element in a child's own life, were the competition, one that children didn't necessarily feel they could best. And things are not so different in the teenage years. Nick, seventeen, says, "My parents text while we eat. I'm used to it. My dad says it is better than his having to be at the office. I say, 'Well, maybe it could just be a short meal.' But my mom, she wants long meals. To get a long meal with a lot of courses, she has to allow the BlackBerry." Things seem at a stalemate.

Children have always competed for their parents' attention, but this generation has experienced something new. Previously, children had to deal with parents being off with work, friends, or each other. Today, children contend with parents who are physically close, tantalizingly so, but mentally elsewhere. Hannah's description of how her mother doesn't look up from her BlackBerry to say hello when she picks her up at school highlights a painful contrast between the woman who goes to the trouble to fetch her daughter and the woman who cannot look up from her screen. Lon says he liked it better when his father had a desktop computer. It meant that he worked from a specific place. Now his father sits next to him on the couch watching a football game but is on his BlackBerry as well. Because they are physically close, his father's turn to the BlackBerry seems particularly excluding.

Miguel, a Hadley senior, says that having his father scroll through his Black-Berry messages during television sports is "stressful" but adds "not the kind that really kills you. More the kind that always bothers you." Miguel says it is hard for him to ask his father to put the BlackBerry away because he himself texts when he is with his father in the car. "He has a son who texts, so why shouldn't he?" But when parents see their children checking their mobile devices and thus feel permission to use their own, the adults are discounting a crucial asymmetry. The multitasking teenagers are just that, teenagers. They want and need adult attention. They are willing to admit that they are often relieved when a parent asks them to put away the phone and sit down to talk. But for parents to make this request—and this no longer goes without saying—they have to put down their phones as well. Sometimes it is children (often in alliance with their mothers) who find a way to insist that dinner time be a time for talking—time away from the smartphone. But habits of shared attention die hard.

One high school senior recalls a time when his father used to sit next to him on the couch, reading. "He read for pleasure and didn't mind being interrupted." But when his father, a doctor, switched from books to his BlackBerry, things became less clear: "He could be playing a game or looking at a patient record, and you would never know. . . . He is in that same BlackBerry zone." It takes work to bring his father out of that zone. When he emerges, he needs time to refocus. "You might ask him a question and he'll say, 'Yeah, one second.' And then he'll finish typing his e-mail or whatever, he'll log off whatever, and he'll say, 'Yeah, I'm sorry, what did you say?'"

It is commonplace to hear children, from the age of eight through the teen years, describe the frustration of trying to get the attention of their multitasking parents. Now, these same children are insecure about having each other's attention. At night, as they sit at computer screens, any messages sent or received share "mind space" with shopping, uploading photos, updating Facebook, watching videos, playing games, and doing homework. One high school senior describes evening "conversation" at his machine: "When I'm IMing, I can be talking to three different people at the same time and listening to music and also looking at a website." During the day, prime time for phone texting, communications happen as teenagers are on their way from one thing to another. Teenagers talk about what they are losing when they text: how someone stands, the tone of their voice, the expression on their face, "the things your eyes and ears tell you," as one eighteen-year-old puts it.

When I first encountered texting, I thought it too telegraphic to be much more than a way to check in. You could use it to confirm an appointment, settle on a restaurant, or say you were home safely. I was wrong. Texting has evolved into a space for confessions, breakups, and declarations of love. There is something to celebrate here: a new, exuberant space for friendship, a way to blow a virtual kiss. But there is a price. All matters—some delicate, some not—are crammed into a medium that quickly communicates a state but is not well suited for opening a dialogue about complexity of feeling. Texting—interrupted by bad reception, incoming calls, and other text messages (not to mention the fact that it all goes on in the presence other people)—can compromise the intimacy it promises. There is a difference, says an eighteen-year-old boy, "between someone laughing and someone *writing* that they're laughing." He says, "My friends are so used to giving their phones all the attention . . . they forget that people are still there to give attention to."

We met Robin, twenty-six, who works as a copywriter in a large and highly competitive advertising agency. She describes the demands of her job as "crushing." She has her BlackBerry with her at all times. She does not put it in her purse; she holds it. At meals, she sets it on the table near her, touching it frequently. At a business lunch, she explains that she needs to leave it on because her job requires her to be "on call" at all times. During lunch, she admits that there is more to the story. Her job certainly requires that she stay in touch. But now, whether or not she is waiting for a message from work, she becomes anxious without her BlackBerry. "If I'm not in touch, I feel almost dizzy. As though something is wrong, something terrible is wrong." The device has become a way to manage anxiety about her parents, her job, and her love life. Even if these don't go quite right, she says, "if I have the BlackBerry in control, I feel that at least everything isn't out of control." But something has gotten out of control. When Robin thinks of stress, she thinks of being without her BlackBerry. But she admits that she thinks of being with her BlackBerry as well.

Robin says that her need for the BlackBerry began with business e-mail, but now she uses it to spend many hours a day on Facebook. She makes no pretense that this is about "business." But Robin is no longer sure it is about pleasure. She describes being increasingly "annoyed" on Facebook. I ask her for an example— one of these moments of annoyance—and Robin begins to talk about her friend Joanne.

Robin and Joanne went to college together in Los Angeles. After graduating, Robin went to Chicago for a first job in publishing; Joanne stayed on the West Coast for graduate school in anthropology. Five years ago, Joanne's dissertation research took her to a village in Thailand. Joanne had e-mail access during her year in the village, and she wrote Robin long, detailed e-mails, five or six pages each. There was a letter every two weeks—a personal journal of Joanne's experience of Thai life. Robin describes them warmly—the letters were "elegant, detailed, poetic." Robin printed out the cherished letters; on occasion she still rereads them. Now Joanne is back in Thailand on a new project, but this time, she posts a biweekly journal to her Facebook page. There has been no falling out between the two women; Joanne has simply chosen a more "efficient" way to get her story out to all her friends. Robin still gets an occasional e-mail. But essentially, what was once a personal letter has turned into a blog.

Robin says she is ashamed of her reaction to Joanne's Facebook postings: "I was jealous of all of the other readers. They are not friends the way I am a friend." Robin understands Joanne's decision to "publish" her journal: "She is reaching more people this way. . . . Some can help in her career." But despite herself, Robin feels abandoned. The all-friend postings do not make her feel close to her friend.

After she tells this story, essentially about a personal loss, Robin adds a post-script that she describes as "not personal. I'm trying to make a general point." She says that when Joanne wrote her letters, they were "from a real person to another real person." They were written to her, in all her particularity. Behind each letter was the history of their long friendship. The new letters on Facebook are generic. For a moment, Robin, the professional writer, allows herself a moment of judgment: "The journal is written to everyone and thus no one. It isn't as good." Robin misses receiving something that was just for her.

SPONTANEITY

In a discussion of online life among seniors at the Fillmore School, Brendan says he is lonely. He attempts humor, describing a typical day as "lost in translation": "My life is about 'I'll send you a quick message, you send me another one in fifteen minutes, an hour, whatever. And then I'll get back to you when I can.'" His humor fades. Texting depresses him. It doesn't make him "feel close," but he is certain that it takes him away from things that might. Brendan wants to see friends in person or have phone conversations in which they are not all rushing off to do something else. Here again, nostalgia circles around attention, commitment, and the aesthetic of doing one thing at a time. Truman, one of Brendan's classmates, thinks his friend is asking too much. Truman says, "Brendan . . . calls me up sometimes, and it's really fun, and I really enjoy it, but it's something I can't really imagine myself doing. . . . Well, it seems like an awkward situation to me, to call someone up just to talk." Truman wants to indulge his friend, but he jokes that Brendan shouldn't "bet on long telephone conversations anytime soon." Truman's remarks require some unpacking. He says he likes the telephone, but he doesn't really. He says conversation is fun, but it's mostly stressful. For Truman, anything other than "a set-up call, a call to make a plan, or tell a location" presumes you are calling someone who has time for you. He is never sure this is the case. So, he worries that this kind of call intrudes. It puts you on the line. You can get hurt.

When young people are insecure, they find ways to manufacture love tests—personal metrics to reassure themselves. These days I hear teenagers measuring degrees of caring by type of communication. An instant message puts you in one window among many. An extended telephone call or a letter—these rare and difficult things—demonstrates full attention. Brad, the Hadley senior taking a break from Facebook, says, "Getting a letter is so special because it is meant only for you. . . . It feels so complimentary, especially nowadays, with people multitasking more and more, for someone to actually go out of their way and give their full attention to something for your sake for five or ten minutes. What is flattering is that they take that amount of time . . . that they're actually giving up that time."

Herb, part of the senior group at Fillmore feels similarly; he and his girlfriend have decided to correspond with letters: "The letter, like, she wrote it, she took her time writing it, and you know it came from her. The e-mail, it's impersonal. Same with a text message, it's impersonal. Anyone, by some chance, someone got her e-mail address, they could've sent it. The fact that you can touch it is really important. . . . E-mails get deleted, but letters get stored in a drawer. It's real; it's tangible. Online, you can't touch the computer screen, but you can touch the letter." His classmate Luis agrees: "There is something about sending a letter. You can use your handwriting. You can decorate a letter. Your handwriting can show where you are." It comes out that he has never received a personal letter. He says, "*I miss those days even though I wasn't alive.*" He goes on, a bit defensively because he fears that his fondness for handwriting might make him seem odd: "Before, you could just feel that way, it was part of the culture. Now, you have to feel like a throwback to something you really didn't grow up with."

Brad says that digital life cheats people out of learning how to read a person's face and "their nuances of feeling." And it cheats people out of what he calls "passively being yourself." It is a curious locution. I come to understand that he means it as shorthand for authenticity. It refers to who you are when you are not "trying," not performing. It refers to who you are when you are in a simple conversation, unplanned. His classmate Miguel likes texting as a "place to hide," but to feel close to someone, you need a more spontaneous medium:

> A phone conversation is so personal because you don't have time to sit there and think about what you're going to say. What you have to say is just going to come out the way it's meant to. If someone sends you a text message, you have a couple of minutes to think about what you're going to say, whereas if you're in a conversation, it'd be a little awkward if you didn't say anything for two minutes, and then you came up with your

answer. . . . That's why I like calls. I'd rather have someone be honest with you. . . . If you call, you're putting yourself out there, but it is also better.

At Fillmore, Grant says of when he used to text, "I end[ed] up feeling too lonely, just typing all day." He has given it up, except for texting his girlfriend. He returns her long text messages with a "k," short for "okay," and then holds off on further communication until he can talk to her on the phone or see her in person. He says, "When someone sends you a text or IM, you don't know *how* they're saying something. They could say something to you, and they could be joking, but they could be serious and you're not really sure."

These young men are asking for time and touch, attention and immediacy. They imagine living with less conscious performance. They are curious about a world where people dealt in the tangible and did one thing at a time. This is ironic. For they belong to a generation that is known, and has been celebrated, for never doing one thing at a time.

Erik Erikson writes that in their search for identity, adolescents need a place of stillness, a place to gather themselves.[2] Psychiatrist Anthony Storr writes of solitude in much the same way. Storr says that in accounts of the creative process, "by far the greater number of new ideas occur during a state of reverie, intermediate between waking and sleeping. . . . It is a state of mind in which ideas and images are allowed to appear and take their course spontaneously . . . the creator need[s] to be able to be passive, to let things happen within the mind."[3] In the digital life, stillness and solitude are hard to come by.

Online we are jarred by the din of the Internet bazaar. Roanne, sixteen, keeps her diary in a paper journal. She says she is too weak to stay focused when she has the Internet to tempt her:

> I can't use the Internet to write in my diary because at any moment I could watch *Desperate Housewives,* or even just a few minutes of it, or *Gossip Girl* or *Glee.* If you want to have an uninterrupted conversation, you might talk to somebody in person. If in person is not an option, then the phone. But there's so many interruptions you can have if you're sitting in front of a computer, because the computer has so many things you could be doing rather than talking to someone.

The physical world is not always a quiet place. There is performance and self-presentation everywhere—at school, in your family, on a date. But when young people describe days of composing and recomposing their digital personae, they accept the reality of this new social milieu, but also insist that online life presents a new kind of "craziness." There are so many sites, games, and worlds. You have to remember the nuances of how you have presented yourself in different places. And, of course, texting demands your attention all the time. "You have no idea," says an exhausted Brad.

THE PERILS OF PERFORMANCE

Brad says, only half jokingly, that he worries about getting "confused" between what he "composes" for his online life and who he "really" is. Not yet confirmed in his identity, it makes him anxious to post things about himself that he doesn't really know are true. It burdens him that the things he says online affect how people treat him in the real. People already relate to him based on things he has said on Facebook. Brad struggles to be more "himself" there, but this is hard. He says that even when he tries to be "honest" on Facebook, he cannot resist the temptation to use the site "to make the right impression." On Facebook, he says, "I write for effect. I sit down and ask, 'If I say this, will it make me sound like I'm too uptight? But if I say this, will it make me sound like I don't care about anything?'" He makes an effort to be "more spontaneous on Facebook . . . to actively say, 'This is who I am, this is what I like, this is what I don't like,'" but he feels that Facebook "perverts" his efforts because self-revelation should be to "another person who cares." For Brad, it loses meaning when it is broadcast as a profile.

The Internet can play a part in constructive identity play, although, as we have seen, it is not so easy to experiment when all rehearsals are archived. But Brad admits that on Facebook he only knows how to play to the crowd. We've seen that he anguishes about the cool bands and the bands that are not so cool. He thinks about the movies he should list as favorites and the ones that will pin him as boring or sexist. There is a chance that admitting he likes the Harry Potter series will be read positively—he'll be seen as someone in touch with the whimsy of his childhood. But more likely, it will make him seem less sexy. Brad points out that in real life, people can see you are cool even if you like some uncool things. In a profile, there is no room for error. You are reduced to a series of right and wrong choices. "Online life," he says, "is about premeditation." Brad sums up his discontents with an old-fashioned word: online life inhibits "authenticity." He wants to experience people directly. When he reads what someone says

about themselves on Facebook, he feels that he is an audience to their performance of cool.

Brad has more than a little of Henry David Thoreau in him. In *Walden*, published in 1854, Thoreau remarks that we are too much in contact with others
and in ways that are random. We cannot respect each other if we "stumble over
one another."[4] He says, we live "thick," unable to acquire value for each other
because there is not enough space between our times together. "Society," writes
Thoreau, "is commonly too cheap."[5] It would be better, he says, to learn or experience something before we join in fellowship with others. We know what
Thoreau did about his opinions. He took his distance. He found communion
with nature and simple objects. He saw old friends and made new ones. All of
these sustained him, but he did not live "thick." In the end, Brad decides to leave
his digital life for his own private Walden. When he wants to see a friend, he
calls, makes a plan, and goes over to visit. He says that life is beginning to feel
more natural. "Humans learn to talk and make eye contact before they learn to
touch-type, so I think it's a more basic, fundamental form of communication,"
he says. Abandoning digital connection, he says, he is "sacrificing three hollow
conversations" in favor of "one really nice social interaction with one person."
He acknowledges that "not doing IM reduces the amount of social interacting
you can do in one day," but doesn't mourn the loss: "Would you rather have
thirty kind-of somewhat-good friends or five really close friends?"

I meet other teenagers, like Brad, who go on self-imposed media "fasts." Some
give up texting, some IM. Because of its centrality to social life, the most decisive
step they can think of is to leave Facebook.[6] Some, like Brad, are exhausted by its
pressure for performance. Some say they find themselves being "cruel"—online
life suppresses healthy inhibitions. Others say they lose touch with their "real"
friends as they spend hours keeping up contacts with the "friended." Some, not
yet many, rebel against the reality that Facebook owns (in the most concrete
terms) the story of their lives. Some believe that the site encourages them to judge
themselves and others in superficial ways. They agonize over what photographs
to post. They digitally alter their Facebook photographs to look more appealing.
But even after so much time, writing profiles and editing photos, the fiction of a
Facebook page is that it is put up with a kind of aristocratic nonchalance. Luis
says, "It's like a girl wearing too much makeup, trying too hard. It's supposed to
look like you didn't care. But no one believes this myth of 'Oh, I just threw some
stuff up on my page. . . . I'm very cool. I have so much else to do.' You see that
they are on their Facebook page all day. Who are they kidding?" His tone turns

wistful: "It must have been nice when you could just discover a person by talking to them." For all of these reasons, dropping out comes as something of a relief.

The terms of these refusals—to find oneself and others more directly and to live a less-mediated life, to move away from performances and toward something that feels more real—suggest the refusals that brought Henry David Thoreau to Walden Pond nearly two centuries before.

WALDEN 2.0

In his essay about his two years of retreat, Thoreau writes, "I went to the woods because I wished to live deliberately, to front only the essential facts of life, and see if I could not learn what it had to teach, and not, when I came to die, discover that I had not lived. I did not wish to live what was not life, living is so dear; nor did I wish to practise resignation, unless it was quite necessary."[7] Thoreau's quest inspires us to ask of our life with technology: Do we live deliberately? Do we turn away from life that is not life? Do we refuse resignation?

Some believe that the new connectivity culture provides a digital Walden. A fifteen-year-old girl describes her phone as her refuge. "My cell phone," she says, "is my only individual zone, just for me." Technology writer Kevin Kelly, the first editor of *Wired*, says that he finds refreshment on the Web. He is replenished in its cool shade: "At times I've entered the web just to get lost. In that lovely surrender, the web swallows my certitude and delivers the unknown. Despite the purposeful design of its human creators, the web is a wilderness. Its boundaries are unknown, unknowable, its mysteries uncountable. The bramble of intertwined ideas, links, documents, and images create an otherness as thick as a jungle. The web smells like life."[8]

But not everyone is as refreshed as Kelly. Brad talks about the "throwaway friendships" of online life. Hannah wonders what she really has to show for the time she has spent hanging out with a small, sarcastic in-crowd and with a best friend who she fears will simply not show up again. It is hard to accept that online friends are not part of your life; yet, they can make themselves disappear just as you can make them vanish. Anxiety about Internet friendships makes people cherish the other kind. The possibility of constant connection makes people value a bit of space. Pattie, fourteen, no longer carries her cell phone. "It feels good," she says, "to have people *not* reach you."

That bit of space could leave room for a child to be a child a bit longer. One of the privileges of childhood is that some of the world is mediated by adults.

Hillary, sixteen, is taking a long break from her cell phone. She doesn't want to be on call, and so she leaves it at home. "I don't like the feeling of being reachable all the time . . . of knowing about everything in real time." For a child—and for this purpose, adolescents are still children—one cost of constant connectivity is that adults lose the ability to act as a buffer against the world. Only a few months before, Hillary was at a party to celebrate the release of a new volume in the Harry Potter series when her father suffered a seizure. She didn't learn about it until she was at home and with family. She was glad for this. Without a cell phone, the bad news waited until there was an adult there to support her, to put it in context. She didn't want to hear it alone, holding a phone.

Hillary is fond of movies but drawn toward "an Amish life minus certain exceptions [these would be the movies] . . . but I wouldn't mind if the Internet went away." She asks, "What could people be doing if they weren't on the Internet?" She answers her own question: "There's piano; there's drawing; there's all these things people could be creating." Hillary talks about how hard it is to keep up "all the different sites you have to keep up," and above all, how time-consuming it is to feed Facebook. These tiring performances leave little space for creativity and reflection: "It really is distracting." There is not much room for what Thoreau meant by a life lived deliberately.

There is nothing more deliberate than the painstaking work of constructing a profile or having a conversation on instant messenger in which one composes and recomposes one's thoughts. And yet, most of the time on the Net, one floats and experiments, follows links, and sends out random feelers. One flips through the photo albums of friends—and then the albums of their friends. One comments on the postings of people one hardly knows. Thoreau complained that people are too quick to share an opinion. Online, social networks instruct us to share whenever there's "something on our mind," no matter how ignorant or ill considered, and then help us broadcast it to the widest possible audience. Every day each of us is bombarded by other people's random thoughts. We start to see such effusions as natural. So, although identity construction on the Net begins in a considered way, with the construction of a profile or an avatar, people can end up feeling that the only deliberate act is the decision to hand oneself over to the Net. After that, one is swept along.

For those so connected, there may be doubts (about life as performance, about losing the nuance of the face-to-face), but there is the pleasure of continual company. For those not connected, there can be an eerie loneliness, even on the streets of one's hometown. Kara, in her fifties, feels that life in her hometown of

Portland, Maine, has emptied out: "Sometimes I walk down the street, and I'm the only person not plugged in. It's like I'm looking for another person who is not plugged in." With nostalgia—which can come with youth or age—for the nod that marks a meeting in shared streets and weather, she adds a bit wistfully, "No one is where they are. They're talking to someone miles away. I miss them. But they are missing out." Nostalgia ensures that certain things stay before us: the things we miss.

There are no simple answers as to whether the Net is a place to be deliberate, to commit to life, and live without resignation. But these are good terms with which to start a conversation. That conversation would have us ask if these are the values by which we want to judge our lives. If they are, and if we are living in a technological culture that does not support them, how can that culture be rebuilt to specifications that respect what we treasure—our sacred spaces. Could we, for example, build a Net that reweights privacy concerns, acknowledging that these, as much as information, are central to democratic life?

The phrase "sacred spaces" became important to me in the 1980s when I studied a cohort of scientists, engineers, and designers newly immersed in simulation. Members of each group held certain aspects of their professional life to be inviolate.[9] These were places they wanted to hold apart from simulation because, in that space, they felt most fully themselves in their discipline. For architects, it was hand drawings. This was where design implicated the body of the architect. This was where architects were engineers, certainly, but they were also artists. This was where the trace of the hand personalized a building. And this was where architects, so often part of large teams, experienced themselves as authors. The most enthusiastic proponents of computer-assisted design defended hand drawing. When their students began to lose the skill, these professors sent them off to drawing class. It was not about rejecting the computer but about making sure that designers came to it with their own values. A sacred space is not a place to hide out. It is a place where we recognize ourselves and our commitments.

When Thoreau considered "where I live and what I live for," he tied together location and values. Where we live doesn't just change how we live; it informs who we become. Most recently, technology promises us lives on the screen. What values, Thoreau would ask, follow from this new location? Immersed in simulation, where do we live, and what do we live for?

necessary conversations

During my earliest days at MIT, I met the idea (at that time altogether novel to me) that part of my job would be to think of ways to keep technology busy. In the fall of 1978, Michael Dertouzos, director of the Laboratory for Computer Science, held a two-day retreat at MIT's Endicott House on the future of personal computers, at the time widely called "home computers." It was clear that "everyday people," as Dertouzos put it, would soon be able to have their own computers. The first of these—the first that could be bought and didn't have to be built—were just coming on the market. But what could people do with them? There was technological potential, but it needed to be put to work. Some of the most brilliant computer scientists in the world—such pioneers of information processing and artificial intelligence as Robert Fano, J. C. R. Licklider, Marvin Minsky, and Seymour Papert—were asked to brainstorm on the question. My notes from this meeting show suggestions on tax preparation and teaching children to program. No one thought that anyone except academics would really want to write on computers. Several people suggested a calendar; others thought that was a dumb idea. There would be games.

Now we know that once computers connected us to each other, once we became tethered to the network, we really didn't need to keep computers busy. *They keep us busy*. It is as though we have become their killer app. As a friend of mine put it in a moment of pique, "We don't do our e-mail; our e-mail does

us." We talk about "spending" hours on e-mail, but we, too, are being spent. Niels
Bohr suggests that the opposite of a "deep truth" is a truth no less profound.[1]
As we contemplate online life, it helps to keep this in mind.

Online, we easily find "company" but are exhausted by the pressures of performance. We enjoy continual connection but rarely have each other's full attention. We can have instant audiences but flatten out what we say to each other
in new reductive genres of abbreviation. We like it that the Web "knows" us, but
this is only possible because we compromise our privacy, leaving electronic
bread crumbs that can be easily exploited, both politically and commercially.
We have many new encounters but may come to experience them as tentative,
to be put "on hold" if better ones come along. Indeed, new encounters need not
be better to get our attention. We are wired to respond positively to their simply
being new. We can work from home, but our work bleeds into our private lives
until we can barely discern the boundaries between them. We like being able to
reach each other almost instantaneously but have to hide our phones to force
ourselves to take a quiet moment.

Overwhelmed by the pace that technology makes possible, we think about how
new, more efficient technologies might help dig us out. But new devices encourage
ever-greater volume and velocity. In this escalation of demands, one of the things
that comes to feel safe is using technology to connect to people at a distance, or
more precisely, to a lot of people from a distance. But even a lot of people from
a distance can turn out to be not enough people at all. We brag about how many
we have "friended" on Facebook, yet Americans say they have fewer friends than
before.[2] When asked in whom they can confide and to whom they turn in an
emergency, more and more say that their only resource is their family.

The ties we form through the Internet are not, in the end, the ties that bind.
But they are the ties that preoccupy. We text each other at family dinners,
while we jog, while we drive, as we push our children on swings in the park. We
don't want to intrude on each other, so instead we constantly intrude on each
other, but not in "real time." When we misplace our mobile devices, we become
anxious—impossible really. We have heard teenagers insist that even when their
cell phones are not on their person, they can feel them vibrate. "I know when
I'm being called," says a sixteen-year-old. "I just do." Sentiments of dependency
echo across generations. "I never am without my cell phone," says a fifty-two-
year-old father. "It is my protection."

In the evening, when sensibilities such as these come together, they are likely
to form what have been called "postfamilial families."[3] Their members are alone

together, each in their own rooms, each on a networked computer or mobile device. We go online because we are busy but end up spending more time with technology and less with each other. We defend connectivity as a way to be close, even as we effectively hide from each other. At the limit, we will settle for the inanimate, if that's what it takes.

Bohr's dictum is equally true in the area of sociable robotics, where things are no less tangled. Roboticists insist that robotic emotions are made up of the same ultimate particles as human ones (because mind is ultimately made of matter), but it is also true that robots' claims to emotion derive from programs designed to get an emotional rise out of us.[4]

Roboticists present, as though it were a first principle, the idea that as our population ages, we simply won't have enough people to take care of our human needs, and so, as a companion, a sociable robot is "better than nothing." But what are our first principles? We know that we warm to machines when they seem to show interest in us, when their affordances speak to our vulnerabilities. But we don't have to say yes to everything that speaks to us in this way. Even if, as adults, we are intrigued by the idea that a sociable robot will distract our aging parents, our children ask, "Don't we have people for these jobs?" We should attend to their hesitations. Sorting all this out will not be easy. But we are at a crossroads—at a time and place to initiate new conversations.

As I was working on this book, I discussed its themes with a former colleague, Richard, who has been left severely disabled by an automobile accident. He is now confined to a wheelchair in his home and needs nearly full-time nursing care. Richard is interested in robots being developed to provide practical help and companionship to people in his situation, but his reaction to the idea is complex. He begins by saying, "Show me a person in my shoes who is looking for a robot, and I'll show you someone who is looking for a person and can't find one," but then he makes the best possible case for robotic helpers when he turns the conversation to *human* cruelty. "Some of the aides and nurses at the rehab center hurt you because they are unskilled, and some hurt you because they mean to. I had both. One of them, she pulled me by the hair. One dragged me by my tubes. A robot would never do that," he says. And then he adds, "But you know, in the end, that person who dragged me by my tubes had a story. I could find out about it. She had a story."

For Richard, being with a person, even an unpleasant, sadistic person, makes him feel that he is still alive. It signifies that his way of being in the world has a certain dignity, even if his activities are radically curtailed. For him, dignity requires a feeling of authenticity, a sense of being connected to the human narrative. It helps sustain him. Although he would not want his life endangered, he prefers the sadist to the robot.

Richard's perspective is a cautionary tale to those who would speak in too-simple terms of purely technical benchmarks for human and machine interactions. We animate robotic creatures by projecting meaning onto them and are thus tempted to speak of their emotions and even their "authenticity." We can do this if we focus on the feelings that robots evoke in us. But too often the unasked question is, What does the robot feel? We know what the robot cannot feel: it cannot feel human empathy or the flow of human connection. Indeed, the robot can feel nothing at all. Do we care? Or does the performance of feeling now suffice? Why would we want to be in conversation with machines that cannot understand or care for us? The question was first raised for me by the ELIZA computer program.[5] What made ELIZA a valued interlocutor? What matters were so private that they could only be discussed with a machine?

Over years and with some reluctance, I came to understand that ELIZA's popularity revealed more than people's willingness to talk to machines; it revealed their reluctance to talk to other people.[6] The idea of an attentive machine provides the fantasy that we may escape from each other. When we say we look forward to computer judges, counselors, teachers, and pastors, we comment on our disappointments with people who have not cared or who have treated us with bias or even abuse. These disappointments begin to make a machine's performance of caring seem like caring enough. We are willing to put aside a program's lack of understanding and, indeed, to work to make it seem to understand more than it does—all to create the fantasy that there is an alternative to people. This is the deeper "ELIZA effect." Trust in ELIZA does not speak to what we think ELIZA will understand but to our lack of trust in the people who might understand.

Kevin Kelly asks, "What does technology want?" and insists that, whatever it is, technology is going to get it. Accepting his premise, what if one of the things technology wants is to exploit our disappointments and emotional vulnerabilities? When this is what technology wants, it wants to be a symptom.

SYMPTOMS AND DREAMS

Wary of each other, the idea of a robot companion brings a sense of control, of welcome substitution. We allow ourselves to be comforted by unrequited love, for there is no robot that can ever love us back. That same wariness marks our networked lives. There, too, we are vulnerable to a desire to control our connections, to titrate our level of availability. Things progress quickly. A lawyer says sensibly, "I can't make it to a client meeting; I'll send notes by e-mail instead." Five steps later, colleagues who work on the same corridor no longer want to see or even telephone each other and explain that "texts are more efficient" or "I'll post something on Facebook."

As we live the flowering of connectivity culture, we dream of sociable robots.[7] Lonely despite our connections, we send ourselves a technological Valentine. If online life is harsh and judgmental, the robot will always be on our side. The idea of a robot companion serves as both symptom and dream. Like all psychological symptoms, it obscures a problem by "solving" it without addressing it. The robot will provide companionship and mask our fears of too-risky intimacies. As dream, robots reveal our wish for relationships we can control.

A symptom carries knowledge that a person fears would be too much to bear. To do its job, a symptom disguises this knowledge so it doesn't have to be faced day to day.[8] So, it is "easier" to feel constantly hungry than to acknowledge that your mother did not nurture you. It is "easier" to be enraged by a long supermarket line than to deal with the feeling that your spouse is not giving you the attention you crave. When technology is a symptom, it disconnects us from our real struggles.

In treatment, symptoms disappear because they become irrelevant. Patients become more interested in looking at what symptoms hide—the ordinary thoughts and experiences of which they are the strangulated expression. So when we look at technology as symptom and dream, we shift our attention away from technology and onto ourselves. As Henry David Thoreau might ask, "Where do we live, and what do we live for?" Kelly writes of technophilia as our natural state: we love our objects and follow where they lead.[9] I would reframe his insight: we love our objects, but enchantment comes with a price.

The psychoanalytic tradition teaches that all creativity has a cost, a caution that applies to psychoanalysis itself.[10] For psychoanalyst Robert Caper, "The transgression in the analytic enterprise is not that we try to make things better; the transgression is that we don't allow ourselves to see its costs and limitations."[11] To make his point Caper revisits the story of Oedipus. As his story is traditionally understood, Oedipus is punished for seeking knowledge—in particular, the knowledge of his parentage. Caper suggests he is punished for something else: his refusal to recognize the limitations of knowledge. A parallel with technology is clear: we transgress not because we try to build the new but because we don't allow ourselves to consider what it disrupts or diminishes. We are not in trouble because of invention but because we think it will solve everything.

A successful analysis disturbs the field in the interest of long-term gain; it learns to repair along the way.[12] One moves forward in a chastened, self-reflective spirit. Acknowledging limits, stopping to make the corrections, doubling back—these are at the heart of the ethic of psychoanalysis. A similar approach to technology frees us from unbending narratives of technological optimism or despair. Consider how it would modulate Kelly's argument about technophilia. Kelly refers to Henry Adams, who in 1900 had a moment of rapture when he first set eyes on forty-foot dynamos. Adams saw them as "symbols of infinity, objects that projected a moral force, much as the early Christians felt the cross."[13] Kelly believes that Adams's desire to be at one with the dynamo foreshadows how Kelly now feels about the Web. As we have seen, Kelly wants to merge with the Web, to find its "lovely surrender." Kelly continues,

> I find myself indebted to the net for its provisions. It is a steadfast bene-
> factor, always there. I caress it with my fidgety fingers; it yields up my de-
> sires, like a lover. . . . I want to remain submerged in its bottomless
> abundance. To stay. To be wrapped in its dreamy embrace. Surrendering
> to the web is like going on aboriginal walkabout. The comforting illogic
> of dreams reigns. In dreamtime you jump from one page, one thought,
> to another. . . . The net's daydreams have touched my own, and stirred
> my heart. If you can honestly love a cat, which can't give you directions
> to a stranger's house, why can't you love the web?[14]

Kelly has a view of connectivity as something that may assuage our deepest fears—of loneliness, loss, and death. This is the rapture. But connectivity also disrupts our attachments to things that have always sustained us—for example, the value we put on face-to-face human connection. Psychoanalysis, with its emphasis on the comedy and tragedy in the arc of human life, can help keep us focused on the specificity of human conversation. Kelly is enthralled by the Web's promise of limitless knowledge, its "bottomless abundance." But the Oedi-pal story reminds us that rapture is costly; it usually means you are overlooking consequences.

Oedipus is also a story about the difference between getting what you want and getting what you think you want. Technology gives us more and more of what we think we want. These days, looking at sociable robots and digitized friends, one might assume that what we want is to be always in touch and never alone, no matter who or what we are in touch with. One might assume that what we want is a preponderance of weak ties, the informal networks that underpin online acquaintanceship. But if we pay attention to the real consequences of what we think we want, we may discover what we really want. We may want some stillness and solitude. As Thoreau put it, we may want to live less "thickly" and wait for more infrequent but meaningful face-to-face encounters. As we put in our many hours of typing—with all fingers or just thumbs—we may discover that we miss the human voice. We may decide that it is fine to play chess with a robot, but that robots are unfit for any conversation about family or friends. A robot might have needs, but to understand desire, one needs language and flesh. We may decide that for these conversations, we must have a person who knows, firsthand, what it means to be born, to have parents and a family, to wish for adult love and perhaps children, and to anticipate death. And, of course, no matter how much "wilderness" Kelly finds on the Web, we are not in a position to let the virtual take us away from our stewardship of nature, the nature that doesn't go away with a power outage.

We let things get away from us. Even now, we are emotionally dependent on online friends and intrigued by robots that, their designers claim, are almost ready to love us.[15] And brave Kevin Kelly says what others are too timid to admit: he is in love with the Web itself. It has become something both erotic and idealized. What are we missing in our lives together that leads us to prefer lives alone together? As I have said, every new technology challenges us, generation after generation, to ask whether it serves our human purposes, something that causes us to reconsider what they are.

In a design seminar, master architect Louis Kahn once asked, "What does a brick want?"[16] In that spirit, if we ask, "What does simulation want?" we know what it wants. It wants—it demands—immersion. But immersed in simulation, it can be hard to remember all that lies beyond it or even to acknowledge that everything is not captured by it. For simulation not only demands immersion but creates a self that prefers simulation. Simulation offers relationships simpler than real life can provide. We become accustomed to the reductions and betrayals that prepare us for life with the robotic.

But being prepared does not mean that we need to take the next step. Sociable robotics puts science into the game of intimacy and the most sensitive moments of children's development. There is no one to tell science what it cannot do, but here one wishes for a referee. Things start innocently: neuroscientists want to study attachment. But things end reductively, with claims that a robot "knows" how to form attachments because it has the algorithms. The dream of today's roboticists is no less than to reverse engineer love. Are we indifferent to whether we are loved by robots or by our own kind?

In Philip K. Dick's classic science fiction story "Do Androids Dream of Electric Sheep" (which most people know through its film adaptation, *Blade Runner*), loving and being loved by a robot seems a good thing. The film's hero, Deckard, is a professional robot hunter in a world where humans and robots look and sound alike. He falls in love with Rachel, an android programmed with human memories and the knowledge that she will "die." I have argued that knowledge of mortality and an experience of the life cycle are what make us uniquely human. This brilliant story asks whether the simulation of these things will suffice.

By the end of the film, we are left to wonder whether Deckard himself may be an android but unaware of his identity. Unable to resolve this question, we cheer for Deckard and Rachel as they escape to whatever time they have remaining—in other words, to the human condition. Decades after the film's release, we are still nowhere near developing its androids. But to me, the message of *Blade Runner* speaks to our current circumstance: long before we have devices that can pass any version of the Turing test, the test will seem beside the point. We will not care if our machines are clever but whether they love us.

Indeed, roboticists want us to know that the point of affective machines is that they will take care of us. This narrative—that we are on our way to being tended by "caring" machines—is now cited as conventional wisdom. We have entered a realm in which conventional wisdom, always inadequate, is dangerously inadequate. That it has become so commonplace reveals our willingness to take the performance of emotion as emotion enough.

EMOTION ENOUGH

When roboticists argue that robots can develop emotions, they begin by asserting the material basis of all thought and take things from there. For example, Rodney Brooks says that a robot could be given a feeling like "sadness" by setting "a number in its computer code." This sadness, for Brooks, would be akin to that felt by humans, for "isn't humans' level of sadness basically a number, too, just a number of the amounts of various neurochemicals circulating in the brain? Why should a robot's numbers be any less authentic than a human's?"[17]

Given my training as a clinician, I tend to object to the relevance of a robot's "numbers" for thinking about emotion because of something humans have that robots don't: a human body and a human life. Living in our bodies sets our human "numbers." Our emotions are tied to a developmental path—from childhood dependence to greater independence—and we experience the traces of our earlier dependencies in later fantasies, wishes, and fears. Brooks speaks of giving the robot the emotion of "sadness." In a few months, I will send my daughter off to college. I'm both sad and thrilled. How would a robot "feel" such things? Why would its "numbers" even "want" to?

Cynthia Breazeal, one of Brooks's former students, takes another tack, arguing that robotic emotions are valid if you take care to consider them as a new category. Cats have cat emotions, and dogs have dog emotions. These differ from each other and from human emotions. We have no problem, says Breazeal, seeing all of these as "genuine" and "authentic." And now, robots will have robot emotions, also in their own category and likewise "genuine" and "authentic." For Breazeal, once you give robotic emotions their own category, there is no need to compare. We should respect emotional robots as "different," just as we respect all diversity.[18] But this argument confuses the authentic with the sui generis. That the robotic performance of emotion might exist in its own category implies nothing about the authenticity of the emotions being performed. And robots do not "have" emotions that we must respect. We build robots to do things that make us feel as though they have emotions. Our responses are their design template.

Whether one debates the question of robotic emotions in terms of materialism or category, we end up in a quandary. Instead of asking whether a robot has emotions, which in the end boils down to how different constituencies define emotion, we should be asking what kind of relationships we want to have with machines. Why do we want robots to perform emotion? I began my career at MIT arguing with Joseph Weizenbaum about whether a computer program might be a valuable dialogue partner. Thirty years later, I find myself debating those who argue, with David Levy, that my daughter might want to marry one.[19]

Simulation is often justified as practice for real-life skills—to become a better pilot, sailor, or race-car driver. But when it comes to human relations, simulation gets us into trouble. Online, in virtual places, simulation turns us into its creatures. But when we step out of our online lives, we may feel suddenly as though in too-bright light. Hank, a law professor in his late thirties, is on the Net for at least twelve hours a day. Stepping out of a computer game is disorienting, but so is stepping out of his e-mail. Leaving the bubble, Hank says, "makes the flat time with my family harder. Like it's taking place in slow motion. I'm short with them." After dinner with his family, Hank is grateful to return to the cool shade of his online life.

Nothing in real life with real people vaguely resembles the environment (controlled yet with always-something-new connections) that Hank finds on the Net. Think of what is implied by his phrase "flat time." Real people have consistency, so if things are going well in our relationships, change is gradual, worked through slowly. In online life, the pace of relationships speeds up. One quickly moves from infatuation to disillusionment and back. And the moment one grows even slightly bored, there is easy access to someone new. One races through e-mail and learns to attend to the "highlights." Subject lines are exaggerated to get attention. In online games, the action often reduces to a pattern of moving from scary to safe and back again. A frightening encounter presents itself. It is dealt with. You regroup, and then there is another. The adrenaline rush is continual; there is no "flat time."

Sometimes people try to make life with others resemble simulation. They try to heighten real-life drama or control those around them. It would be fair to say that such efforts do not often end well. Then, in failure, many are tempted to return to what they do well: living their lives on the screen. If there is an addiction here, it is not to a technology. It is to the habits of mind that technology allows us to practice.

Online, we can lose confidence that we are communicating or cared for. Confused, we may seek solace in even more connection. We may become intolerant of our own company: "I never travel without my BlackBerry," says a fifty-year-old management consultant. She cannot quiet her mind without having things on her mind.

My own study of the networked life has left me thinking about intimacy— about being with people in person, hearing their voices and seeing their faces, trying to know their hearts. And it has left me thinking about solitude—the kind that refreshes and restores. Loneliness is failed solitude.[20] To experience solitude you must be able to summon yourself by yourself; otherwise, you will only know how to be lonely. In raising a daughter in the digital age, I have thought of this very often.

In his history of solitude, Anthony Storr writes about the importance of being able to feel at peace in one's own company.[21] But many find that, trained by the Net, they cannot find solitude even at a lake or beach or on a hike. Stillness makes them anxious. I see the beginnings of a backlash as some young people become disillusioned with social media. There is, too, the renewed interest in yoga, Eastern religions, meditating, and "slowness."

These new practices bear a family resemblance to what I have described as the romantic reaction of the 1980s. Then, people declared that something about their human nature made them unlike any machine ("simulated feeling may be feeling; simulated love is never love"). These days, under the tutelage of imaging technology and neurochemistry, people seem willing to grant their own machine natures. What they rebel against is how we have responded to the affordances of the networked life. Offered continual connectivity, we have said yes. Offered an opportunity to abandon our privacy, so far we have not resisted. And now comes the challenge of a new "species"—sociable robots—whose "emotions" are designed to make us comfortable with them. What are we going to say?

The romantic reaction of the 1980s made a statement about computation as a model of mind; today we struggle with who we have become in the presence of computers. In the 1980s, it was enough to change the way you saw yourself. These days, it is a question of how you live your life. The first manifestations of today's "push back" are tentative experiments to do without the Net. But the Net has become intrinsic to getting an education, getting the news, and getting a job. So, today's second thoughts will require that we actively reshape our lives on the screen. Finding a new balance will be more than a matter of "slowing down." How can we make room for reflection?

QUANDARIES

In arguing for "caring machines," roboticists often make their case by putting things in terms of quandaries. So, they ask, "Do you want your parents and grandparents cared for by robots, or would you rather they not be cared for at all?" And alternatively, "Do you want seniors lonely and bored, or do you want them engaged with a robotic companion?"[22] The forced choice of a quandary, posed over time, threatens to become no quandary at all because we come to accept its framing—in this case, the idea that there is only one choice, between robotic caregivers and loneliness. The widespread use of this particular quandary makes those uncomfortable with robotic companions out to be people who would consign an elderly population to boredom, isolation, and neglect.

There is a rich literature on how to break out of quandary thinking. It suggests that sometimes it helps to turn from the abstract to the concrete.[23] This is what the children in Miss Grant's fifth-grade class did. Caught up in a "for or against" discussion about robot caregivers, they turned away from the dilemma to ask a question ("Don't we have people for these jobs?") that could open up a different conversation. While the children only began that conversation, we, as adults, know where it might go. What about bringing in some new people? What must be done to get them where they are needed? How can we revisit social priorities so that funds are made available? We have the unemployed, the retired, and those currently at war—some of these might be available if there were money to pay them. One place to start would be to elevate elder care above the minimum-wage job that it usually is, often without benefits. The "robots-or-no-one" quandary takes social and political choice out of the picture when it belongs at the center of the picture.

I experienced a moment of reframing during a seminar at MIT that took the role of robots in medicine as its focus. My class considered a robot that could help turn weak or paralyzed patients in their beds for bathing. A robot now on the market is designed as a kind of double spatula: one plate slides under the patient; another is placed on top. The head is supported, and the patient is flipped. The class responded to this technology as though it suggested a dilemma: machines for the elderly or not. So some students insisted that it is inevitable for robots to take over nursing roles (they cited cost, efficiency, and the insufficient numbers of people who want to take the job). Others countered that the elderly deserve the human touch and that anything else is demeaning. The conversation argued absolutes: the inevitable versus the unsupportable.

Into this stalled debate came the voice of a woman in her late twenties whose mother had recently died. She did not buy into the terms of the discussion. Why limit our conversation to no robot or a robotic flipper? Why not imagine a machine that is an extension of the body of one human trying to care lovingly for another? Why not build robotic arms, supported by hydraulic power, into which people could slip their own arms, enhancing their strength? The problem as offered presented her with two unacceptable images: an autonomous machine or a neglected patient. She wanted to have a conversation about how she might have used technology as prosthesis. Had her arms been made stronger, she might have been able to lift her mother when she was ill. She would have welcomed such help. It might have made it possible for her to keep her mother at home during her last weeks. A change of frame embraces technology even as it provides a mother with a daughter's touch.

In the spirit of "break the frame and see something new," philosopher Kwame Anthony Appiah challenges quandary thinking:

> The options are given in the description of the situation. We can call this the *package problem*. In the real world, situations are not bundled together with options. In the real world, the act of framing—the act of describing a situation, and thus of determining that there's a decision to be made— is itself a moral task. It's often *the* moral task. Learning how to recognize what is and isn't an option is part of our ethical development. . . . In life, the challenge is not so much to figure out how best to play the game; the challenge is to figure out what game you're playing.[24]

For Appiah, moral reasoning is best accomplished not by responding to quandaries but by questioning how they are posed, continually reminding ourselves that we are the ones choosing how to frame things.

FORBIDDEN EXPERIMENTS

When the fifth graders considered robot companions for their grandparents and wondered, "Don't we have people for these jobs?" they knew they were asking, "Isn't 'taking care' our parents' job?" And by extension, "Are there people to take care of us if we become 'inconvenient'?" When we consider the robots in our futures, we think through our responsibilities to each other.

Why do we want robots to care for us? I understand the virtues of partnership with a robot in war, space, and medicine. I understand that robots are useful in dangerous working conditions. But why are we so keen on "caring"?[25] To me, it seems transgressive, a "forbidden experiment."[26]

Not everyone sees it this way. Some people consider the development of caring machines as simple common sense. Porter, sixty, recently lost his wife after a long illness. He thinks that if robotic helpers "had been able to do the grunt work, there might have been more time for human nurses to take care of the more personal and emotional things." But often, relationships hinge on these investments of time. We know that the time we spend caring for children, doing the most basic things for them, lays down a crucial substrate.[27] On this ground, children become confident that they are loved no matter what. And we who care for them become confirmed in our capacity to love and care. The ill and the elderly also deserve to be confirmed in this same sense of basic trust. As we provide it, we become more fully human.

The most common justification for the delegation of care to robots focuses on things being "equal" for the person receiving care. This argument is most often used by those who feel that robots are appropriate for people with dementia, who will not "know the difference" between a person and a robot. But we do not really know how impaired people receive the human voice, face, and touch. Providing substitutes for human care may not be "equal" in the least. And again, delegating what was once love's labor changes the person who delegates. When we lose the "burden" of care, we begin to give up on our compact that human beings will care for other human beings. The daughter who wishes for hydraulic arms to lift her bedridden mother wants to keep her close. For the daughter, this last time of caring is among the most important she and her mother will share. If we divest ourselves of such things, we risk being coarsened, reduced. And once you have elder bots and nurse bots, why not nanny bots?

Why would we want a robot as a companion for a child? The relationship of a child to a sociable robot is, as I've said, very different from that of a child to a doll. Children do not try to model themselves on their dolls' expressions. A child projects human expression onto a doll. But a robot babysitter, already envisaged, might seem close enough to human that a child might use it as a model. This raises grave questions. Human beings are capable of infinite combinations of vocal inflection and facial expression. It is from other people that we learn how to listen and bend to each other in conversation. Our eyes "light up" with interest and "darken" with passion or anxiety. We recognize, and are most comfortable with, other people who exhibit this fluidity. We recognize, and are less comfortable with, people—with autism or Asperger's syndrome—who do not exhibit it. The developmental implications of children taking robots as models are unknown, potentially disastrous. Humans need to be surrounded by human touch, faces, and voices. Humans need to be brought up by humans.

Sometimes when I make this point, others counter that even so, robots might do the "simpler" jobs for children, such as feeding them and changing their diapers. But children fed their string beans by a robot will not associate food with human companionship, talk, and relaxation. Eating will become dissociated from emotional nurturance. Children whose diapers are changed by robots will not feel that their bodies are dear to other human beings. Why are we willing to consider such risks?[28]

Some would say that we have already completed a forbidden experiment, using ourselves as subjects with no controls, and the unhappy findings are in: we are connected as we've never been connected before, and we seem to have damaged ourselves in the process. A 2010 analysis of data from over fourteen thousand college students over the past thirty years shows that since the year 2000, young people have reported a dramatic decline in interest in other people. Today's college students are, for example, far less likely to say that it is valuable to try to put oneself in the place of others or to try to understand their feelings.[29] The authors of this study associate students' lack of empathy with the availability of online games and social networking. An online connection can be deeply felt, but you only need to deal with the part of the person you see in your game world or social network. Young people don't seem to feel they need to deal with more, and over time they lose the inclination. One might say that absorbed in those they have "friended," children lose interest in friendship.

These findings confirm the impressions of those psychotherapists—psychiatrists, psychologists, and social workers—who talk to me about the increasing numbers of patients who present in the consulting room as detached from their bodies and seem close to unaware of the most basic courtesies. Purpose-driven, plugged into their media, these patients pay little attention to those around them. In others, they seek what is of use, an echo of that primitive world of "parts." Their detachment is not aggressive. It is as though they just don't see the point.[30]

EARLY DAYS

It is, of course, tempting to talk about all of this in terms of addiction. Adam, who started out playing computer games with people and ends up feeling compelled by a world of bots, certainly uses this language. The addiction metaphor fits a common experience: the more time spent online, the more one wants to spend time online. But however apt the metaphor, we can ill afford the luxury of using it. Talking about addiction subverts our best thinking because it suggests that if there are problems, there is only one solution. To combat addiction, you have to discard the addicting substance. But we are not going to "get rid" of the Internet.

We will not go "cold turkey" or forbid cell phones to our children. We are not going to stop the music or go back to television as the family hearth.

I believe we will find new paths toward each other, but considering ourselves victims of a bad substance is not a good first step. The idea of addiction, with its one solution that we know we won't take, makes us feel hopeless. We have to find a way to live with seductive technology and make it work to our purposes. This is hard and will take work. Simple love of technology is not going to help. Nor is a Luddite impulse.

What I call *realtechnik* suggests that we step back and reassess when we hear triumphalist or apocalyptic narratives about how to live with technology. *Realtechnik* is skeptical about linear progress. It encourages humility, a state of mind in which we are most open to facing problems and reconsidering decisions. It helps us acknowledge costs and recognize the things we hold inviolate. I have said that this way of envisaging our lives with technology is close to the ethic of psychoanalysis. Old-fashioned perhaps, but our times have brought us back to such homilies.

Because we grew up with the Net, we assume that the Net is grown-up. We tend to see it as a technology in its maturity. But in fact, we are in early days. There is time to make the corrections. It is, above all, the young who need to be convinced that when it comes to our networked life, we are still at the beginning of things. I am cautiously optimistic. We have seen young people try to reclaim personal privacy and each other's attention. They crave things as simple as telephone calls made, as one eighteen-year-old puts it, "sitting down and giving each other full attention." Today's young people have a special vulnerability: although always connected, they feel deprived of attention. Some, as children, were pushed on swings while their parents spoke on cell phones.[31] Now, these same parents do their e-mail at the dinner table. Some teenagers coolly compare a dedicated robot with a parent talking to them while doing e-mail, and parents do not always come out ahead. One seventeen-year-old boy says, "A robot would remember everything I said. It might not understand everything, but remembering is a first step. My father, talking to me while on his BlackBerry, he doesn't know what I said, so it is not much use that if he did know, he might understand."

The networked culture is very young. Attendants at its birth, we threw ourselves into its adventure. This is human. But these days, our problems with the Net are becoming too distracting to ignore. At the extreme, we are so enmeshed in our connections that we neglect each other. We don't need to reject or disparage technology. We need to put it in its place. The generation that has grown up with the Net is in a good position to do this, but these young people need help. So as they begin to fight for their right to privacy, we must be their partners. We know how easily information can be politically abused; we have the perspective of history. We have, perhaps, not shared enough about that history with our children. And as we, ourselves enchanted, turned away from them to lose ourselves in our e-mail, we did not sufficiently teach the importance of empathy and attention to what is real.

The narrative of *Alone Together* describes an arc: we expect more from technology and less from each other. This puts us at the still center of a perfect storm. Overwhelmed, we have been drawn to connections that seem low risk and always at hand: Facebook friends, avatars, IRC chat partners. If convenience and control continue to be our priorities, we shall be tempted by sociable robots, where, like gamblers at their slot machines, we are promised excitement programmed in, just enough to keep us in the game. At the robotic moment, we have to be concerned that the simplification and reduction of relationship is no longer something we complain about. It may become what we expect, even desire.

In this book I have referred to our vulnerabilities rather than our needs. Needs imply that we must have something. The idea of being vulnerable leaves a lot of room for choice. There is always room to be less vulnerable, more evolved. We are not stuck. To move forward together—as generations together—we are called upon to embrace the complexity of our situation. We have invented inspiring and enhancing technologies, and yet we have allowed them to diminish us. The prospect of loving, or being loved by, a machine changes what love can be. We know that the young are tempted. They have been brought up to be. Those who have known lifetimes of love can surely offer them more.

When we are at our best, thinking about technology brings us back to questions about what really matters. When I recently travelled to a memorial service for a close friend, the program, on heavy cream-colored card stock, listed the afternoon's speakers, told who would play what music, and displayed photographs of my friend as a young woman and in her prime. Several around me used the program's stiff, protective wings to hide their cell phones as they sent text messages during the service. One of the texting mourners, a woman in her late sixties, came over to chat with me after the service. Matter-of-factly, she offered, "I couldn't stand to sit that long without getting on my phone." The point of the service was to take a moment. This woman had been schooled by a technology she'd had for less than a decade to find this close to impossible.[32] Later,

I discussed the texting with some close friends. Several shrugged. One said, "What are you going to do?"

A shrug is appropriate for a stalemate. That's not where we are. It is too early to have reached such an impasse. Rather, I believe we have reached a point of inflection, where we can see the costs and start to take action. We will begin with very simple things. Some will seem like just reclaiming good manners. Talk to colleagues down the hall, no cell phones at dinner, on the playground, in the car, or in company. There will be more complicated things: to name only one, nascent efforts to reclaim privacy would be supported across the generations. And compassion is due to those of us—and there are many of us—who are so dependent on our devices that we cannot sit still for a funeral service or a lecture or a play. We now know that our brains are rewired every time we use a phone to search or surf or multitask.[33] As we try to reclaim our concentration, we are literally at war with ourselves. Yet, no matter how difficult, it is time to look again toward the virtues of solitude, deliberateness, and living fully in the moment. We have agreed to an experiment in which we are the human subjects. Actually, we have agreed to a series of experiments: robots for children and the elderly, technologies that denigrate and deny privacy, seductive simulations that propose themselves as places to live.[34]

We deserve better. When we remind ourselves that it is we who decide how to keep technology busy, we shall have better.

the letter

I return from Dublin to Boston in September 2009. I have brought my daughter Rebecca to Ireland and helped her to set up her dorm room for a gap year before starting college in New England. I'm one day back from Dublin, and I have already had a lot of contact with Rebecca, all of it very sweet. There are text messages: she forgot a favorite red coat; she wants her green down "puff" jacket and a pink scarf she would like to drape over her bed as a canopy. Could I please mail them to her? I assemble her parcel and send a text: "On the way to the Post Office." I have downloaded Skype and am ready for its unforgiving stare. Yet, even on my first day home, I feel nostalgic. I sit in my basement surrounded by musty boxes, looking for the letters that my mother and I exchanged during my first year in college, the first time I lived away from home. The telephone was expensive. She wrote twice a week. I wrote once a week. I remember our letters as long, emotional, and filled with conflict. We were separating, finding our way toward something new. Forty years later, I find the letters and feel as though I hold her heart in my hands.

As the days pass, I am in regular contact with my daughter on Skype and by text. As though under some generational tutelage, I feel constrained to be charming and brief in our breezy, information-filled encounters. Once, while texting, I am overtaken by a predictable moment in which I experience my mortality. In forty years, what will Rebecca know of her mother's heart as she found her way toward something new?

Now, holding my mother's letters, it is hard to read their brightness and their longing. She wrote them when she was dying and didn't want me to know. Her letters, coded, carried the weight of future letters that would never be written. And once a week, I wrote her a letter, telling my mother what I wanted her to know of my life. In discretion, there were significant omissions. But I shared a lot. She was my touchstone, and I wanted her to understand me. My letters tried to create the space for this conversation.

My daughter's texts and Skype presence leave no space of this kind. Is this breeziness about our relationship, or is it about our media? Through my daughter's senior-class friends—she attended an all-girl's day school—I know a cohort of mothers whose daughters have just left for college or their first year away from home. I talk to them about their experiences and the part that technology is playing.

The "mother narratives" have a certain similarity. They begin with an affirmation of the value of technology: mothers insist that they are more frequently in touch with their daughters than, as one puts it, "I would have ever dared hope." Mothers detail the texts and the Skype calls. A few, only a few, say they get an occasional e-mail. Since Skype has video as well as voice, mothers say they can tell if their daughters are looking well. Everyone is vigilant, worried about swine flu. Several hate that their daughters can see them. The mothers are in their late forties through early sixties, and they are not all happy to be closely observed. "I stopped putting on makeup for Skype," one says. "It was getting ridiculous." Another insists that putting on makeup for Skype is important: "I want her to see me at my best, able to cope. I don't want her to worry."

There is wistfulness in the mothers' accounts. For one, "It's pretty much the old 'news of the week in review,' except it's news of the day. But even with the constant updates, I don't have much of a sense of what is really happening. How she really feels." For another, "Texting makes it easy to lie. You never know where they really are. You never know if they are home. They can be anyplace and text you. Or Skype you on their iPhone. With a landline, you knew they were actually where they were supposed to be." One mother shares my feeling that conversations on Skype are inexplicably superficial. Unlike me, she attributes it to the technical limitations of her Internet connection: "It's like we are shouting at each other in order to be heard. The signal cuts off. I'm shouting at the computer."

And for this mother, things become even more superficial when she and her daughter exchange texts. She says, "I know that some people find it [texting] intimate, but it doesn't seem like a place to get into a long story." To this mother I admit that there is something about Skype that seems so ephemeral that I sometimes take "screenshots" of my daughter during our calls. On Skype you see each other, but you cannot make eye contact. I don't like these screenshots. My daughter has the expression of someone alone. Of course, there is irony in my experience of the digital as ephemeral and in my self-indulgent moment as I imagine my daughter in forty years with no trace of our conversations. Because the digital is only ephemeral if you don't take the trouble to make it permanent.

LIFE CAPTURE

Vannevar Bush, director of the Office of Scientific Research and Development during World War II, was concerned about what would happen once the war was over and scientists could dedicate themselves to civilian life. He wasn't worried about the biologists—they could always work on practical, medical problems—but the physicists needed new direction. In a landmark *Atlantic Monthly* article, "As We May Think," Bush suggested one: the physicists should develop a "memex." This would be "a device in which an individual stores all his books, records, and communications, and which is mechanized so that it may be consulted with exceeding speed and flexibility." It would be, Bush wrote, an "intimate supplement to his memory."[1] Bush dreamed of scientists wearing glasses that could automatically record those things "worthy of the record." He dreamed of annotating all that was captured. In his description of how an individual would make a path through all this data, Bush's narrative captures the essence of a Web search.

In the late 1970s, computer scientist Steve Mann began recording his life in a very different spirit—as an act of resistance. In a world filled with surveillance cameras—on the street, in shopping malls, in banks—Mann wanted to turn cameras against the world. To pursue his project, Mann found a way to wear a computer, keyboard, screen, and radio transmitter on his body. He captured his life and posted it on the Web.[2]

Mann's work was part performance art, part engineering research, and part political statement. Now, his once subversive gesture—documenting a life and putting it on the Web—is almost within everyone's reach. These days, anyone with a smartphone (equipped with a camera and/or video recorder) is close to having a portable archivist. And indeed, many say that when they don't use their mobile phone to document their lives, they feel remiss, guilty for not doing so.

In the mid-1990s, computer pioneer Gordon Bell began a project that would lead him to create a complete life archive. His first steps were to scan books, cards, letters, memos, posters, photographs, and even the logos from his coffee mug and T-shirt collections. Then, he moved on to digitizing home movies, videotaped lectures, and voice recordings. Of course, Bell archived everything he had ever written or read on his computer, from personal e-mails to academic papers. Faced with the question of how to organize and retrieve this data, Bell began to work with his Microsoft colleague Jim Gemmell, and the MyLifeBits project was born. As the system went live, Bell wore voice-recording equipment and a camera programmed to take a new photograph when it sensed (by a change of ambient light) that Bell was with a new person or in a new setting.[3] MyLifeBits recorded Bell's telephone calls, the songs he listened to, and the programs he watched on radio and television. When Bell was at the computer, it recorded the Web pages he visited, the files he opened, the messages he sent and received. It even monitored which windows were in the foreground of his screen at any time and how much mouse and keyboard activity was going on.

Life capture has practical applications. Bell's physician, for example, now has access to a detailed, ongoing record of his patient's life. If Bell doesn't exercise or eats fatty foods, the system knows. But Bell's mind is on posterity. For him, MyLifeBits is a way for people to "tell their life stories to their descendants."[4] His program aspires to be the ultimate tool for life collection.[5] But what of *recollection* in the fully archived life? If technology remembers for us, will we remember less? Will we approach our own lives from a greater distance? Bell talks about how satisfying it is to "get rid" of memories, to get them into the computer. Speaking of photography, Susan Sontag writes that under its influence, "travel becomes a strategy for accumulating photographs."[6] In digital culture, does life become a strategy for establishing an archive?[7] Young people shape their lives to produce an impressive Facebook profile. When we know that everything in our lives is captured, will we begin to live the life that we hope to have archived?

For Bell, a life archive responds to the human desire for a kind of immortality, the ancient fantasy of cheating death. But the experience of building the archive may subvert such intent. We may end up with a life deferred by the business of its own collection. One of life's pleasures is remembering, the good and the bad. Will the fact of the archive convince us that the work of remembering is already done?

When I go to San Francisco to talk with Bell and Gemmell in the summer of 2008, the formal MyLifeBits project is winding down; Bell wears only bits and pieces of his gear to our interview. He turns on a tape recorder. He takes my picture. He has wearied of his hardware. But the two scientists assure me—and I think they have a point—that total recall will be more popular when the technology for documenting your life is less burdensome. In the future there will be no fiddling with cameras and adjusting sound levels. You will be able to wear audio and video recording devices as tiny bits of diamondlike jewelry or, ultimately, as implants.

I am moved by my day with Gordon Bell. We look at his photographs, archived in complex patterns that make it possible to retrieve them by date, subject, and who is in the picture. We look at e-mail archives that span a professional lifetime. But the irony of the visit is that we spend most of our time talking about physical objects: we both love beautiful notebooks, and Bell shows me his Japanese-made journals filled with his elegant sketches of computer circuitry. We talk of physical objects that Bell has saved, things that belonged to his father. At one point, Bell brings out his MIT dissertation written over fifty years ago. It is hand typed. It has the "blueprints" of the circuits he devised—literally, diagrams etched on blue paper. We both touch them with a kind of awe. Now the computer generates such diagrams. But Bell touches the prints with the reverence with which I handle my mother's letters. We are not so ready to let all of this go.

Bell remains an enthusiast of life archiving but admits that it may be having unintended effects. For one thing, he suspects his project may be changing the nature of his memory.[8] Bell describes a lack of curiosity about details of life that he can easily find in his life archive. And he focuses on what the archive makes easily available. So, for example, Bell is mesmerized by a screen saver that draws on his personal archive to display random snapshots. Pictures of long-ago birthdays and family trips trigger waves of nostalgia. But during my visit, Bell tries to use search tools to find a particular photograph that is not coming up on the screen. He pursues one strategy, then another. Nothing works; he loses interest. One senses a new dynamic: when you depend on the computer to remember the past, you focus on whatever past is kept on the computer. And you learn to favor whatever past is easiest to find. My screen saver, my life.

And there are other effects. Bell says he can no longer abide books. He will get one, look at it, but "then I give them away, because they're not in my [computer's] memory. To me they're almost gone."[9] Journalist Clive Thompson, another of Bell's visitors, reflects on this aspect of Bell's experiment. Thompson says, "If it's not in your database, it doesn't exist. That's the sort of eerie philosophical proposition Bell's project raises."[10]

The proposition may not be so philosophical. To a certain degree, we already live it. Consider Washington, D.C., on Inauguration Day in 2009. Arms are held high; cell phones glint in the sun. People are taking pictures of themselves, of strangers, of friends, of the JumboTron plasma screens that will broadcast the ceremony. The event is a celebration of physical presence, but the crowd reaches out to those who are absent. It is important to have images of the day on one's own phone. And it is important to send them along. A photo from the inauguration, or a text, a posting, an e-mail, a Tweet—all validate the sense of being there. It used to be that taking a photograph marked participation—think of all the tourists who wanted to take their own photographs of the *Mona Lisa* as well as photograph themselves with the painting. But these days, the photograph is not enough. Sending implies being. On the inaugural platform, invited guests have cell phones and cameras raised high. The notables who constitute the picture take their own pictures. We are all pressed into the service of technologies of remembrance and validation.[11] As I write in January 2010, a new issue of *The New Yorker* shows a man and woman at the summit of a ski slope. He is using his digital camera; she is on her cell phone.

COLLECTION AND RECOLLECTION

When I learn about how MyLifeBits software will use face-recognition technology to label photographs automatically, I recall childhood times with my mother when she wrote funny things, silly poems, or sentimental inscriptions on the back of family photographs. She liked putting them all together in a big drawer, so that, in a way, picking a photo out of the drawer was like finding a surprise. Moments around the photograph drawer were times of recollection, in laughter and sometimes regret. Bell and Gemmell see photograph labeling as a "pesky" technical problem, something that computers must learn to do. They sum up the issue of labeling by saying that people "don't want to be the librarians of our digital archives—we want the computer to be the librarian."[12] Subtly, attitudes toward one's own life shift. My mother, happily annotating her drawer of snapshots, never saw herself as a librarian.

Bell says that "offloading memories" onto a computer "gives you kind of a feeling of cleanliness." Clean of remembrance? Clean of messy, unreliable associations? Do we want to be "clean" in this way?[13] Marcel Proust mined and reworked his memories—the things that were clear and the things that he felt slipping away—to create *Remembrance of Things Past*. But one never thinks of Proust getting "rid" of memory as he labored in his cork-lined room. For Sigmund Freud, we understand what things mean by what we forget as well as what we remember. Forgetting is motivated; it offers clues about who we are. What Proust struggled to remember is more important than what came easily to him. He found himself in the memories wrenched from the shadows. Artificial remembrance will be the great leveler.

At Microsoft, computer scientist Eric Horvitz is in charge of a project—Life Browser—designed to make MyLifeBits data more user-friendly by giving it shape and pattern. Installed on your computer, Life Browser observes what you attend to—the files you open, the e-mails you answer, the Web searches you return to. It shows who you are based on what you do. You can intervene: for example, you can manually tag as most important, things you do less often. You can say that infrequent calls are to the most important people. But Life Browser will keep coming back at you with what your actual behavior says about your priorities. To demonstrate the program Horvitz tells it, "Go to July Fourth." Life Browser complies with photographs of parades and cookouts. Horvitz says of the program, "It comes to understand your mind, how you organize your memories, by what you choose. It learns to become like you, to help you be a better you."[14]

I think of my mother's photograph drawer, intentionally kept messy. Her Life Browser would have reflected disorder and contradiction, for every time she chose a photograph, she told a different story. Some were true, and some only bore the truth of wishes. Understanding these wishes made my time at the photograph drawer precious to me. In contrast, Gemmell imagines how Life Browser and its artificially intelligent descendants will relieve him of the burden of personal narration: "My dream is I go on vacation and take my pictures and come home and tell the computer, 'Go blog it,' so that my mother can see it. I don't have to do anything; the story is there in the pattern of the images."[15]

Don, twenty-one, a civil engineering student at a West Coast university, wants a life archive. He shoots photographs with his iPhone and uploads them to the Web every night, often a hundred a day. He says that his friends want to see everything he does, so "I put my life on Facebook. I don't like to make choices [among the photographs]. My friends can choose. I just like to have it all up there." There is nothing deliberate in Don's behavior except for its first premise: shoot as much of your life as possible and put it on the Web. Don is confident that "a picture of my life will emerge from, well, all the pictures of my life."

Don hasn't heard of Life Browser but has confidence that it is only a matter of time before he will have access to an artificial intelligence that will be able to see his life "objectively." He welcomes the idea of the documented life, organized by algorithm. The imperfect Facebook archive is only a first step. Rhonda, twenty-six, also uses Facebook to record her life. Her experience is more labored. "Taking and uploading photographs," she says, "feels like a requirement." Rhonda wants to save things on the computer because of a desire to remember ("I'll know exactly what I did") and to forget ("It's all there if I ever need to re-member something. If I put it on the computer, I don't have to think about it anymore"). This is what Gordon Bell calls "clean living"—but with a difference. In Bell's utopian picture, after the saving comes the sifting and savoring. For Rhonda, the practice of saving is an end in itself. Don and Rhonda suggest a world in which technology determines what we remember of the story of our lives. Observing software "learns" our "favorites" to customize what it is impor-tant to remember. Swaddled in our favorites, we miss out on what was in our peripheral vision.

The memex and MyLifeBits both grew out of the idea that technology has developed capacities that should be put to use. There is an implied compact with technology in which we agree not to waste its potential. Kevin Kelly re-frames this understanding in language that gives technology even greater voli-tion: as technology develops, it shows us what it "wants." To live peacefully with technology, we must do our best to accommodate these wants. By this logic, it would seem that right now, one of the things technology "wants" to do is ponder our memories.

A LETTER HOME

I begin drafting this chapter in the late summer of 2009. After a few weeks, my work is interrupted by the Jewish high holy days. On Yom Kippur, the Day of Atonement, there is a special service of mourning for the dead. This is Yiskor. Different synagogues have different practices. In mine, the rabbi delivers a ser-mon just before the service. This year, his comments bring me up short. Things that had seemed complicated now seem clear. The rabbi addresses the impor-tance of talking to the dead. His premise is that we want to, need to, talk to the dead. It is an important, not a maudlin, thing to do. The rabbi suggests that we have four things to say to them: I'm sorry. Thank you. I forgive you. I love you. This is what makes us human, over time, over distance.

When my daughter and I have our first conversation on Skype (Dublin/Boston), I'm in the midst of reviewing my materials on Gordon Bell and the MyLifeBits program. I tell Rebecca I'm writing about the possibility of being able to archive everything we do. I ask her if she would like to have a record of all of her communications during her time in Dublin: e-mails, texts, instant messages, Facebook communications, calls, conversations, searches, pictures of everyone she has met and all the travelling she has done. She thinks about it. After a silence, she finally says, "Well, that's a little pack ratty, creepy." When people are pack rats, the volume of things tends to mean that equal weight is given to every person, conversation, and change of venue. More appealing to her are human acts of remembrance that filter and exclude, that put events into shifting camps of meaning—a scrapbook, a journal. And perhaps, at eighteen, she senses that, for her, archiving might get in the way of living. To live most fully, perhaps we need at least the fiction that we are not archiving. For surely, in the archived life, we begin to live for the record, for how we shall be seen.

As Rebecca and I talk about what has weight for her in her year abroad, I tell her that, prompted by her absence, I have been looking over my freshman-year correspondence with my mother. I ask my daughter if she would like to write me a letter. Since she already sends me regular text messages and we're now on Skype talking about what shoes she should wear to the "Back to the Future" Ball at her Dublin College, she has a genuine moment of puzzlement and says, "I don't know what my subject could be." I appreciate that with the amount of communication we have, it could well seem that all topics have been exhausted. Nevertheless, I say something like, "You could write about your thoughts about being in Ireland, how you feel about it. Things that would mean special things to me." Over time, over distance, through the fishbowl of Skype, Rebecca stares at me from her dorm room and repeats, "Maybe if I could find a subject."

As I talk to Rebecca about the pleasures of my correspondence with my mother, she comments sensibly, "So send me a letter." And so I have.

NOTES

AUTHOR'S NOTE: TURNING POINTS

1. Sherry Turkle, "Inner History," in Sherry Turkle, ed., *The Inner History of Devices* (Cambridge, MA: MIT Press, 2008), 2–29.

2. Sherry Turkle, *The Second Self: Computers and the Human Spirit* (1984; Cambridge, MA: MIT Press, 2005), 2.

3. Sherry Turkle, *Life on the Screen: Identity in the Age of the Internet* (New York: Simon and Schuster, 1995), 13.

4. Roger Entner, "Under-aged Texting: Usage and Actual Cost," Nielsen.com, January 27, 2010, http://blog.nielsen.com/nielsenwire/online_mobile/under-aged-texting -usage-and-actual-cost (accessed May 30, 2010).

INTRODUCTION: ALONE TOGETHER

1. See "What Is Second Life," Second Life, http://secondlife.com/whatis (accessed June 13, 2010).

2. Benedict Carey and John Markoff, "Students, Meet Your New Teacher, Mr. Robot," *New York Times*, July 10, 2010, www.nytimes.com/2010/07/11/science/11 robots.html (accessed July 10, 2010); Anne Tergeson and Miho Inada, "It's Not a Stuffed Animal, It's a $6,000 Medical Device," *Wall Street Journal*, June 21, 2010, http://online .wsj.com/article/SB10001424052748704463504575301051844937276.html (accessed August 10, 2010); Jonathan Fildes, "'Virtual Human' Milo Comes Out to Play at TED in Oxford," *BBC News*, July 13, 2010, www.bbc.co.uk/news/10623423 (accessed July 13, 2010); Amy Harmon, "A Soft Spot for Circuitry: Robot Machines as Companions," *New York Times*, July 4, 2010, www.nytimes.com/2010/07/05/science/05robot.html ?pagewanted=all (accessed July 4, 2010); Emily Veach, "A Robot That Helps You Diet," *Wall Street Journal*, July 20, 2010, http://online.wsj.com/article/SB10001424052 748704682604575369981478383568.html (accessed July 20, 2010).

3. On this, see "The Making of Deep Blue," IBM Research, www.research.ibm .com/deepblue/meet/html/d.3.1.html (accessed June 10, 2010).

4. David L. Levy, *Love and Sex with Robots: The Evolution of Human-Robot Relationships* (New York: Harper Collins, 2007).

5. The book review is Robin Marantz Henig, "Robo Love," *New York Times*, December 2, 2007, www.nytimes.com/2007/12/02/books/review/Henig-t.html (accessed July 21, 2009). The original article about the MIT robot scene is Robin Marantz Henig, "The Real Transformers," *New York Times*, July 29, 2007, www.nytimes.com/ 2007/07/29/magazine/29robots-t.html (accessed July 21, 2009).

6. Levy, *Love and Sex*, 22.

7. On "alterity," the ability to put oneself in the place of another, see Emmanuel Lévinas, *Alterity and Transcendence*, trans. Michael B. Smith (London: Athlone Press, 1999).

8. Sherry Turkle, *The Second Self: Computers and the Human Spirit* (1984; Cambridge, MA: MIT Press, 2005), 183–218.

9. The way here is paved by erotic images of female robots used to sell refrigerators, washing machines, shaving cream, and vodka. See, for example, the campaign for Svedka Vodka (Steve Hall, "Svedka Launches Futuristic, Un-PC Campaign," Andrants.com, September 20, 2005, www.adrants.com/2005/09/svedka-launches-futuristic-unpc .php [accessed September 1, 2009]) and Phillip's shaving system ("Feel the Erotic Union of Man and Shavebot," AdFreak.com, August 21, 2007, http://adweek.blogs.com/ad freak/2007/08/feel-the-erotic.html [accessed September 1, 2009]).

10. Sharon Moshavi, "Putting on the Dog in Japan," *Boston Globe*, June 17, 1999, A1.

11. As preteens, the young women of the first Google generation (born roughly from 1987 to 1993) wore clothing widely referred to as "baby harlot"; they listened to songs about explicit sex well before puberty. Their boomer parents had few ideas about where to draw lines, having spent their own adolescences declaring the lines irrelevant. Boomer parents grew up rejecting parental rules, but knowing that there were rules. One might say it is the job of teenagers to complain about constraints and the job of parents to insist on them, even if the rules are not obeyed. Rules, even unheeded, suggest that twelve to fifteen are not good ages to be emotionally and sexually enmeshed.

Today's teenagers cannot easily articulate any rules about sexual conduct except for those that will keep them "safe." Safety refers to not getting venereal diseases or AIDS. Safety refers to not getting pregnant. And on these matters teenagers are eloquent, unembarrassed, and startlingly well informed. But teenagers are overwhelmed with how unsafe they feel in relationships. A robot to talk to is appealing—even if currently unavailable—as are situations that provide feelings of closeness without emotional demands. I have said that rampant fantasies of vampire lovers (closeness with constraints on sexuality) bear a family resemblance to ideas about robot lovers (sex without intimacy, perfect). And closeness without the possibility of physical intimacy and eroticized encounters that can be switched off in an instant—these are the affordances of online encounters. Online romance expresses the aesthetic of the robotic moment. From a certain perspective, they are a way of preparing for it. On the psychology of adolescents' desire for relationships with constraint, I am indebted to conversations with child and adolescent psychoanalyst Monica Horovitz in August 2009.

12. Commenting on the insatiable desire for robot pets during the 2009 holiday season, a researcher on social trends comments, "A toy trend would be something that reflects the broader society, that tells you where society is going, something society needs." Gerald Celente, founder of the Trends Research Institute, cited in Brad Tuttle, "Toy Craze Explained: A Zhu Zhu Pet Hamster Is Like a 'Viral Infection,'" *Time*, December 9, 2009, http://money.blogs.time.com/2009/12/07/toy-craze-explained-a-zhu-zhu-pet-hamster-is-like-a-viral-infection (accessed December 9, 2009).

13. For classic psychodynamic formulations of the meaning of symptoms, see Sigmund Freud, "The Unconscious," in *The Standard Edition of Sigmund Freud*, ed. and trans. James Strachey et al. (London: Hogarth Press, 1953–1974), 14:159–204; "Introductory Lectures on Psychoanalysis," in *The Standard Edition*, vols. 15 and 16; "From the History of an Infantile Neurosis," in *The Standard Edition*, 17:1–122; "Inhibitions, Symptoms, and Anxiety," in *The Standard Edition*, 20:75–172; and Sigmund Freud and Joseph Breuer, "Studies on Hysteria," in *The Standard Edition*, 2:48–106. For Freud on dreams as wishes, see "The Interpretation of Dreams," in *The Standard Edition*, vol. IV.

14. For an argument about the pleasures of limited worlds in another technological realm, see Natasha Schüll's work on gambling, *Addiction by Design: Machine Gambling in Las Vegas* (Princeton, NJ: Princeton University Press, forthcoming).

15. See, for example, Bill Gates, "A Robot in Every Home," *Scientific American*, January 2007, www.scientificamerican.com/article.cfm?id=a-robot-in-every-home (accessed September 2, 2009).

16. See Sherry Turkle, *Life on the Screen: Identity in the Age of the Internet* (New York: Simon and Schuster, 1995). On life as performance, the classic work is Erving Goffman, *The Presentation of Self in Everyday Life* (Garden City, NY: Doubleday Anchor, 1959).

17. The apt phrase "identity workshop" was coined by my then student Amy Bruckman. See "Identity Workshop: Emergent Social and Psychological Phenomena in Text-Based Virtual Reality" (unpublished essay, Media Lab, Massachusetts Institute of Technology, 1992), www.cc.gatech.edu/~asb/papers (accessed September 2, 2009).

18. Sociologists distinguish between strong ties, those of family and close friendship, and weak ties, the bonds of acquaintanceship that make us comfortable at work and in our communities. Facebook and Twitter, friending rather than friendship—these are worlds of weak ties. Today's technology encourages a celebration of these weak ties as the kind we need in the networked life. The classic work on weak ties is Mark S. Granovetter, "The Strength of Weak Ties," *American Journal of Sociology* 78, no. 6 (May 1973): 1360–1380.

19. See, for example, Matt Richtel, "In Study, Texting Lifts Crash Risk by Large Margin," *New York Times*, July 27, 2009, www.nytimes.com/2009/07/28/technology/28texting.html (accessed September 1, 2009). On the pressure that friends and family members put on drivers who text, see "Driver Texting Now an Issue in Back Seat," *New York Times*, September 9, 2009, www.nytimes.com/2009/09/09/technology/09

distracted.html (accessed September 9, 2009). As I complete this book, Oprah Winfrey has made texting while driving a personal crusade, encouraging people across America to sign an online pledge to not text and drive. See "Oprah's No Phone Zone," Oprah.com, www.oprah.com/packages/no-phone-zone.html (accessed May 30, 2010).

20. The teenage national average as of January 2010 is closer to thirty-five hundred; my affluent, urban neighborhood has a far higher number. Roger Entner, "Under-aged Texting: Usage and Actual Cost," Nielson.com, January 27, 2010, http://blog.nielsen .com/nielsenwire/online_mobile/under-aged-texting-usage-and-actual-cost (accessed May 30, 2010). On texting's impact on teenage life, see Katie Hafner, "Texting May Be Taking Its Toll," *New York Times*, May 25, 2009, www.nytimes.com/2009/05/26/health/ 26teen.html?_r=2&8dpc (accessed July 21, 2009).

21. To find friends in the neighborhood, Loopt for the iPhone is a popular "app."

22. A witty experiment suggests that Facebook "friends" won't even show up when you invite them to a party. Hal Niedzviecki, "Facebook in a Crowd," *New York Times*, October 24, 2008, www.nytimes.com/2008/10/26/magazine/26lives-t.html (accessed July 27, 2010).

23. From Winston Churchill's remarks to the English Architectural Association in 1924, available at the International Centre for Facilities website at www.icf-cebe .com/quotes/quotes.html (accessed August 10, 2010). Churchill's comment is, of course, very similar to the spirit of Marshall McLuhan. See, for example, *Understanding Media: The Extensions of Man* (1964; Cambridge, MA: MIT Press, 1994).

CHAPTER 1: NEAREST NEIGHBORS

1. Weizenbaum had written the program a decade earlier. See Joseph Weizenbaum, "ELIZA—a Computer Program for the Study of Natural Language Communication Between Man and Machine," *Communications of the ACM*, vol. 9, no. 1 (January 1966): 36–45.

2. See Joseph Weizenbaum, *Computer Power and Human Reason: From Judgment to Calculation* (San Francisco: W. H. Freeman, 1976).

3. For whatever kind of companionship, a classical first step is to make robots that are physically identical to people. In America, David Hanson has an Albert Einstein robot that chats about relativity. At the TED conference in February 2009, Hansen discussed his project to create robots with empathy as the "seeds of hope for our future." See http://www.ted.com/talks/david_hanson_robots_that_relate_to_you.html (accessed August 11, 2010) On Hanson, also see Jerome Groopman, "Robots That Care: Advances in Technological Therapy," *The New Yorker*, November 2, 2009, www.newyorker.com/reporting/2009/11/02/091102fa_fact_groopman (accessed November 11, 2009).

These days, you can order a robot clone in your own image (or that of anyone else) from a Japanese department store. The robot clone costs $225,000 and became available in January 2010. See "Dear Santa: I Want a Robot That Looks Just Like Me," Ethics Soup, December 17, 2009, www.ethicsoup.com/2009/12/dear-santa-i-want-a-robot-that-looks-like-me.html (accessed January 12, 2010).

4. Bryan Griggs, "Inventor Unveils $7,000 Talking Sex Robot," CNN, February 1, 2010, www.cnn.com/2010/TECH/02/01/sex.robot/index.html (accessed June 9, 2010).

5. Raymond Kurzweil, *The Singularity Is Near: When Humans Transcend Biology* (New York: Viking, 2005). On radical images of our future, see Joel Garreau, *Radical Evolution: The Promise and Peril of Enhancing Our Minds, Our Bodies—and What It Means to Be Human* (New York: Doubleday, 2005).

6. For my further reflections on computer psychotherapy, see "Taking Things at Interface Value," in Sherry Turkle, *Life on the Screen: Identity in the Age of the Internet* (New York: Simon and Schuster, 1995), 102–124.

7. There is, too, a greater willingness to enter into a relationship with a machine if people think it will make them feel better. On how easy it is to anthropomorphize a computer, see Byron Reeves and Clifford Nass, *The Media Equation: How People Treat Computers, Television and New Media Like Real People and Places* (New York: Cambridge University Press, 1996). See also, on computer psychotherapy, Harold P. Erdman, Marjorie H. Klein, and John H. Greist, "Direct Patient Computer Interviewing," *Journal of Consulting and Clinical Psychology* 53 (1985): 760–773; Kenneth Mark Colby, James B. Watt, and John P. Gilbert, "A Computer Method for Psychotherapy: Preliminary Communication," *Journal of Nervous and Mental Diseases* 142, no. 2 (1966): 148–152; Moshe H. Spero, "Thoughts on Computerized Psychotherapy," *Psychiatry* 41 (1978): 281–282.

8. For my work on early computational objects and the question of aliveness, see Sherry Turkle, *The Second Self: Computers and the Human Spirit* (1984; Cambridge, MA: MIT Press, 2005). That work on aliveness continued with a second generation of computational objects in Turkle, *Life on the Screen*. My inquiry, with an emphasis on children's reasoning rather than their answers, is inspired by Jean Piaget, *The Child's Conception of the World*, trans. Joan Tomlinson and Andrew Tomlinson (Totowa, NJ: Littlefield, Adams, 1960).

9. On the power of the liminal, see, for example, Victor Turner, *The Ritual Process: Structure and Anti-Structure* (Chicago: Aldine, 1969), and *The Forest of Symbols: Aspects of Ndembu Ritual* (1967; Ithaca, NY: Cornell University Press, 1970). See also Mary Douglas, *Purity and Danger: An Analysis of Concepts of Pollution and Taboo* (London: Routledge and Kegan Paul, 1966).

10. Piaget, *The Child's Conception*.

11. Turkle, *The Second Self*, 33–64.

12. Children, in fact, settled on three new formulations. First, when it came to thinking through the aliveness of computational objects, autonomous motion was no longer at the heart of the matter. The question was whether computers had autonomous cognition. Second, they acknowledged that computer toys might have some kind of awareness (particularly of them) without being alive. Consciousness and life were split. Third, computers seemed alive because they could think on their own, but were only "sort of alive" because even though they could think on their own, their histories undermined their autonomy. So an eight-year-old said that Speak & Spell was "sort of alive" but not "really alive" because it had a programmer. "The programmer," he said, "gives it its ideas. So the ideas don't come from the game." These days, sociable

robots, with their autonomous behavior, moods, and faces, seem to take the programmer increasingly out of the picture. And with the formulation "alive enough," children put the robots on a new terrain. As for cognition, it has given way in children's minds to the capacity to show attention, to be part of a relationship of mutual affection.

13. Turkle, *Life on the Screen,* 169.

14. Turkle, *Life on the Screen,* 173–174.

15. The quotation is from a journal entry by Emerson in January 1832. The passage reads in full, "Dreams and beasts are two keys by which we are to find out the secrets of our nature. All mystics use them. They are like comparative anatomy. They are our test objects." See Joel Porte, ed. *Emerson in His Journals* (Cambridge, MA: Belknap Press, 1982), 81.

16. According to psychoanalyst D. W. Winnicott, objects such as teddy bears, baby blankets, or a bit of silk from a first pillow mediate between the infant's earliest bonds with the mother, who is experienced as inseparable from the self, and other people, who will come to be experienced as separate beings. These objects are known as "transitional," and the infant comes to know them both as almost inseparable parts of the self and as the first "not me" possessions. As the child grows, these transitional objects are left behind, but the effects of early encounters with them remain. We see them in the highly charged relationships that people have with later objects and experiences that call forth the feeling of being "at one" with things outside the self. The power of the transitional object is associated with religion, spirituality, the perception of beauty, sexual intimacy, and the sense of connection with nature. And now, the power of the transitional object is associated with computers and, even more dramatically, with sociable robots. On transitional objects, see D. W. Winnicott, *Playing and Reality* (New York: Basic Books, 1971).

17. In the early 1980s, children's notion of people as "emotional machines" seemed to me an unstable category. I anticipated that later generations of children would find other formulations as they learned more about computers. They might, for example, see through the apparent "intelligence" of the machines by developing a greater understanding of how they were created and operated. As a result, children might be less inclined to see computers as kin. However, in only a few years, things moved in a very different direction. Children did not endeavor to make computation more transparent. Like the rest of the culture, they accepted it as opaque, a behaving system. Children taking sociable robots "at interface value" are part of a larger trend. The 1984 introduction of the Macintosh encouraged its users to stay on the surface of things. The Macintosh version of "transparency" stood the traditional meaning of that word on its head. Transparency used to refer to the ability to "open up the hood" and look inside. On a Macintosh it meant double-clicking an icon. In other words, transparency had come to mean being able to make a simulation work without knowing how it worked. The new transparency is what used to be called opacity. For more on this question, see Turkle, *Life on the Screen,* especially 29–43, and Sherry Turkle, *Simulation and Its Discontents* (Cambridge, MA: MIT Press, 2009).

18. Our connections with the virtual intensifies when avatars look, gesture, and move like us; these connections become stronger when we move from the virtual to

the embodied robotic. Computer scientist Cory Kidd studied attachment to a computer program. In one condition the program issued written commands that told study participants what to do. In a second condition, an on-screen avatar issued the same instructions. In a third condition an on-screen robot was used to give the same instructions. The robot engendered the greatest attachment. Cory Kidd, "Human-Robot Interaction" (master's thesis, Massachusetts Institute of Technology, 2003).

19. The Tamagotchi website cautions about unfavorable outcomes: "If you neglect your little cyber creature, your Tamagotchi may grow up to be mean or ugly. How old will your Tamagotchi be when it returns to its home planet? What kind of virtual caretaker will you be?" The packaging on a Tamagotchi makes the agenda clear: "There are a total of 4 hearts on the 'Happy' and 'Hunger' screens and they start out empty. The more hearts that are filled, the better satisfied Tamagotchi is. You must feed or play with Tamagotchi in order to fill the empty hearts. If you keep Tamagotchi full and happy, it will grow into a cute, happy cyberpet. If you neglect Tamagotchi, it will grow into an unattractive alien." The manufacturer of the first Tamagotchi is Bandai. Its website provides clear moral instruction that links nurturance and responsibility. See the Bandai website at www.bandai.com (accessed October 5, 2009).

20. See "Tamagotchi Graveyard," Tamagotchi Dreamworld, http://members.tripod .com/~shesdevilish/grave.html (accessed June 15, 2009).

21. In Japan, a neglected Tamagotchi dies but can be uploaded to a virtual graveyard. In the United States, manufacturers propose gentler resolutions. Some neglected Tamagotchis might become "angels" and return to their home planet. On the Tamagotchis I played with, it was possible to hit a reset button and be presented with another creature.

22. Sigmund Freud, "The Uncanny," in *The Standard Edition of Sigmund Freud,* ed. and trans. James Strachey et al. (London: Hogarth Press, 1953–1974), 17:219–256.

23. Sigmund Freud, "Mourning and Melancholia," in *The Standard Edition,* 14: 237–258.

24. See "Tamagotchi Graveyard."

25. Other writings on the Tamagotchi gravesite include the epitaph for a Tamagotchi named Lacey who lived for ninety-nine years. We know how hard it was for her owner to achieve this result, but he is modest about his efforts: "She wasn't much trouble at all." But even with his considerable accomplishment, he feels her death was due to his neglect: "I slept late on a Sunday and she died." But in the simple expressions of guilt (or perhaps a playing at guilt) are frank admissions of how hard it is to lose someone you love. Mourners say, "I was his mama and he will always love me as I loved him"; "He went everywhere with me. He was a loving and faithful pet"; "I'm sorry and I real[l]y miss you!"; and "God gave him life. I gave him death." Some mourners express their belief in redemption through the generativity of generations. Thus is "Little Guy" memorialized, dead at forty-eight: "I hope you are very happy, Little Guy. I'm currently taking care of your son. I know he's yours because he looks and acts just like you. I'm really sorry I couldn't save you and had you on pause a lot when you were older." See "Tamagotchi Graveyard."

CHAPTER 2: ALIVE ENOUGH

1. The fact that the Furby was so hard to quiet down was evidence of its aliveness. Even adults who knew it was not alive saw it as playing on the boundaries of life. The response of many was to see the Furby as out of control, intolerable, or, as one put it, insane. A online video of an "insane Furby" shows the Furby chatting away, to the increasing consternation of its adult owner. To stop it, he slaps its face, sticks his fingers in its mouth, holds down its ears and eyes, smashes it against a wall, and throws it down a flight of stairs. None of these shuts it down. If anything, its language becomes more manic, more "desperate." Finally comes the solution of taking out the Furby's batteries with a Phillips screwdriver. Now, the quiet Furby is petted. Its owner comments, "That's better." See "Insane Furby," YouTube, March 15, 2007, www.youtube.com/watch?v=g4 Dfg4xJ6Ko (accessed November 11, 2009).

2. These enactments bring theory to ground level. See Donna Haraway, "A Cyborg Manifesto: Science, Technology, and Socialist-Feminism in the Late Twentieth Century," in *Simians, Cyborgs and Women: The Reinvention of Nature* (New York: Routledge, 1991), 149–181, and N. Katherine Hayles, *How We Became Posthuman: Virtual Bodies in Cybernetics, Literature, and Informatics* (Chicago: University of Chicago Press, 1999).

3. Michael Chorost, *Rebuilt: How Becoming Part Computer Made Me More Human* (Boston: Houghton Mifflin, 2005).

4. Here, the Furby acts as what psychoanalyst D. W. Winnicott termed a "transitional object," one where the boundaries between self and object are not clear. See D. W. Winnicott, *Playing and Reality* (New York: Basic Books, 1971).

5. The idea that the Furby had the capacity to learn new words by "listening" to the language around it was persistent. The belief most likely stemmed from the fact that it was possible to have the Furby say certain preprogrammed words or phrases more often by petting it whenever it said them. As a result of this myth, several intelligence agencies banned Furbies from their offices, believing that they were recording devices camouflaged as toys.

6. Children move back and forth between he, she, and it in talking about relational artifacts. Once they make a choice, they do not always stick with it. I report on what children say and, thus, their sentences are sometimes inconsistent.

7. Peter H. Kahn and his colleagues studied online discussion groups that centered on Furbies. For their account, see Batya Friedman, Peter H. Kahn Jr., and Jennifer Hagman, "Hardware Companions? What Online AIBO Discussion Forums Reveal About the Human-Robotic Relationship," in *Proceedings of the Conference on Human Factors in Computing Systems* (New York: ACM Press, 2003), 273–280.

8. The artist Kelly Heaton played on the biological/mechanical tension in the Furby's body by creating a fur coat made entirely from the fur of four hundred "skinned" Furbies, reengineered into a coat for Mrs. Santa Claus. The artwork, titled *Dead Pelt*, was deeply disturbing. It also included a wall of reactive eyes and mouths, taken from Furbies, and a formal anatomical drawing of a Furby. See the Feldman Gallery's Kelly Heaton page at www.feldmangallery.com/pages/artistsrffa/arthea01 .html (accessed August 18, 2009).

9. Baird developed her thought experiment comparing how people would treat a gerbil, a Barbie, and a Furby for a presentation at the Victoria Institute, Gothenburg, Sweden, in 1999.

10. In Turing's paper that argued the existence of intelligence if a machine could not be distinguished from a person, one scenario involved gender. In "Computing Machinery and Intelligence," he suggested an "imitation game": a man and then a computer pose as female, and the interrogator tries to distinguish them from a real woman. See Alan Turing, "Computing Machinery and Intelligence," *Mind* 59, no. 236 (October 1950): 433–460.

11. Antonio Damasio, *The Feeling of What Happens: Body and Emotion in the Making of Consciousness* (New York: Harcourt, 1999). Since emotions are cognitive representations of body states, the body cannot be separated from emotional life, just as emotion cannot be separated from cognition.

12. There are online worlds and communities where people feel comfortable expressing love for Furbies and seriously mourning Tamagotchis. These are places where a deep sense of connection to the robotic are shared. These "sanctioned spaces" play an important part in the development of the robotic moment. When you have company and a community, a sense of intimacy with sociable machines comes to feel natural. Over time, these online places begin to influence the larger community. At the very least, a cohort has grown up thinking that their attitudes toward the inanimate are widely shared.

13. BIT was developed by Brooks and his colleagues at the IS Robotics Corporation. IS Robotics was the precursor to iRobot, which first became well known as the makers of the Roomba robotic vacuum cleaner.

14. Rodney A. Brooks, *Flesh and Machines: How Robots Will Change Us* (New York: Pantheon, 2002), 202.

15. Sherry Turkle, *The Second Self: Computers and the Human Spirit* (1984; Cambridge, MA: MIT Press, 2005), 61.

16. This field has a vast literature. Several works that have influenced my thinking include the early book by Peter D. Kramer, *Listening to Prozac: A Psychiatrist Explores Antidepressants and the Remaking of the Self* (New York: Viking, 1993), and the more recent Margaret Talbot, "Brain Gain: The Underground World of Neuroenhancing Drugs," *The New Yorker*, July 27, 2009, www.newyorker.com/reporting/2009/04/27/090427fa_fact_talbot (accessed July 21, 2009), and Nathan Greenslit, "Depression and Consumption: Psychopharmaceuticals, Branding, and New Identity Practices," *Culture, Medicine, and Psychiatry* 25, no. 4 (2005): 477–502.

CHAPTER 3: TRUE COMPANIONS

1. Three recent works by authors who have influenced my thinking are Jessica Riskin, ed., *Genesis Redux: Essays on the History and Philosophy of Artificial Life* (Chicago: University of Chicago Press, 2007); Gaby Wood, *Edison's Eve: A Magical History of the Quest for Mechanical Life* (New York: Anchor, 2003); and Barbara Johnson, *Persons and Things* (Cambridge, MA: Harvard University Press, 2008). Johnson explores how relations between persons and things can be more fluid while arguing a central ethical tenet: persons should be treated as persons.

2. Norbert Wiener, *God and Golem, Inc.: A Comment on Certain Points Where Cybernetics Impinges on Religion* (Cambridge, MA: MIT Press, 1966).

3. The literature on the negotiation of technology, self, and social world is rich and varied. I have been particularly influenced by the perspectives described in Wiebe Bijker, Thomas P. Hughes, and Trevor Pinch, eds., *The Social Construction of Technological Systems: New Directions in the Sociology and History of Technology* (1987; Cambridge, MA: MIT Press, 1999) and by the work of Karin D. Knorr Cetina and Bruno Latour. See, for example, Karin D. Knorr Cetina, "Sociality with Objects: Social Relations in Postsocial Knowledge Societies," *Theory, Culture and Society* 14, no. 4 (1997): 1–30; Karin D. Knorr Cetina, *Epistemic Cultures: How the Sciences Make Knowledge* (Cambridge, MA: Harvard University Press, 1999); Bruno Latour and Steve Woolgar, *Laboratory Life: The Construction of Scientific Facts* (1979; Princeton, NJ: Princeton University Press, 1986); Bruno Latour, *Science in Action: How to Follow Scientists and Engineers Through Society* (1987; Cambridge, MA: Harvard University Press, 1999); Bruno Latour, *Aramis, or the Love of Technology*, trans. Catherine Porter (1996; Cambridge, MA: Harvard University Press, 2002); and Bruno Latour, *We Have Never Been Modern*, trans. Catherine Porter (1991; Cambridge, MA: Harvard University Press, 2001).

In the specific area of the relationship between people and computational objects, this book is indebted to the work of Sara Kiesler, Lee Sproull, and Clifford Nass and their collaborators. See, for example, Sau-lai Lee and Sara Kiesler, "Human Mental Models of Humanoid Robots" (paper delivered at the International Conference on Robotics and Automation, Barcelona, Spain, April 18–22, 2005); Lee Sproull et al., "When the Interface Is a Face," *Human-Computer Interaction* 11 (1996): 97–124; Sara Kiesler and Lee Sproull, "Social Responses to 'Social' Computers," in *Human Values and the Design of Technology*, ed. Batya Friedman (Stanford, CA: CLSI Publications, 1997); Byron Reeves and Clifford Nass, *The Media Equation: How People Treat Computers, Television and New Media Like Real People and Places* (New York: Cambridge University Press, 1996); Clifford Nass and Scott Brave, *Wired for Speech: How Voice Activates and Advances the Human-Computer Relationship* (Cambridge, MA: MIT Press, 2005); Victoria Groom and Clifford Nass, "Can Robots Be Teammates? Benchmarks and Predictors of Failure in Human-Robot Teams," *Interaction Studies* 8, no. 3 (2008): 483–500; Leila Takayama, Victoria Groom, and Clifford Nass, "I'm Sorry, Dave, I'm Afraid I Won't Do That: Social Aspects of Human-Agent Conflict," *Proceedings of the Conference on Human Factors in Computing Systems* (Boston, MA: ACM Press, 2009), 2209–2108.

4. The object relations tradition in psychoanalytic thought proposes that infants see objects (and people) in terms of their functions. This partial understanding is captured by the phrase "part objects." So, for example, the breast that feeds the hungry infant is the "good breast." The hungry infant unsuccessfully tries to nurse in relation to the "bad breast." By interacting with the world, the child internalizes these external objects, which shape his or her psyche. Infants gradually grow to develop a sense of "whole objects." The internalized objects may not be accurate representations of the outside world, but they are what the child uses as he or she goes forward. D. W. Winnicott reassured mothers that in a "good-enough" facilitating environment, a percep-

tion of part objects eventually transforms into a comprehension of whole objects. This corresponds to the ability to tolerate ambiguity and to see that both the "good" and "bad" breast are part of the same mother. In larger terms, this underpins the ability throughout life to tolerate ambiguous and realistic relationships. See, for example, Melanie Klein, *Love, Guilt and Reparation: And Other Works, 1921–1945*, ed. Roger Money-Kyrle et al. (New York: Free Press, 1975), and D. W. Winnicott, *Playing and Reality* (New York: Basic Books, 1971).

5. Emmanuel Lévinas, *Alterity and Transcendence*, trans. Michael B. Smith (London: Athlone Press, 1999).

6. The Pokémon is a character that appears in a card collection with which Henry plays an elaborate set of war games. He collects character cards in which different creatures from the Pokémon world have different powers. Then, teams of creatures challenge each other. Henry spends a lot of time strategizing about how to maximize his team's powers in his war games. He spends a lot of time thinking about "powers." I thank my research assistant Lauren Klein for her helpful explanations of the Pokémon universe.

7. In the 1980s, the presence of a "programmer" figured in children's conversations about computer toys and games. The physical autonomy of robots seems to make the question of their historical determination fall out of the conversation. This is crucial in people's relating to robots as alive on their own account.

Peter H. Kahn and his colleagues performed a set of experiments that studied how children's attitudes and, crucially, their behavior differed with AIBOs and stuffed doll dogs. When questioned verbally, children reported opinions about AIBO that were similar to their opinions about a stuffed doll dog. But when you look not at what the children say but at what they do, the picture looks very different. Kahn analyzed 2,360 coded interactions. Most dramatically, children playing with AIBO were far more likely to attempt reciprocal behavior (engaging with the robot and expecting it to engage with them in return) than with the stuffed doll dog (683 to 180 occurrences). In the same spirit, half the children in Kahn's study said that both AIBO and the stuffed doll dog could hear, but children actually gave more verbal direction to AIBO (fifty-four occurrences) than to the stuffed doll dog (eleven occurrences). In other words, when children talk about the lifelike qualities of their dolls, children don't believe what they say. They do believe what they say about AIBO.

Similarly, children in Kahn's study were more likely to take action to "animate" the stuffed doll dog (207 occurrences) while they mostly let AIBO animate itself (20 occurrences). Most tellingly, the children were more likely to mistreat the stuffed doll dog than AIBO (184 to 39 occurrences). Relational artifacts, as I stress here, put children on a moral terrain.

Kahn also found evidence that children see AIBO as the "sort of entity with which they could have a meaningful social (human-animal) relationship." This expresses what I have called simultaneous vision: children see relational artifacts as both machine and creature. They both know AIBO is an artifact and treat it as a dog. See Peter H. Kahn Jr. et al., "Robotic Pets in the Lives of Preschool Children," *Interaction Studies:*

Social Behavior and Communication in Biological and Artificial Systems 7, no. 3 (2006): 405–436.

See also a study by Kahn and his colleagues on how people write about AIBO on the Web: Peter H. Kahn Jr., Batya Friedman, and Jennifer Hagman, "Hardware Companions? What Online AIBO Discussion Forums Reveal About the Human-Robotic Relationship," in *Proceedings of the Conference on Human Factors in Computing Systems* (New York: ACM Press, 2003), 273–280.

8. Older children and adults use access to the AIBO programming code more literally to create an AIBO in their image.

9. I note the extensive and growing literature suggesting that computation (including robots) will, in a short period, enable people to be essentially immortal. The best-known writer in this genre is Raymond Kurzweil who posits that within a quarter of a century, computational power will have hit a point he calls the singularity. It is a moment of "take-off," when all bets are off as to what computers will be able to do, think, or accomplish. Among the things that Kurzweil believes will be possible after the singularity is that humans will be able to embed themselves in a chip. They could either take an embodied robotic form or (perhaps until this becomes possible) roam a virtual landscape. This virtual self could evolve into an android self when the technology becomes ready. See Raymond Kurzweil, *The Singularity Is Near: When Humans Transcend Biology* (New York: Viking, 2005).

Kurzweil's work has captured the public imagination. A small sample of attention to the singularity in the media includes "Future Shock—Robots," *Daily Show with Jon Stewart,* August 24, 2006, www.thedailyshow.com/watch/wed-august-23-2006/future -shock-robots (accessed August 10, 2010); the IEEE Spectrum's special issue *The Singularity: A Special Report,* June 3, 2008; James Temple, "Singularity Taps Students' Technology Ideas," *San Francisco Chronicle,* August 28, 2009, www.sfgate.com/cgi -bin/article.cgi?f=/c/a/2009/08/27/BUUQ19EJIL.DTL&type=tech (accessed September 2, 2009); Iain Mackenzie, "Where Tech and Philosophy Collide," *BBC News,* August 12, 2009, http://news.bbc.co.uk/2/hi/technology/8194854.stm (accessed September 2, 2009).

One hears echoes of a "transhuman perspective" (the idea that we will move into a new realm by merging with our machines) in children's asides as they play with AIBO. Matt, nine, reflecting on his AIBO, said, "I think in twenty years from now, if they had the right stuff, they could put a real brain in a robot body." The idea of practicing for an impossible "perfection" comes to mind when I attend meetings of AIBO users. They come together to show off how they have customized their AIBOs. They reprogram and "perfect them." The users I speak to spend as much as fifty to sixty hours a week on their hobby. Some are willing to tell me that they spend more time with their AIBOs than with their families. Yet, AIBO is experienced as relaxation. As one thirty-five-year-old computer technician says, "All of this is less pressure than a real dog. No one will die." Of all the robotic creatures in my studies, the AIBO provoked the most musing about death and the finality of loss.

10. Sherry Turkle, *The Second Self: Computers and the Human Spirit* (1984; Cambridge, MA: MIT Press, 2005), 41.

11. This is a classic use of the defense mechanism known as projective identification, or seeing in someone else what you feel within yourself. So, if a teenager is angry at her prying mother, she may imagine her mother to be hostile. If a wife is angry with an inattentive husband, she may find her husband aggressive. Sometimes these feelings are conscious, but often they are not. Children use play to engage in the projections that can bring unacknowledged feelings to light. "Play is to the child what thinking, planning, and blueprinting are to the adult, a trial universe in which conditions are simplified and methods exploratory, so that past failures can be thought through, expectations tested." See Erik Erikson, *The Erik Erikson Reader*, ed. Robert Coles (New York: W. W. Norton, 2000), 195–196.

12. Interview. July 2000.

13. Hines says that the robot is "designed to engage the owner with conversation rather than lifelike movement." See "Roxxxy Sex Robot [PHOTOS]: World's First 'Robot Girlfriend' Can Do More Than Chat," Huffington Post, January 10, 2010, www.huffingtonpost.com/2010/01/10/roxxxy-sex-robot-photo-wo_n_417976.html ?view=print (accessed January 11, 2010).

14. Paul Edwards, *The Closed World: Computers and the Politics of Discourse in Cold War America* (Cambridge, MA: MIT Press, 1997).

CHAPTER 4: ENCHANTMENT

1. In an interview about the possibilities of a robot babysitter, psychologist Clifford Nass said, "The question is, if robots *could* take care of your children, would you let them? What does it communicate about our society that we're not making child-care a number-one priority?" I later spoke with Nass about his response to the idea of the nanny bot, and he rephrased with greater emphasis: "The first problem with offering children a robotic babysitter is that you would have to explain to the children why you were considering this. Why were there no people around for the child?" See Brandon Keim, "I, Nanny: Robot Babysitters Pose Dilemma," *Wired*, December 18, 2008, www .wired.com/wiredscience/2008/12/babysittingrobo (accessed May 31, 2010).

2. One article puts this in the context of the Japanese work ethic. See Jennifer Van House Hutcheson, "All Work and No Play," MercatorNet, May 31, 2007, www.mer catornet.com/articles/view/all_work_and_no_play (accessed August 20, 2009). Another reports on an elderly couple who hired actors to portray their grandchildren, but then assaulted them as they wished to assault their true grandchildren. See Pencil Louis, "Elderly Yokohama," OurCivilisation.com, www.ourcivilisation.com/smart board/shop/madseng/chap20.htm (accessed August 20, 2009).

3. Chelsea's mother, Grace, fifty-one, explains her position: "Active young people are simply not the right companions for old people and infirm people." She says, "When I bring my daughters to see my mom, I feel guilty. They shouldn't be here. She isn't what she was, even not as good as when they were little. . . . I think it's better they remember her happier, healthier." Grace had seen Paro in my office and now is intrigued with the idea of bringing the robot to her mother. For Grace, her mother's immortality depends on now declaring her no longer the person to remember. It is the case that robots seem "easier" to give as companions to people whom society declares

"nonpersons." Grace is not saying that her mother is a nonperson, but her choice of a robotic companion marks the moment when Grace's mother is no longer the person Grace wants to remember as her mother.

CHAPTER 5: COMPLICITIES

1. Rodney A. Brooks, "The Whole Iguana," in *Robotics Science*, ed. Michael Brady, MIT Press, 1989), 432–456. This was written in response to a two-page challenge by Daniel C. Dennett about lessons to be learned by building a complete system rather than just modules. See Daniel C. Dennett, "Why Not the Whole Iguana?" *Behavioral and Brain Sciences* 1 (1978): 103–104.

2. Kismet is programmed to recognize the word "say" and will repeat the word that follows it. So, children trying to teach Kismet its name would instruct, "Say Kismet," and Kismet would comply, much to their glee. Similarly, children would try to teach Kismet their names by saying, "Say Robert" . . . "Say Evelyn" . . . "Say Mark." Here, too, it was within Kismet's technical ability to comply.

3. Cog and Kismet were both built at the MIT Artificial Intelligence Laboratory. Cog has visual, tactile, and kinesthetic sensory systems and is capable of a variety of social tasks, including visually detecting people and salient objects, orienting to visual targets, pointing to visual targets, differentiating between animate and inanimate movement, and performing simple tasks of imitation. Kismet is a robotic head with five degrees of freedom, an active vision platform, and fourteen degrees of freedom in its display of facial expressions. Though the Kismet head sits disembodied on a platform, it is winsome in appearance. It possesses small, mobile ears made of folded paper, mobile lips made from red rubber tubing, and heavily lidded eyes ringed with false eyelashes. Its behaviors and capabilities are modeled on those of a preverbal infant. Kismet gives the impression of looking into people's eyes and can recognize and generate speech and speech patterns, although to a limited degree.

Much has been written about these two very well-known robots. See Rodney A. Brooks et al., "The Cog Project: Building a Humanoid Robot," in *Computation for Metaphors, Analogy and Agents*, vol. 1562 of *Springer Lecture Notes in Artificial Intelligence*, ed. C. Nehaniv (New York: Springer-Verlag, 1998), and Rodney Brooks, *Flesh and Machines: How Robots Will Change Us* (New York: Pantheon, 2002). Brian Scassellati did his dissertation work on Cog. See Brian Scassellati, *Foundations for a Theory of Mind for a Humanoid Robot* (PhD diss., Massachusetts Institute of Technology, 2001). Scassellati and Cynthia Breazeal worked together during early stages of the Kismet project, which became the foundation of Breazeal's dissertation work. See Cynthia Breazeal and Brian Scassellati, "How to Build Robots That Make Friends and Influence People" (paper presented at the IEEE/RSJ International Conference on Intelligent Robots and Systems, Kyongju, Korea, October 17–21, 1999), in *Proceedings of the IEEE/RSJ International Conference on Intelligent Robots and Systems* (IROS) (1999), 858–863. Cynthia Breazeal and Brian Scassellati, "Infant-like Social Interactions Between a Robot and a Human Caretaker," *Adaptive Behavior* 8 (2000): 49–74; Cynthia Breazeal, "Sociable Machines: Expressive Social Exchange Between Humans and Robots" (PhD diss., Massachusetts Institute of Technology, 2000); and Cynthia Breazeal, *Designing Sociable Robots* (Cambridge, MA: MIT Press, 2002).

4. Cynthia Breazeal discusses the astronaut project in Claudia Dreifus, "A Conversation with Cynthia Breazeal: A Passion to Build a Better Robot, One with Social Skills and a Smile," *New York Times*, June 10, 2003, www.nytimes.com/2003/06/10/science/ conversation-with-cynthia-breazeal-passion-build-better-robot-one-with-social .html?pagewanted=all (accessed September 9, 2009).

5. I cite this student in Sherry Turkle, *The Second Self: Computers and the Human Spirit* (1984; Cambridge, MA: MIT Press, 2005), 271. The full Norbert Weiner citation is "This is an idea with which I have toyed before—that it is conceptually possible for a human being to be sent over a telegraph line." See Norbert Wiener, *God and Golem, Inc.: A Comment on Certain Points Where Cybernetics Impinges on Religion* (Cambridge, MA: MIT Press, 1966), 36.

6. People drawn to sociable robots seem to hit a wall that has come to be known as the "uncanny valley." This phrase is believed to have been coined by Masahiro Mori in "The Uncanny Valley," *Energy* 7, no. 4 (1970): 33–35, An English translation by Karl F. MacDorman and Takashi Minato is available at www.androidscience.com/the uncannyvalley/proceedings2005/uncannyvalley.html (accessed November 14, 2009).

If one plots a graph with humanlike appearance on the x axis and approval of the robot on the y axis, as the robot becomes more lifelike, approval increases until the robot becomes too lifelike, at which point approval plummets into a "valley." When a robot is completely indistinguishable from a human, approval returns. Japanese roboticist Hiroshi Ishiguro thinks he is building realistic androids that are close to climbing out of the uncanny valley. See, for example, Karl F. MacDorman and Hiroshi Ishiguro, "The Uncanny Advantage of Using Androids in Cognitive and Social Science Research," *Interaction Studies* 7, no. 3 (2006): 297–337, and Karl F. MacDorman et al., "Too Real for Comfort: Uncanny Responses to Computer Generated Faces," *Computers in Human Behavior* 25 (2009): 695–710.

Like Ishiguro, roboticist David Hanson aspires to build realistic androids that challenge the notion of the uncanny valley. "We conclude that rendering the social human in all possible detail can help us to better understand social intelligence, both scientifically and artistically." See David Hanson et al., "Upending the Uncanny Valley," Association for the Advancement of Artificial Intelligence, May 11, 2005, www.aaai.org/ Papers/Workshops/2005/WS-05-11/WS05-11-005.pdf (accessed November 14, 2009).

7. A sympathetic reading of the possibilities of deep human-robot connections is represented in Peter H. Kahn Jr. et al., "What Is Human? Toward Psychological Benchmarks in the Field of Human-Robot Interaction," *Interaction Studies* 8, no. 3 (2007): 363–390, and Peter H. Kahn Jr. et al., "Social and Moral Relationships with Robotic Others?" in *Proceedings of the 13th International Workshop on Robot and Human Interactive Communication (RO-MAN '04)* (Piscataway, NJ: Institute of Electrical and Electronics Engineers, 2004), 545–550.

Yet in their 2006 paper "Robotic Pets in the Lives of Preschool Children" (*Interaction Studies: Social Behavior and Communication in Biological and Artificial Systems*

7, no. 3, 405–436), Kahn and his colleagues cite John Searle's 1992 critique of AI in formulating their own. See John Searle, *The Rediscovery of the Mind* (Cambridge, MA: MIT Press, 1992). Kahn concludes, "Although it is an open question our sense is that because computerized robots are formal systems, with syntax but not semantics, they will never be capable of engaging in full social relationships or of engendering full moral development in human beings."

8. The philosopher Emmanuel Lévinas writes that the presence of a face initiates the human ethical compact, which binds us before we know what lies behind a face. The face itself communicates, "You shall not commit murder." We seem to be summoned by the face even if we are looking at the face of a machine, something that cannot be killed. The robot's face certainly announces, as Lévinas puts it, "Thou shalt not abandon me." See Emmanuel Lévinas, "Ethics and the Face," in *Totality and Infinity: An Essay on Exteriority*, trans. Alphonso Lingis (Pittsburgh, PA: Duquesne University Press, 1969), 199. Lévinas notes that the capacity to put ourselves in the place of another, *alterity*, is one of the defining characteristics of the human. I speak of complicity because for a human being to feel that current robots are "others," the human must construct them as capable of alterity. See Emmanuel Lévinas, *Alterity and Transcendence*, trans. Michael Smith (New York: Columbia, 1999).

9. See Sherry Turkle et al., "First Encounters with Kismet and Cog: Children Respond to Relational Artifacts," in *Digital Media: Transformations in Human Communication*, ed. Paul Messaris and Lee Humphreys (New York: Peter Lang Publishing, 2006). I owe a special debt to Jennifer Audley for her contribution to the design and implementation of this study and to Olivia Dasté and Robert Briscoe for work on the analysis of transcripts.

10. Plato, *The Republic*, Book Two: *The Individual, the State, and Education*.

11. J. K. Rowling, *Henry Potter and the Chamber of Secrets* (New York: Scholastic, 1999), 329.

12. One twelve-year-old is troubled to learn about Scassellati's imminent departure. She pleads with him, "But Cog has seen you so much. Won't it miss you? I think Cog sees you as its dad. . . . It is easier for you to teach it than for any of us." She imagines the worst for the robot's future. "What if someone was trying to do something bad to the robot and you won't be there to protect it anymore?"

13. Brian Aldiss, *Supertoys Last All Summer Long and Other Stories of Future Time* (New York: St. Martin, 2001).

14. See Takayuki Kanda et al., "Interactive Humanoid Robots and Androids in Children's Lives," *Children, Youth and Environments* 19, no. 1, (2009): 12–33, www.colorado.edu/journals/cye (accessed July 4, 2009).

CHAPTER 6: LOVE'S LABOR LOST

1. For Paro's Guinness citation, see "Seal-Type Robot 'PARO' to Be Marketed with Best Healing Effect in the World," Paro Robots, January 4, 2005, www.parorobots.com/pdf/pressreleases/PARO%20to%20be%20marketed%202004-9.pdf (accessed July 27, 2010).

2. Publicity films for Paro show older men and women who live with Paro having breakfast with it, watching television with it, taking it to the supermarket and out to dinner. Sometimes Paro is adopted by a couple and sometimes by a younger person who simply does not enjoy living alone. In interviews, people say they are happy to have company that is easier to take care of than a real pet, company that will not die. See the Paro website at www.parorobots.com (accessed August 10, 2010).

3. Over years, I become convinced that in nursing homes, seniors become fascinated by relationships with robots because, among other reasons, they bring to the surface tensions about seniors' autonomy in their institutions. A robot that needs you promotes a fantasy of autonomy: seniors feel competent because something depends on them. Yet, robots can also disrupt fictions of autonomy. They can send the message, "Now you know that you are wholly dependent. Play with this toy. You are like a child." This push and pull makes the robots compelling even as they disturb. I owe this insight to my research assistant William Taggart. See Sherry Turkle et al., "Relational Artifacts with Children and Elders: The Complexities of Cybercompanionship," *Connection Science* 18, no. 4 (December 2006): 347–361, and Cory D. Kidd, William Taggart, and Sherry Turkle, "A Sociable Robot to Encourage Social Interaction Among the Elderly," *Proceedings of the 2006 IEEE International Conference on Robotics and Automation*, Orlando, Florida, May 15–19, 2006.

4. See, for example, Toshiyo Tamura et al., "Is an Entertainment Robot Useful in the Care of Elderly People with Severe Dementia?" *The Journals of Gerontology Series A: Biological Sciences and Medical Sciences* 59 (January 2004): M83–M85, http://biomed. gerontologyjournals.org/cgi/content/full/59/1/M83 (accessed August 15, 2009).

5. Suvendrini Kakuchi, "Robot Lovin'," *Asia Week Magazine Online*, November 9, 2001, www.asiaweek.com/asiaweek/magazine/life/0,8782,182326,00.html (accessed September 15, 2006).

6. It is standard for presentations about the "need" for sociable robots to begin with a slide demonstrating the inability to staff service jobs with people because of demographic trends. This slide is often drawn from the 2002 United Nations Report "UNPD World Population Ageing: 1950–2050," United Nations, www.un.org/esa/population/publications/worldageing19502050 (accessed July 8, 2009). The slides in this report dramatize (particularly for the developed world) that there are more and more older people and fewer and fewer younger people to take care of them. Nothing in my argument disputes the slide. I do question the leap from this slide to the inevitability of robots to care for people.

7. The meeting was sponsored by the Association for the Advancement of Artificial Intelligence. There are were also presentations on relational agents that dwell within the machine—for example, an "affective" health and weight-loss coach, developed by the chair of the symposium, Timothy W. Bickmore. See Timothy W. Bickmore, "Relational Agents: Effecting Change Through Human-Computer Relationships" (PhD diss., Massachusetts Institute of Technology, 2003), and Timothy W. Bickmore and Rosalind W. Picard, "Towards Caring Machines," in *CHI '04 Extended Abstracts on Human Factors in Computing Systems* (New York: ACM Press, 2004). Another presentation discussed a robotic cat, Max, named one of *Time* magazine's "Inventions of

the Year" for 2003. If you stroke Max or call it by name, the robot responds. See "Best Inventions 2003: Lap Cat," Time.com, www.time.com/time/2003/inventions/invcat.html (accessed September 23, 2009). Max is brought to the conference by Elena and Alexander Lubin, who are proponents of "robotherapy." For an overview of their work, see their entry in the *Encyclopedia of Applied Psychology* (Oxford: Elsevier, 2004), 289–293.

8. An estimated 26.2 percent of Americans ages 18 and older (and one in five children) suffer from a diagnosable mental disorder in a given year. National Institutes of Mental Health Statistics, http://www.nimh.nih.gov/health/topics/statistics/index.shtml (accessed August 10, 2010) and "The numbers count: Mental Disorders in America," http://www.nimh.nih.gov/health/publications/the-numbers-count-mental-disorders -in-america/index.shtml (Accessed August 10, 2010).

9. The movement to use robots in therapeutic situations is encouraged by the observed therapeutic potential of pets and the power of a gaze. See, for example, K. Allen et al., "Presence of Human Friends and Pet Dogs As Moderators of Autonomic Responses to Stress in Women," *Journal of Personality and Social Psychology* 61, no. 4 (1991): 582–589, and Michael Argyle and Mark Cook, *Gaze and Mutual Gaze* (Cambridge: Cambridge University Press, 1976).

10. We now have years of experience of people using games and the Internet as places to, in their words, "say what they couldn't say in their real lives." We see people, especially adolescents, using the anonymity and new opportunities of online life to experiment with identity. They try out new things, in safety—never a bad idea. When we see what we do in our lives on the screen, we can learn what we feel we are missing and use this information to enhance our lives in the "real." Over years of study I have learned that the people who do best with their lives on the screen are those who use them as material for self-reflection. This can be the focus of work between a therapist and a patient. In Sherry Turkle, ed., *The Inner History of Devices* (Cambridge, MA: MIT Press, 2008), see especially the contributions of therapists John Hamilton, Kimberlyn Leary, and Marcia Levy-Warren.

11. Cory D. Kidd, "Designing for Long-Term Human-Robot Interaction and Application to Weight Loss" (PhD diss., Massachusetts Institute of Technology, 2008). Rose (and Gordon) are pseudonyms that Kidd provided for his subjects.

12. Cory D. Kidd, "Sociable Robots: The Role of Presence and Task in Human-Robot Interaction" (master's thesis, Massachusetts Institute of Technology, 2003).

13. For an early experiment using a small, plush robot as a trigger for memories, see Marina Bers and Justine Cassell, "Interactive Storytelling Systems for Children: Using Technology to Explore Language and Identity," *Journal of Interactive Learning Research* 9, no. 2 (1999): 603–609.

14. See, for example, Erving Goffman, *The Presentation of Self in Everyday Life* (Garden City, NY: Doubleday Anchor, 1959).

15. The Intel Corporation joins with the Universities of Michigan and Pittsburgh, Carnegie Mellon, and Stanford on the Nursebot project. Nursebot tests a range of ideas for assisting elderly people, such as reminding elderly patients to visit the bathroom, take medicine, drink, or see the doctor; connecting patients with caregivers through the Internet; collecting data and monitoring the well-being of patients; ma-

nipulating objects around the home, such as the refrigerator, washing machine, or microwave; taking over certain social functions such as game playing and simple conversation. For a 2002 film produced by the National Science Foundation on Nursebot, see "Nursebot," YouTube, May 10, 2008, www.youtube.com/watch?v=6T8yhouPolo (accessed August 13, 2009).

16. Chung Chan Lee, "Robot Nurse Escorts and Schmoozes the Elderly," Robots—Because Humans Deserve Better, May 17, 2006, http://i-heart-robots.blogspot .com/2006/03/robot-nurse-escorts-and-schmooze.html (accessed August 13, 2009). This blog's main title is telling: "Because Humans Deserve Better."

17. See Lee, "Robot Nurse Escorts."

18. See comments about "In the Hands of Robots—Japan," YouTube, June 16, 2008, www.youtube.com/watch?v=697FJZnFvJs&NR=1 (accessed August 13, 2009).

19. Amy Harmon, "Discovering a Soft Spot for Circuitry," *New York Times*, July 5, 2010, www.nytimes.com/2010/07/05/science/05robot.html?pagewanted=all (accessed July 5, 2010). The story cites Timothy Hornyak, a student of robots in Japan, on our new challenge to process synthetic emotions.

CHAPTER 7: COMMUNION

1. See "Kismet and Rich," MIT Computer Science and Artificial Intelligence Laboratory, www.ai.mit.edu/projects/sociable/movies/kismet-and-rich.mov (accessed November 14, 2009).

2. I have been fortunate to have colleagues who have both inspired and challenged my readings of complicity and communion. I owe a special debt to Margaret Boden, Linnda R. Caporael, and Lucy Suchman.

For a discussion of the construction of meaning behind what I am terming *complicity* in human-robot interactions, see Margaret Boden, *Mind As Machine: A History of Cognitive Science*, vol. 1 (London: Oxford University Press, 2006). Prior to these constructions of meaning is the general question of why humans anthropomorphize. See, for example, Linnda R. Caporael, "Anthropomorphism and Mechanomorphism: Two Faces of the Human Machine." *Computers in Human Behavior* 2 (1986): 215–34 and Linnda R. Caporael and Cecilia M. Hayes, "Why Anthropomorphize? Folk Psychology and Other Stories," in *Anthropomorphism, Anecdotes, and Animals*, ed. Robert W. Mitchell, Nicholas S. Thompson, and Lyn Miles (Albany: State University of New York, 1997), 59–75. The literature on anthropomorphism is large. I signal two particularly useful volumes: Mitchell, Thompson, and Miles, eds., *Anthropomorphism, Anecdotes, and Animals*, and John Stodart Kennedy, *The New Anthropomorphism* (Cambridge: Cambridge University Press, 1992).

For a critical study of the constructions of meaning in human-robot interactions, see Lucy Suchman, *Human-Machine Reconfigurations: Plans and Situated Actions* (1987; Cambridge: Cambridge University Press, 2007), especially ch. 13. See also Lucy Suchman, "Affiliative Objects," *Organization* 12, no. 2 (2005): 379–399. Suchman and I both participated in panels on computers and society at the Society for the Social Studies of Science (August 2007) and at the Harvard Graduate School of Design (March 2009). At both panels, Suchman eloquently examined human-robot interactions as social constructs. Most recently, Suchman has persuasively argued for a return to "innocence" in how we approach sociable robots, a tonic dialing down of what we are willing to project onto them. See Lucy Suchman, "Subject Objects," accepted for a special issue of *Feminist Theory* devoted to "nonhuman feminisms," edited by Myra Hird and Celia Roberts.

3. On Domo, see Sandra Swanson, "Meet Domo, It Just Wants to Help," *Technology Review* (July/August 2007), www.technologyreview.com/article/18915 (accessed August 6, 2009). Unless otherwise referenced, all citations to Aaron Edsinger are from an interview in March 2007.

4. Swanson, "Meet Domo."

5. A similar experience is reported by Lijin Aryananda, a graduate student at MIT's Computer Science and Artificial Intelligence Laboratory, who has done her thesis work on Mertz, the robot to which Pia Lindman hopes to merge her mind. I spoke with Aryananda in March 2007, as she was about to graduate and take a fellowship in Germany. She said she would miss the robot. Her feelings for it, she said, began with a "technical bonding." She described how she was the person who could interact with the robot best: "I can understand the timing of the robot. I know what sorts of things it can be conditioned to respond to. It would be an understatement to say that I look at this robot and see thirteen degrees of freedom. There is more than that." Aryananda described feeling that the robot was not at its "best" unless she was there, to the point that it almost felt as though she was "letting the robot down" if she was not around. And then, one day, these feelings of "technical missing" became "just missing." She said, "It [Mertz] has been such a big part of your life, its ways of responding are so much a part of the rhythm of your day." For her dissertation work on how people responded to Mertz in a natural environment, see Lijin Aryananda, "A Few Days of a Robot's Life in the Human's World: Toward Incremental Individual Recognition" (PhD diss., Massachusetts Institute of Technology, 2007).

6. Alan Turing, usually credited with inventing the programmable computer, said that intelligence may require the ability to have sensate experience. In 1950, he wrote, "It can also be maintained that it is best to provide the machine with the best sense organs that money can buy, and then teach it to understand and speak English. That process could follow the normal teaching of a child. Things would be pointed out and named, etc." Alan Turing, "Computing Machinery and Intelligence," *Mind* 59, no. 236 (October 1950): 433–460.

7. This branch of artificial intelligence (sometimes called "classical AI") attempts to explicitly represent human knowledge in a declarative form in facts and rules. For an overview of AI and its schools that explores its relations to theories of mind, see Margaret Boden, *Artificial Intelligence and Natural Man* (1981; New York: Basic Books, 1990).

8. Hubert Dreyfus, "Why Computers Must Have Bodies in Order to Be Intelligent," *Review of Metaphysics* 21, no. 1 (September 1967): 13–32. See also Hubert Dreyfus, *What Computers Can't Do: A Critique of Artificial Reason* (New York: Harper & Row, 1972); Hubert Dreyfus with Stuart E. Dreyfus and Tom Athanasiou, *Mind over Machine: The Power of Human Intuition and Expertise in the Era of the Computer* (New York: Free Press, 1986); Hubert Dreyfus with Stuart E. Dreyfus, "Making a Mind Versus Modeling the Brain: Artificial Intelligence Back at a Branchpoint," *Daedalus* 117,

no. 1 (winter 1988): 15–44; Hubert Dreyfus, *What Computers "Still" Can't Do: A Critique of Artificial Reason* (1979; Cambridge, MA: MIT Press, 1992).

For another influential critique of artificial intelligences that stresses the importance of embodiment, see John Searle, "Minds, Brains, and Programs," *Behavioral and Brain Sciences* 3 (1980): 417–424, and "Is the Brain's Mind a Computer Program?" *Scientific American* 262, no. 1 (January 1990): 26–31.

9. Dreyfus, "Why Computers Must Have Bodies."

10. Antonio Damasio, *Descartes' Error: Emotion, Reason, and the Human Brain* (New York: Gosset/Putnam Press, 1994).

11. For an introduction to this phenomenon, see David G. Myers, *Exploring Psychology* (New York: Worth Books, 2005), 392. See also Robert Soussignan, "Duchenne Smile, Emotional Experience, and Autonomic Reactivity: A Test of the Facial Feedback Hypothesis," *Emotion* 1, no. 2 (2002): 52–74, and Randy J. Larsen, Margaret Kasimatis, and Kurt Frey, "Facilitating the Furrowed Brow: An Unobtrusive Test of the Facial Feedback Hypothesis Applied to Unpleasant Affect," *Cognition & Emotion* 6, no. 5 (September 1992): 321–338.

12. See, for example, Stephanie D. Preston and Frans B. M. de Waal, "Empathy: Its Ultimate and Proximate Bases," *Behavioral and Brain Sciences* 25 (2002): 1–72, and Christian Keysers and Valeria Gazzola, "Towards a Unifying Neural Theory of Social Cognition," *Progress in Brain Research* 156 (2006): 379–401.

13. On the New York exhibition of the Lindman/Edsinger/Domo project, see Stephanie Cash, "Pia Lindman at Luxe," *Art in America*, September 2006, http://findarticles.com/p/articles/mi_m1248/is_8_94/ai_n26981348 (accessed September 10, 2009).

14. On fusion with the inanimate, there are other expert witnesses, although their experiences are outside the scope of this book. See, for example, Michael Chorost's *Rebuilt: My Journey Back to the Hearing World* (New York: Mariner Books, 2006), a personal account of receiving a cochlear implant. Other relevant testimony comes from Aimée Mullins, a double amputee who uses prosthetic legs to remake herself. See "Aimee Mullins and Her 12 Pairs of Legs," Ted.com, www.ted.com/talks/aimee_mullins _prosthetic_aesthetics.html (accessed September 11, 2009). In both Chorost's and Mullins's cases, there is evidence that merging with technology results not only in a purely instrumental gain in functionality but in a new prosthetic sensibility as well.

15. Lévinas, Emmanuel, "Ethics and the Face," in *Totality and Infinity: An Essay on Exteriority*, trans. Alphonso Lingis (Pittsburgh, PA: Duquesne University Press, 1969), 197–201.

16. Lindman uses continental philosophy and psychoanalysis as a referent. I see two themes in the work of French psychoanalyst Jacques Lacan as touchstones of her thinking. First, there is always something that cannot be represented, something that Lacan calls "the real." Second, the self is structured by language and society. There is no ego apart from language and society. See Jacques Lacan, *Ecrits: A Selection*, trans. Alan Sheridan (1977; New York: W. W. Norton & Company, 2002), and *The Four Fundamental Concepts of Psychoanalysis*, ed. Jacques-Alain Miller, trans. Alan Sheridan (1973; New York: W. W. Norton & Company, 1998).

17. See, for example, Sherry Turkle, "Authenticity in the Age of Digital Companions," *Interaction Studies* 8, no. 3 (2007): 501–517.

18. The tendency for people to attribute personality, intelligence, and emotion to computational objects has been widely documented in the field of human-computer interaction. Classic experimental studies are reported in Byron Reeves and Clifford Nass, *The Media Equation: How People Treat Computers, Television, and New Media Like Real People and Places* (New York: Cambridge University Press/CSLI, 1996); Clifford Nass et al., "Computers Are Social Actors: A Review of Current Research," in *Human Values and the Design of Computer Technology*, ed. Batya Friedman (Stanford, CA: CSLI Productions, 1997), 137–162; Clifford Nass and Yougmee Moon, "Machines and Mindlessness: Social Response to Computers," *Journal of Social Issues* 56, no. 1 (2000): 81–103. See also Salvatore Parise et al., "Cooperating with Life-like Interface Agents," *Computers in Human Behavior* 15 (1999): 123–142; Lee Sproull et al., "When the Interface Is a Face," *Human-Computer Interaction* 11, no. 2 (June 1996): 97–124; Sara Kiesler and Lee Sproull, "Social Responses to 'Social' Computers," in *Human Values and the Design of Technology*, ed. Batya Friedman (Stanford, CA: CLSI Publications, 1997). A review of research on sociable robotics is T. Fong, I. Nourbakhsh, and K. Dautenhahn, *A Survey of Social Robots* (Pittsburgh, PA: Carnegie Mellon University Robotics Institute, 2002).

19. Nass et al., "Computers Are Social Actors," 138.

20. Nass et al., "Computers Are Social Actors," 158.

21. Nass et al., "Computers Are Social Actors," 138.

22. Rosalind W. Picard, *Affective Computing* (Cambridge, MA: MIT Press, 1997), x.

23. Marvin Minsky, *The Emotion Machine: Commonsense Thinking, Artificial Intelligence, and the Future of the Human Mind* (New York: Simon & Schuster, 2006), 345.

24. See "affective," Thesaurus.com, http://thesaurus.reference.com/browse/affective (accessed July 6, 2009).

25. Raymond Kurzweil believes that it will be possible to download self onto machine. For an overview of his ideas, see Raymond Kurzweil, *The Singularity Is Near: When Humans Transcend Biology* (New York: Viking, 2005).

26. Current research at the MIT Media Lab aspires to such computationally enhanced environments. See, for example, research groups on Fluid Interfaces and Information Ecology, www.media.mit.edu/research-groups/projects (accessed August 14, 2010).

27. Starner discussed his ideas on growing a robot by using sensors embedded in his clothing in a January 2008 interview. See "Wearable Computing Pioneer Thad Starner," Gartner.com, January 29, 2008, www.gartner.com/research/fellows/asset _196289_1176.jsp (accessed April 3, 2010).

28. For an overview of robots in medical settings, focusing on research on Alzheimer's and autism, see Jerome Groopman, "Robots That Care: Advances in Technological Therapy," *The New Yorker*, November 2, 2009, www.newyorker.com/reporting/2009/11/02/091102fa_fact_groopman (accessed November 11, 2009).

29. In Japan, robot babysitters offer lessons, games, and child surveillance as their mothers perform household chores. Androids in the form of sexy women are marketed as receptionists and guides. They are in development to serve as hostesses and elementary school teachers. In a related development in Japan, a lifelike sex doll, anatomically correct and enhanced by sphincter muscles, is publicly marketed and seen as a good way for shut-ins to find pleasure and, more generally, to control the spread of sexually transmitted diseases.

In a new development, there is now a "real," physical vacation resort where Japanese men can spend time with their virtual girlfriends. Although the men check in "alone," the staff is trained to respond to them as though they were in a couple. Daisuke Wakabayashi, "Only in Japan, Real Men Go to a Hotel with Virtual Girlfriends," August 31, 2010, http://online.wsj.com/article/SB10001424052748703632304575451414209 658940.html (accessed September 7, 2010).

CHAPTER 8: ALWAYS ON

1. This chapter expands on themes explored in Sherry Turkle, "Tethering," in *Sensorium: Embodied Experience, Technology, and Contemporary Art*, ed. Caroline Jones (Cambridge, MA: Zone, 2006), 220–226, and "Always-On/Always-on-You: The Tethered Self," in *Handbook of Mobile Communication Studies*, ed. James E. Katz (Cambridge, MA: MIT Press, 2008), 121–138.

2. These statements put me on a contested terrain of what constitutes support and shared celebration. I have interviewed people who say that flurries of virtual condolence and congratulations are sustaining; others say it just reminds them of how alone they are. And this in fact is my thesis: we are confused about when we are alone and when we are together.

3. See "The Guild—Do You Want to Date My Avatar," YouTube, August 17, 2009, www.youtube.com/watch?v=urNyg1ftMIU (accessed January 15, 2010).

4. Internet Relay Chat is a form of real-time Internet text messaging (chat) or synchronous conferencing. It is mainly designed for group communication in discussion forums, called channels, but also allows one-to-one communication via private message as well as chat and data transfers. It is much used during academic conferences, now in addition to Twitter. See, for example, this note on a conference invitation for a conference on media literacy: "Conference attendees are encouraged to bring their laptops, PDAs, netbooks or Twitter enabled phones, so they can participate in on-line social networking that will be part of this year's conference. Directions on how to obtain Internet connectivity and where people will be talking, will be provided in your attendee packet. For those who can not attend, tell them they can backchannel with us on Twitter at #homeinc." See "Conference Program," 2009 Media Literacy Conference, http://ezregister.com/events/536 (accessed October 20, 2009).

5. Hugh Gusterson and Catherine Besteman, eds., *The Insecure American: How We Got Here and What We Should Do About It* (Los Angeles: University of California Press, 2009).

6. See, for example, Robert D. Putnam, *Bowling Alone: The Collapse and Revival of American Community* (New York: Simon and Schuster, 2001); Gusterson and Beste-

man, eds., *The Insecure American*; Theda Skocpol, *Diminished Democracy: From Membership to Management in American Civic Life* (Norman: University of Oklahoma Press, 2003).

7. Sherry Turkle, *Life on the Screen: Identity in the Age of the Internet* (New York: Simon and Schuster, 1995), 182.

8. See "What Is Second Life," Second Life, http://secondlife.com/whatis (accessed June 13, 2010).

9. There is evidence that people experience what they do online as though it happened in the physical real. See, for example, Nick Yee, Jeremy Bailenson, and Nicolas Ducheneaut, "The Proteus Effect: Implications of Transformed Digital Self-representation on Online and Offline Behavior," *Communication Research* 36, no. 2: 285–312. For a video introduction to work in this area by Stanford University's Virtual Human Interaction Laboratory, directed by Jeremy Bailenson, see, "The Avatar Effect," PBS.org, www.pbs.org/wgbh/pages/frontline/digitalnation/virtual-worlds/second-lives/the-avatar-effect.html?play (accessed September 2, 2009).

10. Pete accesses Second Life through an iPhone application known as Spark. It does not bring the entire world to him, but it does enable conversation.

11. Pete insists that Alison does not know of his double life. Over the past twenty years I have had many conversations about virtual infidelity. In the case of women whose husbands are virtually unfaithful, there are sharp differences of opinion. Some think it is preferable to any physical infidelity. Others think it is the worst kind of infidelity, an infidelity that involves not simply sex but talking, considering another, making plans, and building a life.

12. In online life, weak ties—the ties of acquaintanceship—are often celebrated as the best ties of all. For the seminal work on weak ties, see Mark Granovetter, "The Strength of Weak Ties," *American Journal of Sociology* 78, no. 6 (1973): 1360–1380, and "The Strength of Weak Ties: A Network Theory Revisited," *Sociological Theory* 1 (1983): 201–233.

13. Turkle, *Life on the Screen*.

14. This is sometimes referred to as "continuous partial attention," a phrase widely credited to media researcher Linda Stone. See Stone's blog at www.lindastone.net (accessed August 24, 2009).

15. Those who study the boundaries between work and the rest of life suggest that it is helpful to demarcate our changing roles. Sue Campbell Clark, "Work/Family Border Theory: A New Theory of Work/Family Balance," *Human Relations* 53, no. 6 (2000): 747–770; Stephan Desrochers and Leisa D. Sargent, "Work-Family Boundary Ambiguity, Gender and Stress in Dual-Earner Couples" (paper presented at the conference "From 9-to-5 to 24/7: How Workplace Changes Impact Families, Work, and Communities," 2003 BPW/Brandeis University Conference, Orlando, Florida, March 2003); and Michelle Shumate and Janet Fulk, "Boundaries and Role Conflict When Work and Family Are Colocated: A Communication Network and Symbolic Interaction Approach," *Human Relations* 57, no. 1 (2004): 55–74.

16. Media theorist Henry Jenkins is an eloquent spokesperson for the significance of multitasking. See "The Skill of the Future: In a Word 'Multitasking,'" PBS.org,

www.pbs.org/wgbh/pages/frontline/digitalnation/living-faster/split-focus/the-skill-of-the-future.html? (accessed November 16, 2009). His other online interviews on the Digital Nation website beautifully capture a vision of schools bending to new media sensibilities. See "The Tech Fix," PBS.org, www.pbs.org/wgbh/pages/frontline/digital-nation/learning/schools/the-tech-fix.html?play (accessed November 14, 2009), and "Defenders of the Book," PBS.org, www.pbs.org/wgbh/pages/frontline/digitalnation/learning/literacy/defenders-of-the-book.html?play (accessed November 14, 2009).

17. The literature on the downside of multitasking is growing. An influential and much-reported study is Eyal Ophir, Clifford Nass, and Anthony Wagner, "Cognitive Control in Media Multitaskers," *Proceedings of the National Academy of Sciences* 106 (2009): 15583–15587, www.pnas.org/content/106/37/15583 (accessed August 10, 2010). This study found that when people multitask, everything they do is degraded in quality. An excellent work on the general topic is Maggie Jackson, *Distracted: The Erosion of Attention and the Coming Dark Age* (New York: Prometheus, 2008). On the practical downside of thinking that we can do more than one thing at once, see, for example, the nine-part series on the *New York Times* website titled "Driven to Distraction," covering such topics as doing office work while driving at 60 mph, drivers and legislators dismissing cell phone risks, and New York taxi drivers ignoring the ban on cell phone use while driving. "Driven to Distraction," *New York Times*, http://topics.nytimes.com/topics/news/technology/series/driven_to_distraction/index.html (accessed November 14, 2009).

Teenagers routinely drive and text; we know this because their automobile accidents are traced back to texting and cell phone use. A 2009 study of twenty-one teenagers showed them changing speed and weaving in and out of lanes while texting. Eastern Virginia Medical School, "Texting While Driving Can Be Deadly, Study Shows," *ScienceDaily*, May 5, 2009, www.sciencedaily.com/releases/2009/05/090504094434.htm (accessed January 4, 2010). A larger study of nine hundred teenagers in 2007 showed 50 percent of them texted while driving despite the fact that 36 percent of them thought this was dangerous. See Steve Vogel, "Teen Driver Menace: Text-Messaging," Suite101, October 22, 2007, http://parentingteens.suite101.com/article.cfm/teen_driver_menace_textmessaging (accessed January 4, 2009).

Adults also text while driving. Trains collide while conductors text. A plane flies past its destination airport because its pilots are absorbed in a new computer program. In October 2009, pilots attending to their laptop computers—behavior in defiance of safety regulations—were the cause of an aircraft overshooting its Minneapolis destination by 150 miles. "The pilots told the National Transportation Safety Board that they missed their destination because they had taken out their personal laptops in the cockpit, a violation of airline policy, so the first officer, Richard I. Cole, could tutor the captain, Timothy B. Cheney, in a new scheduling system put in place by Delta Air Lines, which acquired Northwest last fall." See Micheline Maynard and Matthew L. Wald, "Off-Course Pilots Cite Computer Distraction," *New York Times*, October 26, 2009, www.nytimes.com/2009/10/27/us/27plane.html?_r=1 (accessed November 16, 2009).

18. In practical terms, what works best is to remind students that media literacy is about knowing when not to use technology as well as how to use it. I am optimistic

that over time, we will make better use of technology in the classroom and we will be less afraid to turn it off when that is what makes sense pedagogically.

19. Melissa Mazmanian, "Some Thoughts on BlackBerries" (unpublished memo, Massachusetts Institute of Technology, 2005). See also Melissa Mazmanian, Wanda Orlikowski, and Joanne Yates, "Ubiquitous E-mail: Individual Experiences and Organizational Consequences of BlackBerry Use," *Proceedings of the 65th Annual Meeting of the Academy of Management*, Atlanta, Georgia, August 2006, http://seeit.mit.edu/ Publications/BlackBerry_AoM.pdf (accessed August 24, 2009).

20. The first book club selection by Arianna Huffington for the *Huffington Post's* book club was Carl Honoré's *In Praise of Slowness: How a Worldwide Movement Is Challenging the Cult of Speed* (New York: HarperCollins, 2004).

21. Diana B. Gant and Sara Kiesler, "Blurring the Boundaries: Cell Phones, Mobility and the Line Between Work and Personal Life," in *Wireless World: Social and Interactional Aspects of the Mobile Age*, ed. N. G. R. H. Brown (New York: Springer, 2001).

22. Donna Haraway, "A Cyborg Manifesto," in *Simians, Cyborgs and Women: The Reinvention of Nature* (New York; Routledge, 1991), 149–181.

23. Thomas R. Herzog et al., "Reflection and Attentional Recovery As Distinctive Benefits of Restorative Environments," *Journal of Environmental Psychology* 17 (1997): 165–170. See also Stephen Kaplan, "The Restorative Benefits of Nature: Toward an Integrative Framework," *Journal of Environmental Psychology* 15 (1995): 169–182.

24. I studied teenagers from a wide range of economic, social, and ethnic backgrounds. They attended seven different schools: two private boys preparatory schools, one in an urban center (Fillmore) and one in a rural setting (Hadley), one urban private girls school (Richelieu), an urban Catholic coeducational high school (Silver Academy), a private urban coeducational high school (Cranston), and two public high schools, one suburban (Roosevelt) and one urban (Branscomb). All students, from wealthy to disadvantaged, had cell phones with texting capability. Class distinctions showed themselves not in whether students possessed a phone but in what kind of contract they had with their providers. Teenagers with fewer resources, such as Julia in the following chapter, tended to have plans that constrained who they could text for free. Free texts are most usually for people on the same network. Ever resourceful, students with restricted plans try to get their friends to sign up with their cell providers. We shall see that teenagers don't care much about who they can call. I often hear, "I never use my calling minutes." On teenagers and digital culture, see Mizuko Ito et. al., *Hanging Out, Messing Around, and Geeking Out: Kids Learning and Living with New Media* (Cambridge, MA: MIT Press, 2010) and Danah Boyd, "Why Youth (Heart) Social Network Sites: The Role of Networked Publics in Teenage Social Life," MacArthur Foundation Series on Digital Learning—Youth, Identity, and Digital Media, ed. Davind Buckingham (Cambridge, MA: Mit Press 2007), 119–142.

CHAPTER 9: GROWING UP TETHERED

1. Carol Gilligan, *In a Different Voice: Psychological Theory and Women's Development* (1982; Cambridge, MA: Harvard University Press, 1993).

2. Erik Erikson, *Identity and the Life Cycle* (1952; New York: W. W. Norton, 1980) and *Childhood and Society* (New York: Norton, 1950).

3. In Julia's world, e-mail is considered "slow" and rarely used because texting has greater immediacy.

4. It is so common to see teenagers (and others) attending to their mobiles rather than what is around them, that it was possible for a fake news story to gain traction in Britain. Taken up by the media, the story went out that there was a trial program to pad lampposts in major cities. Although it was a hoax, I fell for it when it was presented online as news. In fact, in the year prior to the hoax, one in five Britons did walk into a lamppost or other obstruction while attending to a mobile device. This is not surprising because research reported that "62 per cent of Britons concentrate so hard on their mobile phone when texting they lose peripheral vision." See Charlie Sorrel, "Padded Lampposts Cause Fuss in London," *Wired*, March 10, 2008, www.wired .com/gadgetlab/2008/03/padded-lamposts (accessed October 5, 2009).

5. New communications technology makes it easier to serve up people as slivers of self, providing a sense that to get what you need from others you have multiple and inexhaustible options. On the psychology that needs these "slivers," see Paul H. Ornstein, ed., *The Search for Self: Selected Writings of Heinz Kohut (1950–1978)*, vol. 2 (New York: International Universities Press, 1978).

6. David Riesman, Nathan Glazer, and Reuel Denney, *The Lonely Crowd: A Study of the Changing American Character* (1950; New Haven, CT: Yale University Press, 2001).

7. Orenstein, *The Search for Self*. For an earlier work, of a very different time, that linked cultural change and narcissistic personality style, see Christopher Lasch, *The Culture of Narcissism* (New York: Norton, 1979). Lasch said that "pathology represents a heightened version of normality." This formulation is helpful in thinking about the "normal" self in a tethered society and those who suffer more acutely from its discontents. From a psychodynamic perspective, we all suffer from the same things, some of us more acutely than others.

8. See Erik Erikson, *Identity and the Life Cycle* and *Childhood and Society*; *Young Man Luther: A Study in Psychoanalysis and History* (New York: W. W. Norton and Company, 1958).

9. Robert Jay Lifton, *The Protean Self: Human Resilience in an Age of Fragmentation* (New York: Basic Books, 1993).

10. Lifton shared this story at a meeting of the Wellfleet Seminar in October 2009, an annual gathering that began as a forum for Erikson and his students as they turned their attention to psychohistory.

11. The performances of everyday life—playing the roles of father, mother, child, wife, husband, life partner, worker—also provide "a bit of stress." There is room for considerable debate about how much online life really shares with our performances of self in "real life." Some look to the sociology of "self-presentation" to argue that online and off, we are always onstage. Erving Goffman, *The Presentation of Self in Everyday Life* (Garden City, NY: Doubleday Anchor, 1959).

CHAPTER 10: NO NEED TO CALL

1. In the object relations tradition of psychoanalysis, an object is that which one relates to. Usually, objects are people, especially a significant person who is the object or
target of another's feelings or intentions. A whole object is a person in his or her entirety.
It is common in development for people to internalize part objects, representations of
others that are not the whole person. Online life provides an environment that makes
it easier for people to relate to part objects. This puts relationships at risk. On object
relations theory, see, for example, Stephen A. Mitchell and Margaret J. Black, *Freud and
Beyond: A History of Modern Psychoanalytic Thought* (New York: Basic Books, 1995).

2. See Stefana Broadbent, "How the Internet Enables Intimacy," Ted.com,
www.ted.com/talks/stefana_broadbent_how_the_internet_enables_intimacy.html
(accessed August 8, 2010). According to Broadbent, 80 percent of calls on cell phones
are made to four people, 80 percent of Skype calls are made to two people, and most
Facebook exchanges are with four to six people.

3. This mother is being destructive to her relationship with her daughter. Research
shows that people use the phone in ways that surely undermine relationships with
adult partners as well. In one striking finding, according to Dan Schulman, CEO of
cell operator Virgin Mobile, one in five people will interrupt sex to answer their phone.
David Kirkpatrick, "Do You Answer Your Cellphone During Sex?" *Fortune*, August
28 2006, http://money.cnn.com/2006/08/25/technology/fastforward_kirkpatrick.for-
tune/index.htm (accessed November 11, 2009).

4. See Amanda Lenhart et al., "Teens and Mobile Phones," The Pew Foundation,
April 20, 2010, www.pewinternet.org/Reports/2010/Teens-and-Mobile-Phones.aspx
?r=1 (accessed August 10, 2010).

5. See "What Is Second Life," Second Life, http://secondlife.com/whatis (accessed
June 13, 2010).

6. Erik Erikson, *Childhood and Society* (New York: Norton, 1950).

7. To use the psychoanalyst Philip Bromberg's language, finding fluidity of self in
online life enables us to "stand in the spaces between realities and still not los[e] any
of them . . . the capacity to feel like one self while being many." See Philip Bromberg,
"Shadow and Substance: A Relational Perspective on Clinical Process," *Psychoanalytic
Psychology* 10 (1993): 166. In AI pioneer Marvin Minsky's language, cycling through
online personae reveals different aspects of a "society of mind," a computational notion
of identity as distributed and heterogeneous. Identity, from the Latin *idem*, has been
typically used to refer to the sameness between two qualities. On the Internet, however,
one can be many and usually is. See Marvin Minsky, *Society of Mind* (New York: Basic
Books, 1987).

8. Nielsen recently found that children send eight text messages for every phone
call they make or receive. See Anna-Jane Grossman, "I Hate the Phone," Huffington
Post, October 14, 2009, www.huffingtonpost.com/anna-jane-grossman/i-hate-the-
phone_b_320108.html (accessed October 17, 2009).

9. "Number of US Facebook Users over 35 Nearly Doubles in Last 60 Days," Inside
Facebook, March 25, 2009, www.insidefacebook.com/2009/03/25/number-of-us-face-
book-users-over-35-nearly-doubles-in-last-60-days (accessed October 19, 2009).

10. Dan is aware of his withdrawal, but a new generation takes machine-mediated communication simply as the nature of things. Two young girls, ten and twelve, trapped inside a storm drain turned to Facebook for help instead of calling the police. They used their mobile phones to update their Facebook statuses. Even with their lives at risk, these girls saw Facebook as their portal to the world. Firefighters eventually rescued the pair after being contacted by one of their male school friends, who had been online and saw they were trapped. The news report read as follows:

The drama happened near Adelaide, Australia. Firefighter Glenn Benham, who took part in the rescue, said, "These girls were able to access Facebook on their mobile phones so they could have called the emergency services. It seems absolutely crazy but they updated their status rather than call us directly. We could have come to their rescue much faster than relying on someone else being online, then replying to them, then calling us. It is a worrying development. Young people should realize it's better to contact us directly. Luckily they are safe and well. It's awful to think what could have happened because of the delay.'"

See "Girls Trapped in Storm Drain Use Facebook to Call for Help," *Daily Mail*, September 8, 2009, www.dailymail.co.uk/news/worldnews/article-1211909/Girls-trapped -storm-drain-use-Facebook-help-instead-phoning-emergency-services.html#ixzz0T9 iWpeNR (accessed October 6, 2009).

11. This paraphrases a line from Sonnet 73: "Consum'd with that which it was nour- ish'd by . . . "

12. The author of a recent blog post titled "I Hate the Phone" would not call Trey old-school, but nor would she want to call him. Anna-Jane Grossman admits to growing up loving her pink princess phone, answering machine, and long, drawn-out conversations with friends she had just seen at school. Now she hates the phone: "I feel an inexplicable kind of dread when I hear a phone ring, even when the caller ID displays the number of someone I like. . . . My dislike for the phone probably first started to grow when I began using Instant Messenger. Perhaps phone-talking is a skill that one has to practice, and the more IMing I've done, the more my skills have dwindled to the level of a modern day 13-year-old who never has touched a landline. . . . I don't even listen to my [phone] messages any more: They get transcribed automatically and then are sent to me via e-mail or text." The author was introduced to Skype and sees its virtues; she also sees the ways in which it undermines conversation: "It occurs to me that if there's one thing that'll become obsolete because of video-chatting, it's not phones: it's natural flowing conversations with people far away." See Grossman, "I Hate the Phone."

In my experience with Skype, pauses seem long and awkward, and it is an effort not to look bored. Peggy Ornstein makes this point in "The Overextended Family," *New York Times Magazine*, June 25, 2009, www.nytimes.com/2009/06/28/magazine/28fob-wwln-t.html (accessed October 17, 2009). Ornstein characterizes Skype as providing "too much information," something that derails intimacy: "Suddenly I understood why slumber-party confessions always came after lights were out, why children tend to admit the juicy stuff to the back of your head while you're driving, why psychoanalysts stay out of a patient's sightline."

CHAPTER 11: REDUCTION AND BETRAYAL

1. Seth Schiesel, "All Together Now: Play the Game, Mom," *New York Times*, September 1, 2009, www.nytimes.com/2009/09/06/arts/television/06schi.html (accessed December 13, 2009).

2. Amy Bruckman. "Identity Workshop: Emergent Social and Psychological Phenomena in Text-Based Virtual Reality" (unpublished essay, Media Lab, Massachusetts Institute of Technology, 1992), www.cc.gatech.edu/~asb/papers (accessed September 2, 2009).

3. For rich material on the "boundary work" between self and avatar, see Adam Boellstorff, *Coming of Age in Second Life: An Anthropologist Explores the Virtually Human* (Princeton, NJ: Princeton University Press, 2008), and T. L. Taylor, *Play Between Worlds: Exploring Online Game Culture* (Cambridge, MA: MIT Press, 2006). See also Sherry Turkle, *Life on the Screen: Identity in the Age of the Internet* (New York: Simon and Schuster, 1995).

4. This is true whether they are in the text-based multiuser domains, or MUDs, of the early 1990s (such as Lambda Moo), in the visually rich massively multiplayer online role-playing games of the end of that decade (Everquest and Ultima II), or in today's cinemalike virtual worlds, such as World of Warcraft or Second Life.

5. Victor Turner, *The Ritual Process: Structure and Anti-Structure* (Chicago: Aldine, 1969).

6. The work coming out of Stanford University's virtual reality laboratory presents compelling evidence that if you are, for example, tall in virtual reality, you will feel more assertive in meetings that follow online sessions. See, for example, J. N. Bailenson, J. A. Fox, and J. Binney, "Virtual Experiences, Physical Behaviors: The Effect of Presence on Imitation of an Eating Avatar," *PRESENCE: Teleoperators and Virtual Environments* 18, no. 4: 294–303, and J. A. Fox and J. N. Bailenson, "Virtual Self-modeling: The Effects of Vicarious Reinforcement and Identification on Exercise Behaviors," *Media Psychology* 12 (2009): 1–25.

7. Turkle, *Life on the Screen*.

8. The Loebner Prize Competition also awards a prize to the person who is most obviously a person, the person who is *least* confused with an artificial intelligence. See Charles Platt, "What's It Mean to Be Human, Anyway?" *Wired*, May 1995, www.wired.com/wired/archive/3.04/turing_pr.html (accessed May 31, 2010).

9. Mihaly Csíkszentmihalyi, *Beyond Boredom and Anxiety* (San Francisco: Jossey-Bass, 2000 [1st ed. 1975]), and Natasha Schüll, *Addiction by Design: Machine Gambling in Las Vegas* (Princeton, NJ: Princeton University Press, forthcoming).

10. Mihaly Csíkszentmihalyi, *Flow: The Psychology of Optimal Experience* (New York: Harper & and Row, 1990).

11. With too much volume, of course, e-mail becomes too stressful to be relaxing. But "doing e-mail," no matter how onerous, can put one in the zone.

12. Natasha Schüll, *Addiction by Design*. On the issue of unreal choices, Schüll refers to the work of psychologist Barry Schwartz, *The Paradox of Choice: Why More Is Less* (New York: Harper Collins, 2003).

13. Sherry Turkle, *The Second Self: Computers and the Human Spirit* (1984; Cambridge, MA: MIT Press, 2005), see, especially, "Video Games and Computer Holding Power," 65–90.

14. Washington State University neuroscientist Jaak Panksepp describes a compelled behavior he calls the "seeking drive." When humans (indeed, all mammals) receive stimulation to the lateral hypothalamus (this happens every time we hear the ping of a new e-mail or hit return to start a Google search), we are caught in a loop "where each stimulation evoke[s] a reinvigorated search strategy." See Jaak Panksepp, *Affective Neuroscience: The Foundations of Human and Animal Emotions* (Oxford: Oxford University Press, 1998), 151. The implication is that search provokes search; seeking provokes seeking. Panksepp says that when we get thrilled about the world of ideas, about making intellectual connections, about divining meaning, it is the seeking circuits that are firing.

In an article in *Slate*, Emily Yoffe, reviews the relationship between our digital lives and how the brain experiences pleasure. She says:

Actually all our electronic communication devices—e-mail, Facebook feeds, texts, Twitter—are feeding the same drive as our searches. . . . If the rewards come unpredictably—as e-mail, texts, updates do—we get even more carried away. No wonder we call it a "CrackBerry." . . .

[Psychologist Kent] Berridge says the "ding" announcing a new e-mail or the vibration that signals the arrival of a text message serves as a reward cue for us. And when we respond, we get a little piece of news (Twitter, anyone?), making us want more. These information nuggets may be as uniquely potent for humans as a Froot Loop to a rat. When you give a rat a minuscule dose of sugar, it engenders "a panting appetite," Berridge says—a powerful and not necessarily pleasant state.

See Emily Yoffe, "Seeking How the Brain Hard-Wires Us to Love Google, Twitter, and Texting. And Why That's Dangerous," *Slate*, August 12, 2009, www.slate .com/id/2224932/pagenum/all/#p2 (accessed September 25, 2009). See also Nicholas Carr, "Is Google Making Us Stupid?" *The Atlantic*, July–August 2008, www.theatlantic.com/doc/200807/google (accessed November 20, 2009), and Kent C. Berridge and Terry E. Robinson, "What Is the Role of Dopamine in Reward: Hedonic Impact, Reward Learning, or Incentive Salience?" *Brain Research Reviews* 28 (1998): 309–369.

CHAPTER 12: TRUE CONFESSIONS

1. The PostSecret site is run by Frank Warren. See http://postsecret.blogspot.com (accessed August 22, 2009). For his views on the positive aspects of venting through confessional sites, see Tom Ashcroft's *On Point* interview with Frank Warren, "Baring Secrets Online," WBUR, June 10, 2009, www.onpointradio.org/2009/06/secret-sharers (accessed August 2, 2010). See also Michele Norris's *All Things Considered* interview with Frank Warren, "Postcards Feature Secret Messages from Strangers," NPR, March 30, 2005, www.npr.org/templates/story/story.php?storyId=4568035 (accessed August 2, 2010).

2. See www.postsecret.com (accessed August 4, 2010).

3. As throughout this book, I have disguised the details of this case and all others I cite.

4. Ashley Fantz, "Forgive Us Father; We'd Rather Go Online," CNN.com, March 13, 2008, www.cnn.com/2008/LIVING/wayoflife/03/13/online.confessions/index.html (accessed August 22, 2009).

5. The exceptions are significant: if at the earliest ages you were not nurtured—you often cried and were not fed—the vulnerability/nurturance expectation can be broken. Erik Erikson calls the positive laying down of expectations "basic trust." See *Childhood and Society* (New York: Norton, 1950), 247–250.

6. This is the defense mechanism of "projective identification." Instead of facing our own issues, we see them in others. There, they can be safely attacked. Insecure about her own appearance, a wife criticizes her husband's weight; the habitually angry see a hostile world.

7. The Reverend Bobby Gruenwald of the Oklahoma-based LifeChurch.tv, an evangelical consortium of thirteen churches affiliated with the online confessional MySecret.tv, is one of those who argues that our notion of "community" should include online congregations. In the first year it was open, about thirty thousand people posted "secrets" on the MySecret website. The posts are linked to categories, which include lusting, cheating, stealing, and bestiality. When the site was featured on America Online's homepage, it got over 1.3 million hits in a single day. Confessional sites like MySecret do not track IP addresses, which could identify those who post. This means that if someone posts a confession of a criminal nature, the site managers cannot do much about it. So, online, we read about people admitting to murder (these are often interpreted as soldiers writing about the experience of war) and enjoying child pornography: "A recent message on ivescrewedup.com reads, 'I have killed four people. One of them was a 17 year old boy.'" See Fantz, "Forgive Us Father."

8. Ray Oldenberg. *The Great Good Place: Cafés, Coffee Ships, Community Centers, Beauty Parlors, General Stores, Bars, Hangouts, and How They Get You Through the Day* (New York: Paragon House, 1989). On virtual environments as communities, see Howard Rheingold, *The Virtual Community: Homesteading on the Electronic Frontier* (Reading, MA: Addison Wesley, 1993).

9. There is, too, the word "world." Sociologist William Bainbridge, a student of World of Warcraft, takes its title seriously and talks of the game as a world. See William Bainbridge, *The Warcraft Civilization: Social Science in a Virtual World* (Cambridge, MA: MIT Press, 2010). For an interchange on the game as a "world," or perhaps a "neighborhood," see Tom Ashcroft's *On Point* interview with William Bainbridge, "Deep in the 'World of Warcraft,'" WBUR, March 30, 2010, www.onpointradio.org /2010/03/warcraft-civilization (accessed August 10, 2010).

CHAPTER 13: ANXIETY

1. "No More Teachers? No More Books? Higher Education in the Networked Age," A Centennial Panel on Information Technology, Harvard University, Cambridge, Massachusetts, November 16, 2009.

2. There are at least three displacements in the passage from the book to the online text. First, there is the displacement of personal and idiosyncratic associations. Second, online links that are there to be double-clicked are determined by what the author of the text thinks it is important for you to be exposed to. Third, even if one accepts that such links are convenient, they are easily bypassed when reading is interrupted by an incoming e-mail or other online distractions.

3. In *The Year of Magical Thinking* (New York: Alfred A. Knopf, 2005), a memoir about the year after her husband's death, Joan Didion describes how material objects became charged with meaning. So, for example, Didion cannot bring herself to throw away her husband's shoes because she is convinced that he may need them. This same magical thinking is associated both with religious devotion and the "illness" of mourning. With time, Freud believed, the true object, the lost husband, comes to have a full internal representation. See Sigmund Freud, "Mourning and Melancholia," in *The Standard Edition of Sigmund Freud*, ed. and trans. James Strachey et al. (London: Hogarth Press, 1953–1974), 14: 237–258.

4. At many summer camps, there are rules that campers should not have cell phones, which are routinely "confiscated" at the start of camp. Children now tell me that parents give them two phones: one to "turn in" on the first day of camp and a second to keep for calling home.

5. In October 2005, ABC News called the term "in vogue." See "Do 'Helicopter Moms' Do More Harm Than Good?" ABCNews.com, October 21, 2005, http://abc news.go.com/2020/Health/story?id=1237868&page=1 (accessed April 7, 2004).

6. In 2004, the Pentagon canceled its so-called LifeLog project, an ambitious effort to build a database tracking a person's entire existence: phone calls made, TV shows watched, magazines read, plane tickets bought, e-mails sent and received. It was then partially revived nine months later. See Noah Schachtman, "Pentagon Revives Memory Project," *Wired News*, www.wired.com/politics/security/news/2004/09/6491 (accessed August 4, 2010). Backers of the project saw it as a near-perfect digital memory. Civil libertarians argued that it could become the ultimate invasive profiling tool. Such objections, of course, are undermined if people make agreements with private services (for instance, Facebook and Google) in which they sacrifice rights to their data in return for services on the system. When one agrees to such terms of service, the questions become, How transparent are the privacy settings on the service, and how easy is it for people to choose the privacy options they wish to have? Facebook has been the focus of much of the public discussion of these matters, centering on whether the "default" is more privacy or less. So, for example, in 2007, Facebook users turned against Beacon, a service that posted information about users' purchases both to Facebook and other sites. More than fifty thousand users signed an online petition in protest, and Beacon became an application that users had to "opt into." By 2009, it was shut down completely, and Facebook agreed to use $9.5 million to start a foundation dedicated to questions of online privacy. See "Facebook Shuts Down Beacon Marketing Tool," *CBC News*, September 21, 2009, www.cbc.ca/technology/story/2009/09/21/tech-internet-facebook-beacon.html (accessed October 15, 2009).

In spring 2010, Facebook's privacy policies again became front-page news. See Jenna Wortham, "Facebook Glitch Brings New Privacy Worries," *New York Times*, May 5, 2010, www.nytimes.com/2010/05/06/technology/internet/06facebook.html ?scp=2&sq=wortham%20facebook&st=cse (accessed May 10, 2010), and Miguel Helft and Jenna Wortham, "Facebook Bows to Pressure over Privacy," *New York Times*, May 27, 2010, www.nytimes.com/2010/05/27/technology/27facebook.html (accessed May 29, 2010). This conversation will surely continue.

7. Miguel Helft, "Anger Leads to Apology from Google About Buzz," *New York Times*, February 14, 2010, www.nytimes.com/2010/02/15/technology/internet/15google.html (accessed May 29, 2010).

8. The corporate world has certainly behaved as though transparency about privacy policy is not necessarily in its best interest. When Facebook has been open about how much user data it feels it owns, users have not been happy. The corporate reality, however, is on the public record. An anonymous Facebook employee disclosed that the company saves "all the data on all of our servers, every hour of every day." "At least two people," the employee said, "have been fired" for spying on accounts. Cited in Stephen Burt, "Always On," *London Review of Books* 32, no. 11 (June 10, 2010): 21–22.

9. Polly Sprenger, "Sun on Privacy: Get over It," *Wired News*, January 26, 1999, www.wired.com/politics/law/news/1999/01/17538 (accessed August 4, 2010).

10. Eric Schmidt made the first remark about controlling behavior rather than worrying about privacy to CNBC. The video is available at Ryan Tate, "Google CEO: Secrets Are for Filthy People," Gawker, December 4, 2009, http://gawker.com/5419271/google-ceo-secrets-are-for-filthy-people (accessed June 5, 2010). His remark about name changing was made to the *Wall Street Journal*. Holman W. Jenkins Jr., "Google and the Search for the Future," online.wsj.com/article/SB10001420527487049011045 75423294099527212.html (accessed September 3, 2010).

11. On the issue of computational metaphors being taken as reality, see Harry R. Lewis (with Hal Abelson and Ken Ledeen), *Blown to Bits: Your Life, Liberty, and Happiness After the Digital Explosion* (New York: Pearson, 2006), ch. 3.

12. Robert Jay Lifton, "Protean Man," *Archives of General Psychiatry* 24 (1971): 298–304, and Robert Jay Lifton, *The Protean Self: Human Resilience in an Age of Fragmentation* (New York: Basic Books, 1993). See also Sherry Turkle, *Life on the Screen: Identity in the Age of the Internet* (New York: Simon and Schuster, 1995).

13. Michel Foucault, *Discipline and Punish: The Birth of the Prison*, trans. Alan Sheridan (1979; New York: Vintage Books, 1995).

14. Foucault, *Discipline and Punish*, 195–228. Here is one example of Foucault on the relationship between remembrance and the constitution of a new kind of self: "First, to bring out a certain number of historical facts which are often glossed over when posing this problem of writing, we must look into the famous question of the hypomnemata.... Now, in fact, hypomnemata has a very precise meaning. It is a copybook, a notebook. Precisely this type of notebook was coming into vogue in Plato's time for personal and administrative use. This new technology was as disrupting as the introduction of the computer into private life today. It seems to me the question of writing and the self must be posed in terms of the technical and material framework in which it arose.... What seems remarkable to me is that these new instruments

were immediately used for the constitution of a permanent relationship to oneself—one must manage oneself as a governor manages the governed, as a head of an enterprise manages his enterprise, a head of household manages his household."

See Paul Rabinow, "An Interview with Michel Foucault," in *The Foucault Reader*, ed. Paul Rabinow (New York: Pantheon, 1984), 363–365.

CHAPTER 14: THE NOSTALGIA OF THE YOUNG

1. This recalls how French psychoanalyst Jacques Lacan talks about the analytic encounter. The offer to listen creates a demand to be heard. "In short, I have succeeded in doing what in the field of ordinary commerce people would dearly like to be able to do with such ease: with supply, I have created demand." See Jacques Lacan, "The Direction of the Treatment and the Principles of Its Power," *Ecrits: A Selection*, trans. Alan Sheridan (New York: W.W. Norton, 1977), 254. For a discussion of Lacan and the "intransitive demand," see Sherry Turkle, *Psychoanalytic Politic: Jacques Lacan and Freud's French Revolution* (1978; New York: Guilford Press, 1992), 85.

2. David Andersen, "Erik H. Erikson's Challenge to Modernity" (PhD diss., Bowling Green State University, 1993). After writing this chapter and the next, I found Alan Lightman's elegant essay, "Prisoner of the Wired World," which evokes many of the themes I treat here. In *A Sense of the Mysterious: Science and the Human Spirit* (New York: Pantheon, 2005), 183–208.

3. Anthony Storr, *Solitude: A Return to the Self* (New York: Random House, 1988), 198.

4. Henry David Thoreau, "Where I Lived and What I Lived For," in *Walden* (1854; New York: American Renaissance Books, 2009), 47. I thank Erikson biographer Lawrence J. Friedman for his insights on Erikson and "stillness."

5. Thoreau, "Where I Lived," 47.

6. Katy Hafner, "To Deal with Obsession, Some Defriend Facebook," *New York Times*, December 20, 2009, www.nytimes.com/2009/12/21/technology/internet/21facebook.html?_r=1 (accessed January 6, 2009).

7. Thoreau, "Where I Lived," 47.

8. Kevin Kelly, "Technophilia," The Technium, June 8, 2009, www.kk.org/thetechnium/archives/2009/06/technophilia.php (accessed December 9, 2009).

9. See Sherry Turkle, "Simulation and Its Discontents," in Sherry Turkle, *Simulation and Its Discontents* (Cambridge, MA: MIT Press, 2009), 3–84.

CONCLUSION: NECESSARY CONVERSATIONS

1. Bohr says, "It is the hallmark of any deep truth that its negation is also a deep truth" (as quoted in Max Delbrück, *Mind from Matter: An Essay on Evolutionary Epistemology* [Palo Alto, CA: Blackwell Scientific Publications, 1986], 167).

2. One study comparing data from 1985 and 2004 found that the mean number of people with whom Americans can discuss matters important to them dropped by nearly one-third, from 2.94 people in 1985 to 2.08 in 2004. Researchers also found that the number of people who said they had no one with whom to discuss such matters more than doubled, to nearly 25 percent. The survey found that both family and nonfamily confidants dropped, with the loss greatest in nonfamily connections. Miller

McPherson, Lynn Smith-Lovin, and Matthew E. Brashears, "Social Isolation in America: Changes in Core Discussion Networks over Two Decades," *American Sociological Review* 71 (June 2006): 353–375.

3. Barry Wellman and Bernie Hogan (with Kristen Berg et al.), "Connected Lives: The Project," in *Networked Neighborhoods*, ed. Patrick Purcell (London: Springer-Verlag, 2006), 161–216.

4. Moving past the philosophical, there are contradictions on the ground: a "huggable" robot is a responsive teddy bear that makes it possible for a grandmother in Detroit to send a squeeze to her grandson in Cambridge, Massachusetts. The grandmother hears and sees her grandson through the eyes and ears of the bear, and the robot communicates her caress. All well and good. But when videoconferences and hugs mediated by teddy bears keep grandparents from making several-thousand-mile treks to see their grandchildren in person (and there is already evidence that they do), children will be denied something precious: the starchy feel of a grandmother's apron, the smell of her perfume up close, and the taste of her cooking. Amy Harmon, "Grandma's on the Computer Screen," *New York Times*, November 26, 2008, www.nytimes.com/2008/11/27/us/27minicam.htm?pagewanted=all (accessed December 11, 2009). On the "Huggable" project, see http://robotic.media.mit.edu/projects/robots/huggable/overview/overview.html (accessed April 5, 2010).

5. On ELIZA, see Joseph Weizenbaum, *Computer Power and Human Reason: From Judgment to Calculation* (San Francisco: Freeman, 1976); Sherry Turkle, *The Second Self: Computers and the Human Spirit* (1984; Cambridge, MA: MIT Press, 2005); Sherry Turkle, *Life on the Screen: Identity in the Age of the Internet* (New York: Simon and Schuster, 1995).

6. People who feel that psychotherapists are dismissive or disrespectful may also prefer to have computers as counselors. An MIT administrative assistant says to me: "When you go to a psychoanalyst, well, you're already going to a robot."

7. In fact, we have two robotic dreams. In one, we imagine the robots as perfect companions. In another, we join with them to become new selves. This second scenario itself has two variants. In a first, we evolve. We assimilate robotic parts until there is no "us" and "them." In the short term, we feel smarter and healthier. In the long term, we become immortal. In the second variant, there is a decisive turn, a moment of "singularity" in which computing power is so vast that people essentially become one with machines. For a critique of what he calls "cybernetic totalism," see Jaron Lanier, "One Half a Manifesto," www.edge.org/3rd-culture/lanier-pl.html (accessed August 3, 2010) and *You Are Not a Gadget: A Manifesto* (New York: Knopf, 2010).

8. Psychoanalysis sees truth in the symptom. But it is a truth that has not been given free expression. You don't want to get rid of these truths for they are "signs that something has disconnected a significant experience from the mass of other, non-symptomatic significant experiences. The aim of psychoanalysis is to restore the broken connection, thereby converting the distorted, disconnected experience (the symptom) into an ordinary, connected one." See Robert Caper, *Building Out into the Dark: Theory and Observation in Science and Psychoanalysis* (New York: Routledge, 2009), 90.

9. Kevin Kelly, "Technophilia," The Technium, June 8, 2009, www.kk.org/thetechnium/archives/2009/06/technophilia.php (accessed December 9, 2009).

10. Caper, *Building Out into the Dark*, 93.

11. Personal communication, October 2008.

12. Caper says, "We tolerate the plague of our neurotic symptoms because we fear that discovering the truths they simultaneously rest on and cover over will lead to our destruction." And further, an interpretation, like a new technology, "always poses a danger. . . . The danger consists not in the analysts' search for truth, and not even in the fact that his interpretations are inevitably flawed, but in his not recognizing that this is so." See Caper, *Building Out into the Dark*, 91, 94.

13. Henry Adams, "The Dynamo and the Virgin," in *The Education of Henry Adams: An Autobiography* (Boston: Massachusetts Historical Society, 1918), 380.

14. Kelly, "Technophilia."

15. One roboticist who makes quite extravagant claims about our futures is David Hanson. For videos and progress reports, see Hanson Robotics at www.hansonrobot-ics.com (accessed December 11, 2009). And, of course, there is David Levy's book on the future of robot affections, *Love and Sex with Robots: The Evolution of Human-Robot Relationships* (New York: Harper Collins, 2007).

16. This is a paraphrase. The exact citation is, "When you want to give something presence, you have to consult nature and that is where design comes in. If you think of brick, for instance, you say to brick, 'What do you want, brick?' And brick says to you, 'I'd like an arch.' And if you say to brick, 'Look, arches are expensive and I can use a concrete lintel over you, what do you think of that, brick?' And brick says to you, 'I'd like an arch.'" See Nathaniel Kahn, *My Architect: A Son's Journey* (New Yorker Films, 2003).

17. Rodney Brooks, cited in "MIT: 'Creating a Robot So Alive You Feel Bad About Switching It Off'—a Galaxy Classic," The Daily Galaxy, December 24, 2009, www.dailygalaxy.com/my_weblog/2009/12/there-is-ongoing-debate-about-what-constitutes-life-synthetic-bacteria-for-example-are-created-by-man-and-yet-also-alive.html (accessed June 4, 2010).

18. Cynthia Breazeal and Rodney Brooks both make the point that robot emotions do not have to be like human ones. They should be judged on their own merits. See Cynthia Breazeal and Rodney Brooks (2005). "Robot Emotion: A Functional Perspective," in J.-M. Fellous and M. Arbib (eds.) *Who Needs Emotions: The Brain Meets the Robot*, MIT Press. 271–310. Breazeal insists that "the question for robots is not, 'Will they ever have human emotions?' Dogs don't have human emotions, either, but we all agree they have genuine emotions. The question is, 'What are the emotions that are genuine for the robot?'" Breazeal talks about Kismet as a synthetic being and expects that it will be "given the same respect and consideration that you would to any living thing." WNPR, "Morning Edition," April 9, 2001, www.npr.org/programs/morning/features/2001/apr/010409.kismet.html (accessed August 12, 2010). See also Susan K. Lewis, "Friendly Robots," Nova, www.pbs.org/wgbh/nova/tech/friendly-robots.html and Robin Marantz Henig, "The Real Transformers," *New York Times*, July 29, 2007, www.nytimes.com/2007/07/29/magazine/29robots-t.html (accessed September 3, 2010).

19. There is much talk these days of a "robot bill of rights." As robots become more complex, there is a movement to have formal rules for how we treat artificial sentience. Robot rights are the subject of parliamentary inquiry in the United Kingdom. In South

Korea, where the government plans to put a sociable robot into every home by 2020, there are plans to draw up legal guidelines on how they must be treated. The focus of these efforts is on protecting the robots. But as early as the mid-1990s, people abused virtual creatures called "norns," tormenting them until they became psychotic, beating their virtual heads against virtual walls. Popular Web videos show even as simple a robotic toy as Hasbro's Elmo Live being doused with gas and set on fire, his red fur turning to charcoal as he writhes in what looks like pain. I have watched the abuse of Tamagotchis, Furbies, My Real Babies, and Paros. The "robot rights movement" is all about not hurting the robots. My concern is that when we torture sociable robots that we believe to have "states of mind," we damage ourselves.

Daniel Roth, "Do Humanlike Machines Deserve Human Rights," *Wired Magazine*, January 19, 2009, www.wired.com/culture/culturereviews/magazine/17–02/st_essay (accessed June 4, 2010).

20. For drawing my attention to what he calls "*formes frustes* of feeling," I thank my colleague Cambridge psychiatrist and psychoanalyst Dr. David Mann, who has reformulated an entire range of unpleasant affects (for example, envy, greed, resentment) in an as-yet-unpublished essay, "Failures of Feeling" (2009).

21. Anthony Storr, *Solitude: A Return to the Self* (New York: Random House, 1988).

22. Quandaries have become a classic way of thinking about moral dilemmas. See, for example, Marc Hauser, *Moral Minds: How Nature Designed Our Universal Sense of Right and Wrong* (New York: Ecco, 2006). Some of the most common quandaries involve trolley cars and the certainty of death. A typical scenario has you driving a trolley car with five workers ahead of you on the track. Doing nothing will kill all five. You can swerve onto a track on which there is only one worker. Do you act to kill one person rather than five? Then, the scenario may be shifted so you are on a bridge, observing the trolley cars. There is a fat man standing beside you. Do you push him onto the track to stop the trolley, thus saving the five people? And so it goes.

23. Traditional psychology was constructed based on experiments done only with men and through theories that only took into account male development. During the first and second world wars, psychological tests were standardized for the male soldiers with whom they were developed. End of story. Psychologists came to see male responses as "normal" ones. The behaviors, attitudes, and patterns of relationship exhibited by most men became the norm for "people." Psychologist Carol Gilligan's 1982 *In a Different Voice* is an example of work that broke this frame. Gilligan portrays the canonical (and stereotypically "male") view of moral reasoning and then points out that it constitutes only one way in which people make moral decisions. The canonical pattern looks at moral choices in terms of abstract principles. Another, equally evolved moral voice relies on concrete situations and relationships. For example, see Gilligan's treatment of "Amy and Heinz" in *In a Different Voice: Psychological Theory and Women's Development* (Cambridge, MA: Harvard University Press, 1993), 26–28, 30. The "robots-or-nothing" thinking about elder care frames a dilemma that begs for a contextual approach; this is what the fifth graders in Miss Grant's class brought to the table.

We hear another moment of reframing when seventeen-year-old Nick tries to find a way to get his father to put away his BlackBerry during family dinners. Recall that in Nick's home, family dinners are long. His mother takes pride in her beautifully prepared meals with many courses. Nick suggests shorter meals. His parents argue principles: the priority of work versus that of a meal prepared with love. Nick focuses on relationship. The family needs family time. How can they provide that for each other? Nick suggests a shorter meal with no phones.

24. Anthony Appiah, *Experiments in Ethics* (Cambridge, MA: Harvard University Press, 2008), 196–197. Appiah is writing about "trolley car" quandaries, but he could be writing about the "robots-or-nothing" problem.

25. Here I note the work on using robots as a therapeutic tool with people on the autism spectrum. Robots do not overwhelm them as people may. The predictability of robots is comforting. The question remains whether these robots can serve as transitions to relationships with people. I cotaught a course at MIT on robotics and autism with Rosalind Picard and Cynthia Breazeal. Roboticists are of course gratified to feel that they can contribute to therapy in this area; the jury is still out on whether nonhuman faces get us ready for human ones. For a discussion that focuses on the work of roboticist Maja Matarić in this area, see Jerome Groopman, "Robots That Care: Advances in Technological Therapy," *The New Yorker*, November 2, 2009, www.newyorker .com/reporting/2009/11/02/091102fa_fact_groopman (accessed November 11, 2009).

26. This phrase is drawn from Roger Shattuck's book on the "Wild Child" of Aveyron. *The Forbidden Experiment* (New York: Farrar, Strauss, and Giroux, 1980).

27. "Basic trust" is Erik Erikson's phrase; see *Childhood and Society* (New York: Norton, 1950) and *Identity and the Life Cycle* (1952; New York: Norton, 1980).

28. At MIT, the question of risk strikes most of my students as odd. They assume, along with roboticist David Hanson, that eventually robots "will evolve into socially intelligent beings, capable of love and earning a place in the extended human family." See Groopman, "Robots That Care."

29. A University of Michigan study found that today's college students have less empathy than those of the 1980s or 1990s. Today's generation scored about 40 percent lower in empathy than their counterparts did twenty or thirty years ago. Sara Konrath, a researcher at the University of Michigan's Institute for Social Research, conducted, with University of Michigan graduate student Edward O'Brien and undergraduate student Courtney Hsing, a meta-analysis that looked at data on empathy, combining the results of seventy-two different studies of American college students conducted between 1979 and 2009. Compared to college students of the late 1970s, the study found, college students today are less likely to agree with statements such as "I sometimes try to understand my friends better by imagining how things look from their perspective" and "I often have tender, concerned feelings for people less fortunate than me." See "Empathy: College Students Don't Have As Much As They Used To," EurekAlert! May 28, 2010, www.eurekalert.org/pub_releases/2010–05/uom-ecs052610 .php (accessed June 4, 2010).

30. I thank my psychotherapist colleagues for ongoing conversations on these matters. In particular I acknowledge the adolescent psychiatrist John Hamilton and the panels on "Adolescence in Cyberspace" on which we have collaborated at the Annual Meetings of the American Academy of Child and Adolescent Psychiatry in October

2004 and October 2008; the participants in the MIT working group, "Whither Psycho-analysis in Digital Culture" Initiative on Technology and Self, 2003–2004; and partic-ipants at the Washington Institute for Psychoanalysis's "New Directions" Conference, April 30, 2010.

31. Maggie Jackson, *Distracted: The Erosion of Attention and the Coming Dark Age* (New York: Prometheus, 2008).

32. Matt Richtel, "Hooked on Gadgets and Paying a Mental Price," *New York Times*, July 7, 2010, http://community.nytimes.com/comments/www.nytimes.com/2010/06/07/technology/07brain.html?sort=oldest&offset=2 (accessed July 7, 2010).

33. Nicholas Carr, *The Shallows: What the Internet Is Doing to Our Brains* (New York: W. W. Norton and Company, 2010). Here, the argument is that online activities—surfing, searching, jumping from e-mail to text—actually change the nature of the brain. The more time we spend online, the more we are incapable of quiet reverie, not because of habits of mind but because of a rewiring of our circuitry. This area of research is, happily, getting more and more public attention. See Matt Richtel, "Your Brain on Computers: Outdoor and Out of Reach, Studying the Brain," *New York Times*, August 16, 2010, www.nytimes.com/2010/08/16/technology/16brain.html (accessed August 16, 2010).

34. Of course, one of my concerns is that the moment to summon ourselves to ac-tion might pass. We are at a point at which, when robots are proposed as companions for the elderly or as babysitters, we can still have a conversation that challenges these ideas. We still remember why they are problematic. I am concerned that in twenty years, one may simply boast, "I'm leaving my kid with the nanny bot." After the cost of purchase, it will be free and reliable. It will contact you if there is any deviation from the plan you have left for your child—be these deviations in your child's temperature or in a range of acceptable behaviors. I vividly remember leading an MIT seminar in 2001, one that was part of a celebration at the release of Steven Spielberg's *A.I.: Arti-ficial Intelligence*, when for the first time, I was the only person in a room of thirty who did not see any issue at all with the prospect of a computer psychotherapist. Mo-ments when big steps with technology seem problematic have a way of passing.

EPILOGUE: THE LETTER

1. Vannevar Bush, "As We May Think," *Atlantic Monthly* (July 1945): 101–106, www.theatlantic.com/doc/194507/bush (accessed November 20, 2009).

2. See Steve Mann (with Hal Niedzviecki), *Digital Destiny and Human Possibility in the Age of the Wearable Computer* (New York: Random House, 2001).

3. C. Gordon Bell and Jim Gemmell, "A Digital Life," *Scientific American* 296, no. 3 (March 2007): 58–65, http://sciam.com/print_version.cfm?articleID=CC50D7BF-E7F2–99DF-34DA5FF0B0A22B50 (accessed August 7, 2007). The My Life Bits web-site is http://research.microsoft.com/en-us/projects/mylifebits (accessed July 30, 2010). Bell and Gemmell published a book-length discussion of this project, *Total Recall: How the E-Memory Revolution Will Change Everything* (New York: Dutton, 2009).

4. Bell and Gemmell, "A Digital Life."

5. Thompson notes of his 2007 visit, "MyLifeBits records his telephone calls and archives every picture—up to 1,000 a day—snapped by his automatic 'SenseCam,' that

device slung around his neck. He has even stowed his entire past: The massive stacks of documents from his 47-year computer career, first as a millionaire executive then as a government Internet bureaucrat, have been hoovered up and scanned in. The last time he counted, MyLifeBits had more than 101,000 emails, almost 15,000 Word and PDF documents, 99,000 Web pages, and 44,000 pictures." See Clive Thompson, "A Head for Detail," *Fast Company*, December 19, 2007, www.fastcompany.com/magazine/110/head-for-detail.html (accessed October 1, 2009).

6. Susan Sontag, *On Photography* (New York: Dell, 1978), 9.

7. Bell and Gemmell discuss the burdens of having a digital shadow. They anticipate that other people captured in one's sights may need to be pixilated so as not to invade their privacy, data will have to be stored "offshore" to protect it from loss and/or illegal seizure, and there is danger posed by "identity thieves, gossipmongers, or an authoritarian state." The fact that these three are grouped together as problems to be solved technically illustrates the power of the fantasy of total life capture. For after all, the potential damage from gossipmongers and an authoritarian state are not commensurate. They surely cannot be dealt with by the same technical maneuvers. Yet the fantasy is potent. Bell and Gemmell admit that despite all problems, "for us the excitement outweighs the fear." See Bell and Gemmell, "A Digital Life."

8. Indeed, with far less "remembrance technology," many of us wonder if Google is "making us stupid" because it is always easier to search than remember. The originator of this memorable phrase is Nicholas Carr, "Is Google Making Us Stupid?" *The Atlantic*, July/August 2008, http://www.theatlantic.com/magazine/archive/2008/07/is-google-making-us-stupid/6868/ (accessed August 12, 2010).

9. Thompson, "A Head for Detail."

10. Thompson, "A Head for Detail."

11. Obama himself fought hard and famously to keep his BlackBerry, arguing that he counts on this digital device to make sure that the "bubble" of his office does not separate him from the "real" world. Obama kept his BlackBerry, but in March 2009, the Vatican asked the Catholic bishops of Italy to request that their flocks give up texting, social-networking websites, and computer games for Lent, or at least on Fridays. Pope Benedict has warned Catholics not to "substitute virtual friendship" for real human relationships. On his YouTube site, the pope warned of "obsessive" use of mobile phones and computers, which "may isolate individuals from real social interaction while also disrupting the patterns of rest, silence, and reflection that are necessary for healthy human development." The *London Times* reports that "even Pope Benedict . . . experienced the distractions of obsessive texting" when President Nicolas Sarkozy of France was flagged for rudeness when he checked his mobile device for text messages during a personal audience with the pontiff. See Richard Owen, "Thou Shalt Not Text until Easter, Italians Told," *The Times*, March 3, 2009 (accessed July 30, 2010).

12. See Sherry Turkle, "Reading the *Inner History of Devices*," in Sherry Turkle, ed., *The Inner History of Devices* (Cambridge, MA: MIT Press, 2008).

13. Technology and remembrance is a growing discipline. In addition to *Cyborg*, Steve Mann has written extensively about computation and remembrance. See, for example, "Wearable Computing: Toward Humanistic Intelligence," *Intelligent Systems*

16, no. 3 (May–June 2001): 10–15. From 1996 on, Thad Starner, who like Steve Mann was a member of the MIT cyborg group, worked on the Remembrance Agent, a tool that would sit on your computer desktop (or now, your mobile device) and not only record what you were doing but make suggestions about what you might be interested in looking at next. See Bradley J. Rhodes and Thad Starner, "Remembrance Agent: A Continuously Running Personal Information Retrieval System," *Proceedings of the First International Conference on the Practical Application of Intelligent Agents and Multi Agent Technology* (PAAM '96),487–495, 487–495, www.bradleyrhodes.com/Papers/remembrance.html (accessed December 14, 2009).

Albert Frigo's "Storing, Indexing and Retrieving My Autobiography," presented at the 2004 Workshop on Memory and the Sharing of Experience in Vienna, Austria, describes a device to take pictures of what comes into his hand. He comments on the implications: "The objects I photograph, while used, represent single specific activities that from a more general perspective can visualize how, throughout my life, my intentions, my desires, my sorrows have mutated. The objects become my emblems, the code through which the whole of me can be reconstructed, interpreted." See Albert Frigo, "Storing, Indexing and Retrieving My Autobiography," Nishida & Sumi Lab, www.ii.ist.i.kyoto-u.ac.jp/~sumi/pervasive04/program/Frigo.pdf (accessed November 2009). For a sense of the field's current ambitions, see the Memories for Life project at www.memoriesforlife.org (accessed July 30, 2010) and the Reality Mining group at MIT and the Santa Fe Institute at http://reality.media.mit.edu/about.php (accessed December 14, 2009).

William C. Cheng, Leana Golubchik, and David G. Kay write about the politics of remembrance. They anticipate a future in which we will all wear self-monitoring and recording devices. They discuss the danger that state authority will presume that when behaving lawfully, people will be wearing the device. Not wearing the device will be taken as indicative of guilt. Yet, even given this dark scenario, they conclude with the claim that, essentially, the train has left the station: "We believe that systems like Total Recall will get built, they will have valuable uses, and they will radically change our notions of privacy. Even though there is reason to be skeptical that there will be any meaningful legal protection for the privacy status quo, we believe that useful technologies are largely inevitable, that they often bring social changes with them, and that we will inevitably both suffer and benefit from their consequences." See William C. Cheng, Leana Golubchik, and David G. Kay, "Total Recall: Are Privacy Changes Inevitable?" (paper presented at Capture, Archiving, and Retrieval of Personal Experiences [CARPE] workshop, New York, October 15, 2004), http://bourbon.usc.edu/iml/recall/papers/carpe2k4-pub.pdf (accessed December 14, 2009).

14. Alec Wilkinson, "Remember This?" *The New Yorker*, May 28, 2007, 38–44, www.newyorker.com/reporting/2007/05/28/070528fa_fact_wilkinson (accessed November 20, 2009).

15. Wilkinson, "Remember This?"

INDEX

Credit: Peter Urban

Sherry Turkle is the Abby Rockefeller Mauzé Professor of the Social Studies of Science and Technology at MIT, the founder and director of the MIT Initiative on Technology and Self, and a licensed clinical psychologist. She is the author of *The Second Self* and *Life on the Screen,* with which *Alone Together* forms a trilogy, and most recently *Reclaiming Conversation.* Turkle is a recipient of the Harvard Centennial Medal and is a Fellow of the American Academy of Arts and Sciences.